R072323

The Role of Computers in Manufacturing Processes

THE ROLE OF COMPUTERS IN MANUFACTURING PROCESSES

Gideon Halevi

Director CAM/CAD R&D Center, IMI
Tel Aviv, Israel

A Wiley-Interscience Publication
JOHN WILEY & SONS
New York • Chichester • Brisbane • Toronto

TS
176
H33

Copyright © 1980 by John Wiley & Sons, Inc.

All rights reserved. Published simultaneously in Canada.

Reproduction or translation of any part of this work beyond that permitted by Sections 107 or 108 of the 1976 United States Copyright Act without the permission of the copyright owner is unlawful. Requests for permission or further information should be addressed to the Permissions Department, John Wiley & Sons, Inc.

Library of Congress Cataloging in Publication Data

Halevi, Gideon, 1928–
 The role of computers in manufacturing processes.

 Includes index.
 1. Production engineering—Data processing.
I. Title.

TS176.H33   658.5'0028'54   80-11378
ISBN 0-471-04383-4

Printed in the United States of America

10 9 8 7 6 5 4 3 2 1

To Malka, Liat, Ilan

Preface

The role of computers in manufacturing is growing rapidly and achieving widespread industrial recognition and acceptance. Their application has led to a reevaluation of the analytic approach to manufacturing processes, while at the same time contributing to its advancement. The formulation of this approach has been initiated in the individual phases of the manufacturing process. For example, mathematicians have made remarkable progress in the development of sequencing and scheduling theory; economists have developed the area of inventory management and control; computer personnel have created a new and flexible data base organization; and management scientists have developed the concept of requirement planning and costing. However, progress has not been as advanced in the engineering phase (the heart of manufacturing) due to the large number of interdependent variables, which makes analytical treatment difficult. This field remains an "art" that is heavily dependent on skill, experience, and intuition.

Fortunately (and as was to be expected), this "missing link" has not prevented implementation of the advances in the other phases. The combination of these separate phases into a chain of activities, with each phase "giving and taking" data from the previous and following ones, has led to the present-day technology of integrated manufacturing systems.

The integrated system merely supplies administrative services to the engineering phase of manufacturing, which then uses this information as the basic data for the other phases of the system. This is why the system can only be as good as its basic (engineering) data.

In the past few years, my research has been devoted to these fundamental fields (i.e., the engineering phases of manufacturing) in an attempt to replace "art" by "science," that is, to replace intuition by computation, while turning skill and experience into formulas.

The rapid growth of the capabilities of digital computers together with the increased computational power that exists today have made it possible for us to meet our research targets. An engineering decision requiring thousands of computations was impractical in the past, and one had to rely on intuition and experience. Today, however, a computer may be employed to evaluate all alternatives and thus reach an optimum decision.

The first step was to establish a systematic approach to the field of process planning and to prepare a generative-type computer program for the implementation of a particular process (in our case, metal cutting in turning operations). This generative process planning (GPP) computer program has three main sections:

1. Pattern recognition (geometric modeling)—a special type of data organization that enables the computer to "see" the part and make mathematical manipulations on it.
2. Metal cutting technology.
3. A specially developed mathematical technique to solve the decision problem concerning the optimum, practical sequence of operations and specific machine to be used.

The next step was to evaluate the performance of this approach as the "missing link," that is, to transform the treatment of the engineering phase in the integrated manufacturing system from an administrative one into a functional one.

The evaluation indicated that the approach limited the flexibility and inherent dynamics of GPP. It became apparent that since different professions are involved in the various phases of the manufacturing process, many artificial constraints are imposed on the system, thereby limiting the flexibility and productivity obtained. We concluded that it would be more advantageous to extend our work in two directions. First, the mathematical section of the GPP program (which results in the optimum processing of a component) was extended to cover capacity and production planning. This resulted in an optimum process that took into account the time phase and a given product mix, thereby replacing existing sequencing and scheduling problems. Second, the pattern recognition section was extended to cover and to aid in the fields of engineering design, standardization, purchasing, inventory control, and requirement planning.

Clearly, the impact of our work called for a reassessment of the existing manufacturing process. This resulted in its replacement by a new, all-embracing, computer-oriented technology—Hal-Technology (Hal in Hebrew means all-embracing)—that views the manufacturing process as a single unique system.

This approach is now ready for widespread application to computer-aided manufacturing and computer-aided design (CAM/CAD). Several stepping-stones in the evolution of Hal-Technology have already been described in the literature, but this book represents its first comprehensive description. Most of the material is appearing in book form for the first time, while much of it has never been published in any form.

The major purpose of this book is to present, in a practical and descriptive form, the philosophies of Hal and to demonstrate its advantages over existing methods. To achieve this purpose, the book is divided into three parts: In Part I, an extensive overview of current methods in manufacturing technology is presented; Part II discusses the GPP system and its modules; and Part III presents an in-depth discussion of Hal, detailing its capabilities and advantages over current methods across all phases of the production process. Practical problems and their solutions are given in order to illustrate Hal-Technology.

This book will be of use to all those involved in manufacturing. It will provide the practitioner with up-to-date information on all the current techniques besides Hal, as well as with sufficient data to make an independent decision on which of these techniques to adopt. We hope, of course, that it will be Hal.

I would like to express my gratitude to Myra Bank, Rachel Lahav, and Benyamin Bobach for their special review efforts. Credit should be given to CAM-I, Inc., the CIRP, NSF, and to the others engaged in production research, whose work and publications were very helpful in the preparation of this book. Finally, I wish to acknowledge the generosity of the International Business Machines Corporation in permitting the reproduction of excerpts from their various publications (in particular, from the COPICS series).

GIDEON HALEVI

Tel Aviv
January 1980

Contents

**Part I ASPECTS OF MANUFACTURING TECHNOLOGY—
A STATE OF THE ART REVIEW**   **1**

1 Introduction   **3**

    1.1   The Evolution of Computer Applications in Industry   7

2 The Integrated Manufacturing System   **13**

    2.1   The Manufacturing Process   14
    2.2   Manufacturing and Industrial Management   21
    2.3   Computerized Manufacturing System Design Concepts   24
    2.4   The Integrated Manufacturing System   32
    2.5   Design for Reliability of Computer Systems—General   37
    2.6   Design for Reliability—The Integrated Manufacturing System   43

3 Engineering in the Manufacturing Process   **47**

    3.1   Product Design—Engineering Design   48
    3.2   Performance Evaluation and Computation   52
    3.3   Standardization   64
    3.4   Design for Production   67
    3.5   Process Planning   72
    3.6   Group-Technology   76
    3.7   CAM/CAD   86

4 Engineering Data Control — 95

  4.1 Classification and Coding — 96
  4.2 Product Definition—Bill of Material — 104
  4.3 Organizing—Bill of Material — 113
  4.4 Manufacturing Routing — 121

5 Master Production Planning — 125

  5.1 Customer Orders — 125
  5.2 Forecasting — 129
  5.3 Master Production Schedule — 135
  5.4 Management Control and Finance Planning — 143

6 Requirement Planning — 150

  6.1 Manufacturing Activity Planning — 151
  6.2 Manual versus Computerized Systems — 155
  6.3 Requirement Planning Technique — 162
  6.4 Pegging — 169
  6.5 Discipline Needed for Requirement Planning — 173

7 Inventory Management and Control — 179

  7.1 Inventory Objectives — 180
  7.2 Inventory System Technique—General — 183
  7.3 Design for Reliability of the Inventory System — 192
  7.4 The Inventory System as a Management Control Tool — 195

8 Capacity Planning and Order Release — 199

  8.1 Capacity Planning Objectives — 200
  8.2 Capacity Planning Terminology — 203
  8.3 Capacity Planning Technique — 206
  8.4 Order Release — 217

9 Shop-Floor Control — 220

  9.1 Short-Term (Daily) Capacity Planning — 221
  9.2 Dispatching Rules — 226
  9.3 Job Recording — 234

Contents   xiii

    9.4   Job Recording Design for Reliability   240
    9.5   Control Reports   247

10 Cost Planning Control   252

    10.1   Conversion Cost   253
    10.2   Cost Planning (Standard Cost)   259
    10.3   Actual Cost   264
    10.4   Cost Control   268

Part II GENERATIVE PROCESS PLANNING (GPP)   275

11 Generative Process Planning—Prerequisite   277

    11.1   State-of-the-Art   278
    11.2   Economics of the Basic Turning Operation   284
    11.3   Process Planning   289
    11.4   Retrieval-Type Process Planning Programs   294
    11.5   Generative Process Planning   299
    11.6   Summary   303

12 Generative Process Planning—Engineering   305

    12.1   Basic Cutting Conditions Equation   306
    12.2   Parameters Analysis   310
    12.3   Operation Establishment   323
    12.4   Chucking Type and Location   329
    12.5   Summary   335

13 Generative Process Planning—Mathematics   338

    13.1   Definition of the Mathematical Problem   339
    13.2   Mathematical Methods Review   341
    13.3   Constructing the Machine–Operation Matrix   343
    13.4   Preliminary Machine Selection   348
    13.5   Operation Arrangement in Matrix   352
    13.6   Matrix Solution   353
    13.7   Summary   364

14 Generative Process Planning—Summary — 366

14.1 Generative Process Planning—Example Demonstration — 367

Part III Hal-Technology — 391

15 Hal—Technology Concepts — 393

15.1 Hal Concepts — 395
15.2 Hal-Technology System Architecture — 397
15.3 Part Description System — 400
15.4 Hal Benefits — 414

16 Hal—in Engineering — 421

16.1 Product Design—Preview — 422
16.2 Design of Parts and Products — 427
16.3 Other Design Features of Hal — 438
16.4 Process Planning under Hal — 441

17 Hal—in Production — 448

17.1 Adaptive Production-Process Planning — 449
17.2 Forced Process Planning — 454
17.3 Stock Utilization Features — 457
17.4 Hal-Master Production Planning — 460

18 Hal—in Industrial Management — 483

18.1 Plant Performance Measurements — 484
18.2 Facility Planning — 487

Index — 495

The Role of Computers in Manufacturing Processes

PART I
ASPECTS OF MANUFACTURING TECHNOLOGY—A STATE-OF-THE-ART REVIEW

Chapter One
INTRODUCTION

The manufacturing process is a chain of activities aimed at meeting a set of objectives defined by management. These objectives are mainly concerned with the production of tangible goods, and the activities are, as a rule, specified by the engineering profession.

In order to specify the required activities, the engineer must consider many parameters and alternatives, and reach decisions as to how, what, where, and when. The more alternatives and parameters considered, the better the design finally devised; however, the decision-making process will take more time. Since the manufacturing process is a dynamic one, conditions are constantly changing and decisions have to be made within a short space of time. It is often preferable to have a decision at hand at the right moment rather than to seek the optimum decision without any set time limit. The computer is a tool that can be employed to narrow the gap between the conflicting demands of "time" and "decision." This instrument can store and manipulate a large quantity of a data in a short period of time. Therefore, with the aid of a computer, many decisions that for practical reasons were previously based on intuition can today be based on a decision-making technique, thereby improving the quality of the decision. Figure 1.1 shows the quality of decision as a function of the time required to reach the decision with and without the aid of a computer.

The computer as a problem-solving tool in the manufacturing process is no different than the other tools that industry has adapted and utilized so well to improve productivity. Introducing a new tool into manufacturing practice usually has an effect far beyond the limited sphere of its immediate function.

For example, the introduction of the conveyor belt into assembly lines solved the problem of transferring work pieces from one work station to another; however, although it made the machine loading and scheduling problem obsolete, it also introduced the problem of line balancing. An-

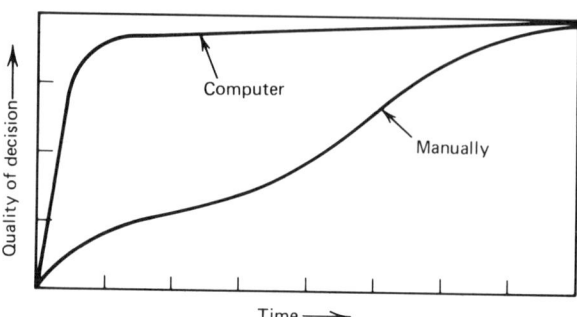

Figure 1.1 Quality of decision versus time.

other example: the use of electric motors as the driving power for machines solved the problem of power transmission and led to the disappearance of drive shafts and belts from industry; at the same time it introduced completely new possibilities with respect to plant layout. A third example: the use of servomechanism and control circuits led to the development of a new type of automatic, numerical control (NC) machine that solved *man–machine problems;* however, new problems of setup and maintenance arose, solutions for which have been found.

Thus, the use of a new tool merely to solve an immediate problem or to serve an immediate need without examining its effect on the overall manufacturing process and without adjusting the procedures accordingly is a waste and misuse of it.

Today computers are used in many industrial activities. They are being utilized more and more to perform the controlling and logistic functions of machines. Numerical control machines, industrial robots, process controls, automatic warehouses, inspection machines, and so forth, are available today. Computers make feasible the automatic communication between machines that points the way toward an automatic factory.

In engineering, the field of computer-aided-design (CAD) is being developed. Several software and hardware systems are available. The aim of these systems is to assist the designer with daily routine jobs. Many engineering-supported computer programs, such as finite element analysis, thermal flow analysis, automatic gear design, hydraulic system design, sculptural surfaces, aerodynamic computations, electronic board layout, and various scientific applications, are available today. Although its field is not well defined, CAM has recently become very fashionable.

In the production phase of the manufacturing process, there probably is not a single activity that does not utilize a computer. The company log book, bookkeeping, inventory, purchasing, and personnel records have long been managed by computers. Computers are performing requirement and capacity planning as well as employee vacation balances and accident and sick leave statistics.

However, the utilization of computers in the engineering and production field (and as machine members), is carried out independently. Each field is being separately developed (i.e., application approach), and thus uses its own logic, which is usually the same logic and method used to perform the task manually. The computer merely speeds up performance, and relieves the person from tedious computations and recordings. Not only are the engineering and production phases carried out independently, but also each separate phase within this field is being developed independently of the other phases. For example, computer-aided-design, the graphic programs and the finite-element analysis programs are two distinct sets of programs that usually cannot communicate between themselves (i.e., share the same data form) unless a complicated interface program is instituted that allows such communication.

The utmost coordination between CAD and CAM is that the part programmer (for a numerical control machine) uses the part display on the CAD graphic terminal. There is absolutely no interdependence between CAD, CAM, and the production phase.

The introduction of computers into industry was an evolutionary process, as described in Section 1.1. It began with the application approach, followed by the second stage of data entry integration. The third stage, distributed processing, still exists today and many favor it for obvious reasons. The fourth, currently the state of the art, is the integrated-manufacturing-system (IMS). The need to have all phases of the manufacturing process working in coordination and in communication with one another has been recognized. One established method, which has been quite successful, is integration. Its success may be attributed to the fact that there is a continuity of thoughts and activities in the production phase—the activities are foreseeable and therefore may be planned and integrated. The problems and activities in the engineering phase, that is, product design and process planning, are quite different by nature. Their integration into the manufacturing system has not been successful.

The term "integrated-manufacturing-system" is used in this book to describe the objectives and not necessarily the reality. Some writers refer to this system as "production-information-system" (PIS), consciously leaving the engineering phase outside the system. Others might enlarge the scope of the system and instead of information system, they refer to it as a "production-information and control-system" (PICS). IBM recently used the term "communication-oriented-production-information-and-control-system" (COPICS); other corporations might use different names.

IMS is by no means a breakthrough or change in technology. It uses the same logic and production theories as older systems, but does it by computer rather than manually. This results in more accurate and up-to-date information. Because of the speed of computers, it is possible to prepare a Gantt-chart, for example, every day, but the logic and need for one remains unchanged. The main drawback of IMS is its inability to integrate the engineering phase as well as the production phase. It thus becomes primarily an information system, rather than a working and decision-making system, which is a role that computers should take today.

The IMS does not dwell on technological factors. It is concerned with carrying out, in an economic manner, the engineering decisions as stated in the bill-of-material, routing files, and the available facilities. When using the IMS for facility planning was tried, the result was only a clerical summation of machine load as a result of the routing; the system treats the facilities as numbers and not as machines having a specific technical capability. The IMS is concerned with utilizing machine time, but it does not have the technical tools to tackle manufacturing lead time.

Group technology (GT), is devised to tackle the problem of manufacturing lead time rather than machine time. GT is concerned with facilities, jigs and fixtures, machine layout, and saving duplicate jobs in engineering. GT does not interfere, and does not have any engineering design solution besides the advice of "do not reinvent the wheel," and do use the same solution to as many problems as possible. Most of the ingredients for a breakthrough in manufacturing technology are available, and they function satisfactorily in a limited-application stand-alone approach.

The all-embracing computer-oriented technology combines all the individual developments: computer-aided-design, engineering support computer programs, computer-aided manufacturing, group technology, integrated-manufacturing-system (and computer controlled machines) into one system, called Hal-technology. Many agree with the idea that such a system is required; however, they are sceptical about the feasibility of constructing such a system. They point out the difficulties and the vast amount of effort that it will take to construct the system. Part III of this book describes the Hal-technology. It can be seen in two ways—either as a solution method to the construction of the all-embracing technology or as a hypothetical exercise. It describes a new way of thinking and points out new manufacturing capabilities. Even if the reader does not agree or accept the proposed solution method, he or she will be convinced that the effort in developing an all-embracing system is worthwhile.

Part I of this book introduces and discusses the IMS. This chapter is mainly concerned with the evolution of computer applications in industry that eventually led to the IMS. Chapter 2 outlines the overall system, while each of the subsequent chapters describes separate activities and their interrelationships. The weaker points of IMS will also be pointed out.

1.1 The Evolution of Computer Applications in Industry

The computer industry is a dynamic one. Every three to five years a new family of computers is introduced, each more powerful and less expensive than the previous one. The task-performance capabilities of the computer are continuously increasing, and the leveling off to a steady state is not yet in sight. Likewise, the potential applications of the computer are also increasing and changing form. A review of these possible applications must necessarily include a consideration of the technological development and cost of the hardware.

First Stage—The Application Approach

Initially, computers applications were designed to solve a specific problem, or to supply specific information to a specific department or a person. This approach is known today as the application approach. The purchasing planner, for example, established a need in terms of the specific computer reports required and contacted the Data Processing Department. In doing so, the purchasing planner was forced to draw boundaries around the system and with the assistance of data processing personnel usually had to define the logic and interdependence of data. Stage one represented a stand-alone application and thus did not significantly affect manufacturing.

This approach to data processing was very successful and actually provided the fundamental base upon which the modern technique was built. It introduced the computer to industry personnel and made it possible for the basic elements of data processing and computers to be learned on a day-to-day practical basis. It gave personnel confidence in this tool and the will to expand its use. The success of this approach can be attributed to the fact that the users themselves were actively involved in the design of the system. It was their data and their reports. If the report was not available in time, it meant that they had not delivered the necessary data on time and thus had noone to blame but themselves. This was also true with respect to the content of the data. The users furnished the data and the software logic. Thus, it was as much their responsibility as that of the data processing personnel to produce a good reliable report on time. It was their data, not company data. They would get the credit by being able to provide top management with updated reports on the subjects under their responsibility, with their own interpretation.

The data required for manufacturing purposes or by top management are usually a combination of data from several stand-alone applications. However, such information is not necessarily synchronized. Preparatory work that involves comparing data, investigating any differences, deciding which data are correct and relevant, and only then using them is required. In stage one this preparatory work was usually carried out manually.

This stage and approach, which was essential in the development of data processing, has been misjudged by many professionals in retrospect. They forget the technology, hardware, and general state-of-the-art of data processing in those days. More importantly, they forget the objectives of these stand-alone applications. An inventory control application that was designed with the objectives of keeping track of stock balance, furnishing data for costing, spotting dead stock and slow-moving items, and giving data to the balance sheet cannot be of any significant use to production planning. However, this does not mean that it is a bad system. It is an excellent system if it performs its intended function and successfully meets its predetermined objectives; boundaries must be set somewhere.

Second Stage—Data Entry Integration

As the computer gained use in more and more applications, it was noticed that the boundaries of the separate applications were tangent or overlapping and that captured data could be transferred to more than one application. Thus, inventory transactions could be used both as receiving feedback data for purchasing and as shipping feedback data for customer orders, while job recording could not only be used for computation of wages, but also as feedback to production planning and as basic data for labor costing. The interest in transferring data from one application to another was mutual. The application user could save time and effort by not having to write down input forms for the computer, while data processing saved keypunching work. This stage represented the beginning of

1.1 The Evolution of Computer Applications in Industry

integration, albeit only on a functional level. It was a cross-functional stage in which each function had its own files and programs. The integration was merely in the capture of data and not in its actual utilization. It provided better and more efficient service to the specific function, but not to the company and higher-level management.

The implementation of this stage was not as successful as that of the first stage because three integration problems arose. The first problem involved the updating of data and the control of the reporting date. At this stage each function had become dependent on the primary function—the function that initially collected the data. Purchasing required inventory transactions in order to update order status and thus had to wait for inventory processing to be completed before it could start, the personnel file had to be updated before the job recording application was processed, and so on. The users no longer controlled the reporting dates for their applications.

The second problem was related to the data. Each function viewed the data differently; thus, a piece of information that was vital to one application may not have been at all necessary for the other application. For example, the purchase order number, vendor's name, and item costs are irrelevant to inventory control personnel, and in the application approach the inventory control personnel will ignore this information. The above mentioned information is mandatory to purchasing and bookkeeping. The information appears on the shipping documents, and therefore, in the data entry integration approach, inventory control personnel must record it.

Data processing was the matchmaker between the applications. Data processing personnel had to convince the primary user to expand the input so that it would be of use to other applications as well. Interdependence of applications increased, and it was not so easy to convince inventory users that they could not obtain a report because a particular piece of information that they did not even require (one of the expanded input) had been rejected in the validity test of the input data. It was also not easy to convince purchasing that they could not have their report because the storage location in the inventory transaction was found to be incorrect.

All the users generously shared their own data and were even willing to expand and pass on information that they themselves did not require. However, they did not anticipate that their own applications would suffer. No longer controlling the input or the reporting dates, they became passive and frustrated and had noone to blame but data processing. However, data processing was helpless, since they processed—not generated—data. They could be blamed for one thing only, namely, being matchmaker. In order to ease the minds of the users, data processing was tempted to loosen the connections between applications. This could be done by passing on data from one application to another even if they were not absolutely free of errors, and the users could check the validity of the data by their own criteria. This appeared to be a possible solution, but soon enough it was observed that despite the use of a common input, the common output data were different. Thus, orders accepted by inventory would still be open orders in purchasing, while customer orders that had been shipped and recorded in inventory could still be open orders in the customer orders application.

This situation had also existed in stage one of computer use. However, everyone was aware that it might occur and had acted accordingly. The situation was not as bad in the first stage because there was one single person responsible for the reliability of the data. In the second stage noone was directly responsible and, consequently, data processing was blamed, although they could do nothing about it. This situation could have been prevented by resisting the temptation to loosen the interapplication connection. However, probably no one would have appreciated this, and data processing would have been blamed for stubbornness, inflexibility, and being unrealistic.

The third problem involved the special reports. By its nature, the second stage was an application-oriented, not system-oriented, use of the computer. Management knew that although many applications and much data were in the computer, certain managerial problems posed to data processing could not be solved. However, management was still unjustifiably disappointed when this occurred, even though the systems were doing exactly what they were designed to. They were not designed to solve managerial problems, so there was actually no reason to be disappointed. Management played a passive role, not becoming involved in data processing applications. The single specific function or user was the one to define the scope and boundaries of the application. Management was one among many users of data processing. Most of the data needed to solve the problem were probably present somewhere in the files of the existing applications, but additional data unique to management alone did not exist. Someone would have had to volunteer to furnish this information that they themselves did not require for their own computer application, information that might even be used against them. Information is power, and volunteering information meant voluntarily surrendering some of the power that one possessed by virtue of their function in the company. As a rule, noone could refuse the boss, but the information might be imperfect. However, it was data processing that would be blamed for not supplying the requested report of the desired reliability.

1.1 The Evolution of Computer Applications in Industry

Third Stage—Distributed Processing

Two distinct solutions were proposed for the problems in the second stage. On the one hand, the users learned about minicomputers and became in favor of distributed (decentralized) processing. They gave strong official reasons for abandoning the main company computer, such as poor response to user development needs, inadequate computer services, lack of understanding of the users' need by the Data Processing Department, unreliable data processing computer equipment, and the existence of less expensive, more efficient, small computers. They were, in fact, advocating a return to the first stage of computer development. Top management, on the other hand, saw the solution in a data base, integrated system, and they also had very strong arguments to justify their solution. Management required information that could be generated only from a company-wide pool that included data from the narrower applications located further down in the organizational hierarchy. Management said that it was more feasible and much more efficient to use professional programmers and run the operation from a data base rather than from the separate data files of the specific applications. This would prevent redundancy of data, the reports of all functions would be synchronized in time, and the entire company would be using the same data.

The development of data processing as a manufacturing tool had shown throughout its history that management either did not understand the decision that had to be made at a given point in time or that it could not or would not stand up to the division managers. Thus the third stage in the development of data processing was distributed processing. Some say that this stage was inevitable, since noone is ever willing to learn from the experience and mistakes of others. Stage three involved a return to the stage one application approach, which had all the ingredients of success. Minicomputers were introduced into several departments or divisions of the corporation, thus satisfying the users.

This system began growing with respect to memory size, storage space, printing speed, and unfortunately, cost. Upon reaching a certain size, it possessed all the faults of a centralized Data Processing Department, with subusers tending to have their own minicomputer. The dispersion of computers in this uncontrolled, uncoordinated, disconnected fashion resulted in a dramatic rise of computing cost for the corporation. Duplicate, incompatible data resulted, the large computer became crippled, and top management could not obtain the data processing services it required. One corporation has reported that by 1975 they had 35 small computers. This rose to 102 in 1976 and to 150 by mid-1977.

Fourth Stage—Integrated Manufacturing System

Upon realizing the situation, top management will act and stage four will take place. Stage three is not technological evolution, but rather a "noise" or "disturbance" that comes about due to personality conflicts, uncooperative managers, and lack of knowledge or willingness on the part of top management to learn from and cope with data processing problems.

Stage four is the system integration approach. It recognizes that all manufacturing activities are tied together and are using the same information. Thus information should be available when needed, synchronized, and ready for use without any preparatory work. Such information can provide a flexible basis for production planning and enable the use of a computer in complex manufacturing activities. The organization of information should be on a company-wide basis with the objectives of:

- Eliminating the duplication of data and therefore inconsistencies by representing each data item only once.
- Making the data available to each user in the form requested.
- Developing techniques by which the basic data can be maintained in a current and accurate condition so that information can be provided on demand.

The IMS recognizes the problems of the specific users in stage two and offers them control over their own applications. It also offers these users remote job entry and intelligent terminals by which they may operate the central computer as if it were theirs alone. However, it imposes on the user certain limitations that are required to preserve the integrated system frame for the benefit of the corporation and all levels of management.

For this stage to be successful, it must be understood that system integration is not a data processing slogan or technique. It is, above all, a management technique in which it is realized that the manufacturing process is neither a set of independent systems nor several sets of integrated systems, but one, logical, overall system. This approach can increase productivity, reduce cost and overhead, and meet management objectives. The system can be implemented only by the use of a computer and integrated data base system, it also being required that management take an active role in its construction and implementation. Without management involvement, the chances of success are next to zero, and it is not advisable to begin implementing stage four. It is preferable to return to stage two or even stage one.

Chapter Two
The Integrated Manufacturing System

The integrated manufacturing system (IMS) is a system that recognizes and supplies computer services to each phase of the manufacturing cycle independently while at the same time maintaining a data base that serves as a single source of data for all company activities and applications. Basic data are maintained in a current and accurate condition so that information can be provided on demand.

The scope of the IMS is considered in the first two sections of this chapter. The third section discusses the concepts upon which the system is based, while the fourth section describes the system itself. The last two sections are devoted to the topic of system reliability. Techniques for IMS implementation will be discussed in detail in subsequent chapters.

The IMS must encompass almost all of the activities of the industrial enterprise, but no single profession has been trained to handle a system of such scope. Data processing personnel are qualified to handle such computer-related technical problems as data base organization, but they are not qualified to handle the application aspects, and neither are mechanical, industrial, or production engineers. Probably a new profession is required.

It is very difficult to implement a data base IMS. It is a system that involves both materials and people. Therefore, the active involvement of management is mandatory for the successful implementation of the IMS. In addition, IMS reliability is a necessary requirement, since errors could result in irreversible damage. The subject of reliability is usually neglected, and the literature seldom bothers to elaborate on it. However, since the concept of reliability is as fundamental as an understanding of a system itself, the concluding two sections of this chapter are devoted to this subject.

2.1 The Manufacturing Process

The manufacturing process begins with a set of objectives. These objectives, which are set by management and such nonengineering functions of the organization as sales, marketing, and finance, could be:

- To develop and produce products.
- To produce parts or products designed by the customer.
- To reproduce items that have been manufactured in the past.

The manufacturing environment can differ with respect to:

- Size of plant.
- Type of industry.
- Type of production.

However, the fundamental principles of the manufacturing process are the same for all manufacturing concerns, and thus a general cycle can be formulated. Since each mode of manufacturing is subject to different specific problems, the emphasis on any particular phase of the cycle will vary accordingly. For example, in mass production or continuous-type production, emphasis should be placed on process planning and methods. Auxiliary production aids such as special tools and machinery, jigs and fixtures, and automatic inspection devices should be designed and used. On the other hand, material requirement planning and scheduling usually need not be very sophisticated, since the rate of production is limited or fixed by the initial design. In production of the large-scale job-shop type, where many orders are to be produced in small quantities and only once, emphasis should be placed on scheduling; as a rule, there is not much sense in developing auxiliary production aids or in devoting too much time to methods. Furthermore, in a small-size job-shop specializing in the production of parts designed by customers, the engineering design phase may not be required at all, while in a goods-type industry, emphasis should be placed on the engineering design phase.

Figure 2.1 shows that the manufacturing cycle can be divided into several main phases. Each phase consists of a continuous chain of activities, as discussed in the following sections.

Engineering Design

The purpose of this phase is to transform the objective into a detailed set of engineering ideas, concepts, and specifications. Engineering design

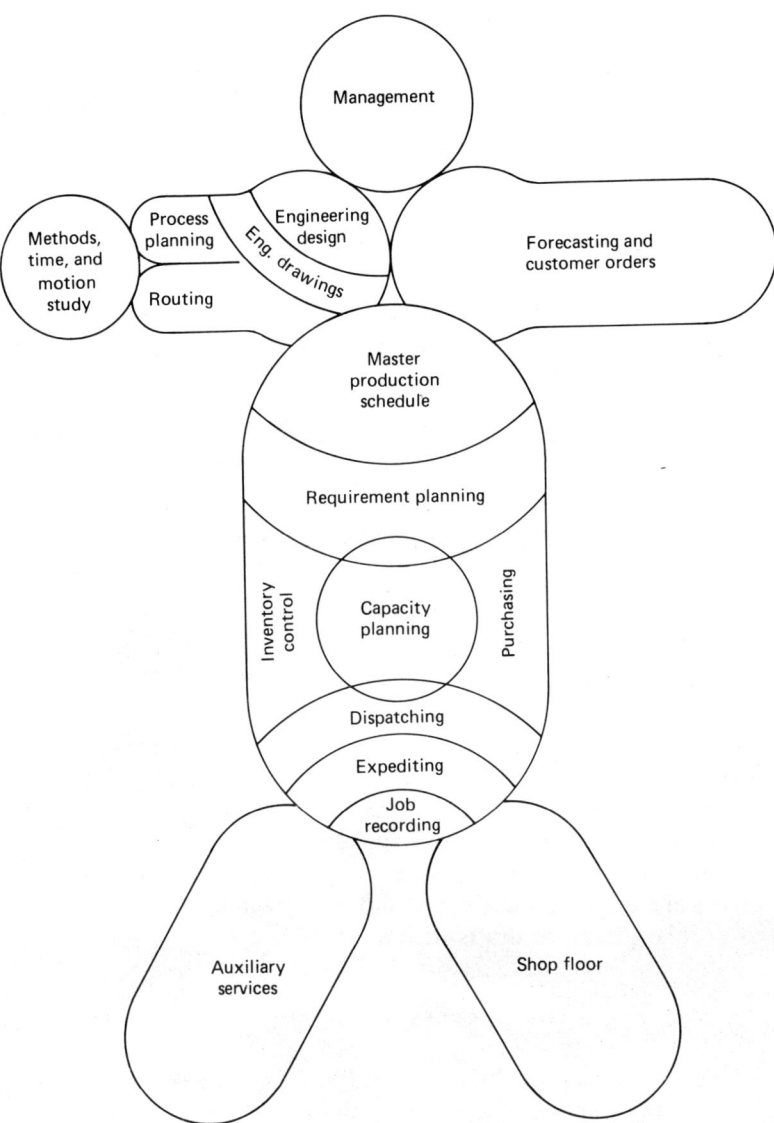

Figure 2.1 The manufacturing cycle.

theories are employed, the objective is translated into engineering specifications, and the engineering task is defined. Thus it is an innovation process. Many ideas and concepts will be formulated and analyzed, and the best conceptual solution will be determined. This conceptual solution will define separate engineering tasks of lower level (and so on) until the last detail of the design is decided upon.

The optimization criteria for decisions made in this phase are for the most part engineering considerations: weight, size, stability, durability, ease of operation, ease of maintenance, noise, cost, and so on. Some of these criteria conflict with each other, and thus the decision will often be a compromise. However, the designer's primary criterion in making a decision is to meet the design objectives. This is the designer's most important responsibility, since errors in production are not as critical as errors in design. To be on the safe side, the designer will tend to incorporate as many safety factors as possible.

The personnel involved in this phase are engineers and designers who are skilled in solving the relevant engineering problems.

Engineering Drawings

In this phase the design decisions reached in the engineering design phase are transformed into a set of detailed engineering drawings and part lists. It is an editing process, constrained by the explicit rules and grammar of engineering language, namely, drawings.

The decisions required in this stage are concerned with layout, the number of projections required, and (in some cases) the assignment of the noncritical dimensions. The optimization criteria for decisions made in this phase are clarity, readability, and unambiguity. The personnel involved in this phase are draftsmen.

Process Planning

In this phase the process that transforms raw material into the form specified by the engineering drawings is defined. This task should be carried out separately for each part and assembly of the product. This phase is basically analogous to the engineering design phase, but here the nature of the objective is different.

Process planning is a decision-making task for which the primary optimization criterion is to meet the specifications given in the engineering drawings. The secondary criteria are cost and time with respect to the constraints set by company facilities, tooling, know-how, quantity required, and machine load balancing. Some of these constraints are variable or semifixed; hence, the optimum solution obtained will be valid only with respect to those conditions considered in making the decisions.

The personnel involved in this phase are engineers and technicians or foremen with long experience in a manufacturing shop.

Methods, Time, and Motion Study

The economical way of performing the operations specified in the process planning phase and time standards are established in this phase. It is a decision-making and computational process, and the optimization criteria are cost and time.

The personnel involved in this phase are industrial engineers and technicians who are familiar with shop practice, personnel, and company standards.

Routing

The flow of work in the plant is prescribed in this phase. It is an editing process constrained by the data of the previous phases and taking into consideration plant layout, storage locations, and the material handling system. The optimization criteria for decisions made in this phase are clarity, readability, and unambiguity.

The personnel involved in this phase are industrial engineers and technicians who are familiar with shop practice and procedures.

Forecasting and Customer Orders

The purpose of this phase is to link sales and management strategy to manufacturing. It represents the driving force behind the manufacturing process that begins with orders and ends with deliveries. The specific type of industry and management policy adopted will determine the particular mode of operation: by confirmation of customer orders, by forecasting, or by both.

Decisions made in this phase are mostly governed by economics and business factors not within the scope of the manufacturing function. However, manufacturing provides much of the data required to arrive at optimum decisions. The manufacturing cycle regards this phase as an objective to be accomplished.

The personnel involved in this phase might include economists, business administrators, and mathematicians.

Master Production Schedule

The master production schedule transforms the manufacturing objectives of quantity and due date for the final product, which are assigned by the nonengineering functions of the organization, into an engineering production plan. The decisions in this phase depend either on the forecast or on confirmed customer orders, and the optimization criteria are meeting due dates, minimum level of work-in-process, and plant load balance. These criteria are subject to the constraint of plant capacity and to the constraints set in the routing phase.

The master production schedule is a long-range plan. Decisions concerning lot size, make or buy, addition of facilities, overtime work and shifts, and confirm or change due dates are made until the objectives can be met.

The personnel involved in this phase are production engineers with operations research background.

Requirement Planning

The purpose of this phase is to plan the manufacturing and purchasing activities necessary in order to meet the targets of the master production schedule. A quantity and a due date are set for each part of the final product.

The decisions in this phase are confined to the demands of the master production schedule, and the optimization criteria are meeting due dates, minimum level of inventory and work-in-process, and department load balance. The parameters are on-hand inventory, in-process orders, and on-order quantities.

The personnel involved in this phase are production engineers and mathematicians who have specialized in operations research.

Capacity Planning

The goal here is to transform the manufacturing requirement, as set forth in the requirement planning phase, into a detailed machine loading plan for each machine or group of machines in the plant. It is a scheduling and sequencing task. The decisions in this phase are confined to the demands of the requirement planning phase, and the optimization criteria are capacity balancing, meeting due dates, minimum level of work-in-process, and manufacturing lead time. The parameters are plant capacity, tooling, on-hand materials, and employees.

The personnel involved in this phase are production engineers, operations research analysts, and mathematicians who have specialized in queuing theory, sequencing, and scheduling.

Dispatching Order Release

This phase serves as a link between production planning and execution. It initiates the productive activities by the issuance of orders to the shop floor according to the program formulated in the capacity planning phase. Although it is mainly an execution task, as a result of shop dynamics immediate decisions concerning required changes may become necessary. The primary optimization criterion in this phase is to supply sufficient work to each work station in the plant and department. The secondary criteria are meeting due dates, minimum level of work-in-process and inventory, and any other parameter specified by the dispatching rules used in the particular plant.

Expediting (Follow-up, Plant Monitoring, and Control)

Expediting is used to ensure that the execution of job orders released to the shop by the dispatching phase will keep as closely as possible to the plan. Although it is mainly an execution task, even with good planning unforeseen interruptions may occur, and thus decisions regarding the appropriate course of action would be required.

The primary optimization criteria are meeting production plans and scheduling, while the secondary criteria are coordination between production and supporting activities (inspection, material handling, and maintenance) and the amount of time operators spend waiting for work. In other words, the goals of optimization can be defined as minimum manufacturing lead time and work-in-process.

Job Recording

Job recording supplies data concerning worker activity to the expediting phase and links manufacturing to costing, personnel, salary, incentive plans, and general management. It is not a decision-making task but rather is clerical in nature, being based on company procedure.

Purchasing

The purpose of this phase is to obtain the required quantity of supplies of the specified quality at the right time; it also serves as a link between the manufacturing data and management functions.

Basically, the purchasing phase can be regarded as a manufacturing department where job orders are issued and items ordered are supplied. However, the decisions that procurement personnel have to make are of a different nature than those made by foremen in the shop. The decisions in purchasing concern selection of a supplier subject to the optimization criteria of quality, quantity, delivery date, and cost. These optimization criteria may conflict with each other (e.g., cost versus quality or delivery date versus quantity), and procurement personnel must find the best compromise, taking into account the constraints of the requirement planning and master production schedule phases, namely, quantity, quality, and time.

The personnel involved in this phase are economists and business administrators.

Inventory Control

The purpose of this phase is to keep track of the quantity of material and number of items that should be and that are present in inventory at any given moment; it also supplies data required by the other phases of the manufacturing cycle and links manufacturing to costing, bookkeeping, and general management.

Inventory control is a clerical execution task based on company procedure. The decisions in this phase are usually confined to choosing the procedure to be applied in any given case.

Shop Floor

The actual manufacturing takes place on the shop floor. In all previous phases, personnel dealt with documents, information, and paper. In this phase workers deal with material and produce products. The shop-floor foremen are responsible for the quantity and quality of items produced and for keeping the workers busy. Their decisions will be based on these criteria.

Auxiliary Services

We should not fail to mention the supporting functions that are essential to the manufacturing industry, namely, inspection, material handling, and maintenance. Each of these three functions has its own responsibility, and this responsibility serves as the primary criteria when decisions have to be made.

2.2 Manufacturing and Industrial Management

A typical organization chart of an industrial enterprise is presented in Figure 2.2. Manufacturing constitutes only one function, albeit a dominant one, since it controls the daily activities of the other functions. However, it represents only one aspect of the activities of industrial management. The IMS must consider all the activities of all the functions in the enterprise.

The objectives of management are to ensure:

- Implementation of the policy adopted by the owners or board of directors.
- Optimum return on investment.
- Efficient utilization of personnel, machines, and money.

In other words, industry must make a profit. Therefore, the optimization criteria for management decisions must be cost, capital tied down in production, and profit; these are collectively referred to as the finance criterion.

In designing an IMS, management objectives should be considered the dominant factor. It should be borne in mind that each phase in the manufacturing cycle has its own objectives and its own optimization

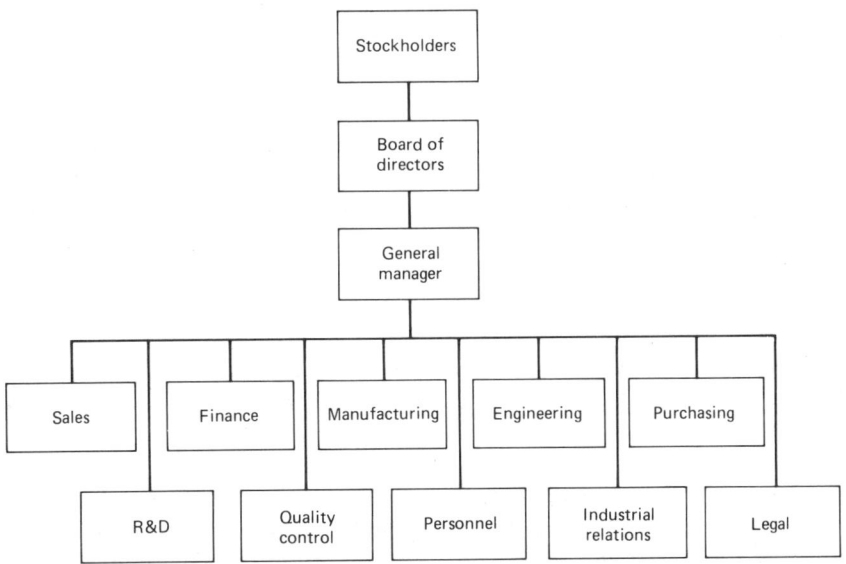

Figure 2.2 Typical organization chart of an industrial enterprise.

criteria. Even if each phase functions optimally, this does not necessarily guarantee overall optimum success with respect to management objectives. Figure 2.3 demonstrates this point.

The manufacturing cycle is a one-way chain of activities, each link having a specific task to perform and the previous link being regarded as a constraint. Thus, for example, master production schedulers accept the quantities and due dates from the customer orders or forecast as a requirement and the routing and bill of material as fixed data; they do not question these data and their planning must comply with them. Process planners accept the product design and its bill of material without question; in fact, they do not even consider the product as a whole, but, rather, the production of each part is regarded as a separate task. Only if problems are encountered in defining the process for a particular part will they turn to the product designer and suggest or ask for a change in design.

Therefore, the chain of activities that comprises the manufacturing cycle is considered as a series of independent elements having individual probabilities of achieving a criterion. The probability of success of any link is independent of every other link with which it is functionally associated. Thus the overall probability of the chain optimally achieving a particular criterion is

$$P_j = P_{j1} \times P_{j2} \times P_{j3} \times \cdots \times P_{jn} = \prod_{i=1}^{i=n} P_{ji}$$

where P_j = the overall chain probability of achieving criterion j
P_{ji} = the probability of achieving criterion j in link i.

Figure 2.3 shows the seven phases of the manufacturing cycle that are subject to the finance criterion; the other phases need not be considered. If we assume an 80% probability of achieving the finance criterion in each of these seven phases, then the overall probability of the finance criterion being optimally achieved in the manufacturing cycle will be $0.8^7 \times 100\%$ = 21%. It should be noted that the ability to predict each probability P_{ji} is difficult at best; therefore, discussion is qualitative only.

Management must have tools (controls) in order to achieve its own objectives. One of these tools is the budget. However, the budget is based on engineering data, the bill of material, and the routing. Thus it can only assure that the outcome will not be worse than the planning; it cannot improve the planning.

Another tool is the organization of the company. A company is organized with respect to key functions, each function being concerned with a different aspect of the operation and fighting for its point of view. For example, sales would want to be able to promise early delivery and competitive prices and thus would favor a high level of inventory and low

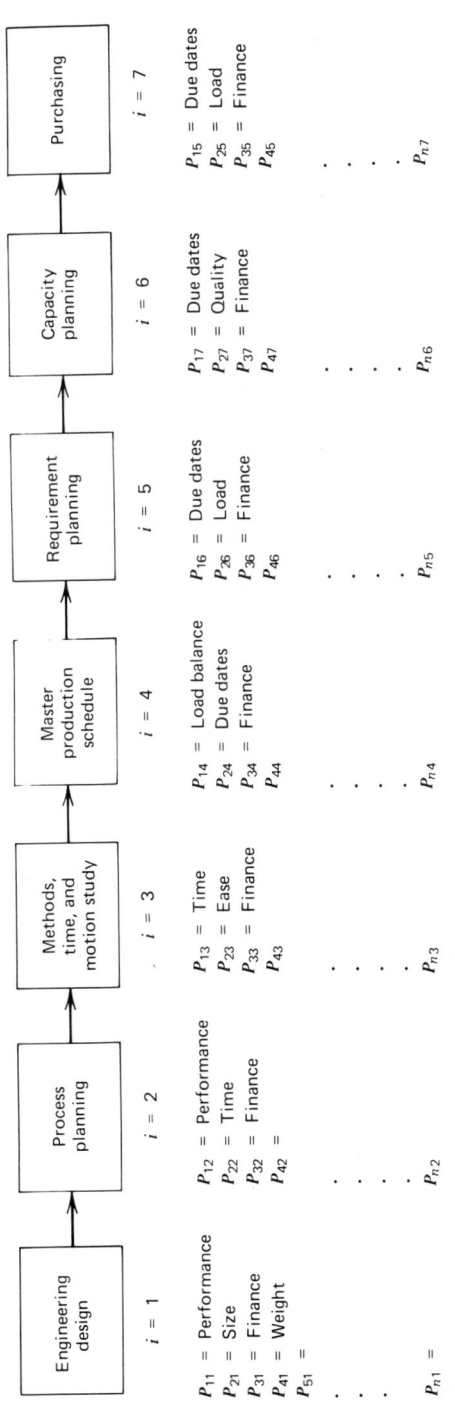

Figure 2.3 Probability of achieving a criterion.

cost of production; finance would prefer a minimum amount of capital tied down in production and thus would favor a low level of inventory and short lead time production; finally, the production manager would emphasize that all work stations have jobs and thus would favor a high level of in-process inventory and long lead times. Only if each of the functions stands up for its own interests will a good balance in the overall operation of the plant be reached.

Value engineering is another tool that management can use to examine and improve the various manufacturing activities. It will usually be used when providing a particular product in response to market demands. Although value engineering is an important tool, it is seldom employed as part of the manufacturing cycle. If it were to become a standard part of the manufacturing cycle, it would be part of the "Establishment" and probably cease to serve its original purpose.

Standardization and simplification are additional tools that can be used to improve the financial aspects of manufacturing. However, they are administrative measures without actual control.

Another approach that management may take is to focus on profit opportunities rather than on efficiency.

Management must have tools in order to exercise control over operations. Manual tools are not perfect, and, consequently, the use of the computer can assist management in performing its task in an improved manner.

2.3 Computerized Manufacturing System Design Concepts

The integrated manufacturing system is based on:

- General data processing concepts.
- Specific manufacturing concepts.

The following general data processing concepts are self-explanatory:

- The system should be management-oriented and not data processing-oriented.
- The system should be adaptable to change, responsive, and economical.
- The system should be reliable.
- The system should reduce paperwork.
- The system should be realistic and consider the environment in which it operates.

2.3 Computerized Manufacturing System Design Concepts

The specific manufacturing concepts require explanation, and thus a brief discussion will precede the statement of each:

1. A computer, like any other machine, has both capabilities and limitations. Industry understands how to utilize machinery to increase productivity, to perform dangerous jobs, and to ease the burden on the worker. Industry personnel are well versed in the operation of available equipment and know how to draw the line that distinguishes between jobs where a person is superior to a machine and those where the machine is superior to the person. For example, a lift track is a machine designed to move loads from one location to another. No reasonable person would lift and move heavy loads manually when a lift track is available; for this task the machine is superior. However, no one would expect the machine—the lift track—to determine either the optimum path of movement between locations or the height to which the load should be lifted. For this task a human is superior.

Machines in general have eliminated the need for manual labor, while at the same time increasing the power with which various jobs can be performed. Advances in technology have allowed control circuits or microprocessors to become part of the machines, thus permitting them to regulate activities formerly governed by human intelligence. The initial step in this direction was the introduction of limit switches and sensors of different kinds with relay control circuits. As the logic becomes more complex, the relays are replaced by solid-state gate elements. Finally, when many sensors are required and a complicated interrelationship logic exists, the gate elements are replaced by a computer.

This type of intelligence has been accepted by industry and is widely used to perform specific tasks. It is agreed that the machine is superior to humans in performing a routine decision job where little intelligence is required and where decisions are based on known parameters and a known algorithm. In these cases the machine will be faster and more reliable than humans, while relieving them of such boring and monotonous work as watching dials and adjusting valves in a chemical plant.

In all these applications the computer is used as a job-performing tool and not merely for data collection and display. There is no reason why the same logic should not be applied to the manufacturing system. Hence, the *first concept* is: *In the manufacturing process the computer should have the role of performing tasks and not merely constitute an information center.*

2. The manufacturing process is a defined quantity, and the task of each link in the chain of activities is known. Which tasks should be performed by computer and which by workers should be decided on the basis of worker–machine task-performance superiority.

A computer is basically a high-speed calculating machine. With today's modern computers, it takes about one-millionth of a second to perform an arithmetic manipulation. A computer can store billions of bits of information and have access to any of them in a tenth of a second; it can be programmed to solve any problem that can be formulated as a mathematical expression or as an algorithm; finally, it can correlate and use any number of parameters and variables desired and in a very short period of time solve the problem posed. With respect to these characteristics there is no doubt that the machine—the computer—is superior to the human mind.

Additional intelligence can be introduced into the computer by means of human-written software, which enables the computer to perform a series of manipulations in any sequence, depending on predefined conditions and rules. The computer will select the right formula for the appropriate set of parameters; it will perform iterations and simulations of any conditions as long as a human can prepare software. If the problem is well defined and the solution can be formulated, a programmer can write a computer program. The software and intelligence of a computer can only be as good as the programmer. The computer does not solve problems or perform tasks, but rather the person behind it. Just as the driver and not the car is blamed for an accident, the same applies to the computer. It enables people to outline the logic and formulas (algorithms) required for the solution of a particular problem and relieves them of the laborious and tedious task of performing the actual manipulations. The computer can do it faster and more reliably. It will repeat the computations precisely, even to the point that the same errors will constantly reappear.

This person–machine (person–computer) combination is a powerful one where both contribute their superior aspects relative to the other for the benefit of the task at hand. The human being does not have to make the same computations and the same decisions over and over again. It is boring to the human mind, and, since it is monotonous work, a high probability of errors exists. Hence, the *second concept* may be stated as follows: *Let humans define the strategy for a solution, and let the machine perform it precisely.*

3. The manufacturing process is a decision-making one. The essential characteristics of a decision situation are:

- An objective.
- Optimization criteria.
- Alternatives.

By an objective, it is simply meant that in order to make a decision, one must have a desired goal.

2.3 Computerized Manufacturing System Design Concepts

Optimization criteria provide the scale that one must have in order to measure the accomplishment of the objective. Besides the primary objective, there are usually many relevant factors (economic, personnel, technical, etc.) that are of secondary importance. It is easy to decide what course of action to take in order to satisfy the primary objective, since many solutions will meet this requirement. However, it will be the secondary objectives—namely, the optimization criteria—that determine the quality of a decision.

Finally, in order to make a decision, more than one course of action must be available (obviously, if there is only one course of action available, there is no choice and thus no decision to be made). The more alternatives considered, the better the quality of decision reached.

A decision is a compromise between the various optimization criteria. Usually, no alternative will satisfy the demands of all the criteria; the decision is in selecting the most desirable (least undesirable) alternative.

The compromise or decision can be made either by employing mathematical techniques or by human intuition. The prerequisite of the mathematical technique is that the criterion function, the constraints, and the controllable parameters be defined quantitatively and formulated.

Included among the available mathematical techniques are optimization theory, differential calculus, Lagrangian multipliers, dual variables, and linear programming. In situations of partial knowledge and uncertainty, the probability and statistical techniques of expected value, variance, standard deviation, truth tables, correlation regressions, and so on can be used. In a variety of value dimension situations, one can employ decision theory and operations research techniques of minimax, the Monte Carlo method, traveling salesman problems, branch and bound methods, and so on. Thus there is quite a variety of mathematical techniques that satisfy the prerequisite.

Some problems and parameters are based not on law, logic, or reason, but on a person's belief or hope. Thus personal relationships between key people in the plant or prejudice on the part of the boss can also represent constraints. Other parameters might be only qualitatively known and thus not incorporable in any algorithm. In this case the decision will be made manually, that is, by intuition. In other cases, even though the problem satisfies the mathematical technique prerequisite, its formulation and solution may still be unfeasible. Intuition would again have to be employed in order to make a decision.

Decisions made by computation with sound mathematical backing are of a better quality than decisions arrived at by intuition. A computer can perform a known series of computations better and faster than any person. Hence, the *third concept* is: *Whenever possible a computer should be employed in decision-making.*

4. In the manufacturing process, the objectives and optimization criteria are defined for each phase of the activities. Alternative courses of action should be defined for each task separately. The alternatives in the engineering design, process planning, and methods phases are innovative in nature and depend on the creativity of the person performing the task. Some improved creative techniques and idea-stimulating methods, such as brainstorming, fantasy, analogy, and systematic search for ideas, have been developed; however, all of them are qualitative and do not involve the mechanical generation of ideas or alternatives. A human is essential here and superior to any machine or algorithm known today. Therefore, the *fourth concept* is: *Tasks and decisions in the engineering design, process planning, and methods, time, and motion study phases cannot be performed by a computer alone and unattended.*

The computer can be of assistance to designers in the mechanical and computational aspects of their job. An interactive, conversational-mode use of computers has been receiving increased attention under the name of computer-aided design (CAD). It is a very helpful time-saver, but it is still the human who plays first fiddle. (See Section 3.7.)

5. The alternative courses of action in the subsequent production phases of the manufacturing process are of a different nature. They are combinable where the population is finite and known. Due to the nature of the problems, a huge number of alternatives can be constructed (in contrast to the limited number of alternatives in the engineering phases).

Actually, the decision-making problem in the production phases is not to generate alternatives, but to evaluate many alternatives within a reasonable period of time. Obviously, this will take longer if done manually. A computer is capable of generating the alternatives and evaluating them far better than humans. Interruptions in production (e.g., a machine breakdown or missing material) pose no problem other than a new situation that requires the appropriate alternatives to be generated and evaluated. Hence, the *fifth concept* is: *A computer should be used to make decisions in the production phases of the manufacturing process.*

6. The conflict between management objectives and the optimization criteria for the production phases has been discussed in Section 2.2. Industry as a whole should work toward management objectives, but the logic of the manufacturing cycle does not assist in meeting these objectives. Many phases, each working toward a different goal, but linked in series in such a way that each accepts and treats the outcome of the previous phase as its input or constraint, will produce a low overall probability of optimally achieving the finance criterion with respect to the total operation.

2.3 Computerized Manufacturing System Design Concepts

This overall probability is equal to the product of the individual probabilities from each phase of the serial connection. The greater the number of phases, the lower the probability. Many tools that management may use in order to overcome this conflict have been mentioned. A theoretical approach from the mathematical point of view involves reducing the number of phases in series. In this case, even if the combined phases have a lower overall probability of meeting the finance criterion, the overall system optimality will be higher.

An example will demonstrate this point. Assume that there are five phases in series, three of them with 80% probability and two with 90% probability of optimally achieving the finance criterion. The overall probability will be

$$P_j = 0.8 \times 0.8 \times 0.8 \times 0.9 \times 0.9 = 0.41472$$

If these five phases can be reduced to two phases in series, one with 70% probability and the other with 80% probability, then the overall probability will be

$$P_j = 0.7 \times 0.8 = 0.56$$

The overall probability has risen from 41% to 56%.

A practical way to reduce the number of phases in series is by combining the objectives of a few phases into one phase. This may be accomplished by integration. Hence, the *sixth concept* is: *An integrated manufacturing control system should be used.*

The integration can be carried out only by use of a computer, since the scope is too wide for manual integration.

7. The integration can include all those phases where decision by computer is superior to manual decisions. In present-day technology, production phases may be integrated and treated as a single phase with a strategy defined by a human translated into algorithm form and solved by a computer. According to the fourth concept, the engineering phases—namely, engineering design, process planning, and methods cannot be included in the integration and will be performed manually. The *seventh concept* follows from this: *The outcome of the engineering phases, bill of material, and routing will be the starting point of the integrated manufacturing control system.*

Part III of this book will introduce Hal-Technology and demonstrate how the engineering phases can be included in the integrated system.

8. All functions in industry require data and information in order to perform their tasks. Many functions often require the same information. For example, the foreman must know what his workers are doing and

what operations have been completed; the expediter must know what operations have been completed in order to move these parts to the next machining center; the capacity planner requires this information in order to replan the production schedule for the following period; cost personnel must be in possession of this information and also know the time it took to accomplish the task; the incentive system must have the same information in order to compute the incentive pay earned by each employee; wage and salary personnel must know the time each employee worked in order to prepare the salaries; and, finally, the controller requires this information in order to control direct and indirect work.

It is extremely important that all of the functions work with the same data, that is, the same language. The bill of material specifies the code (name) for the raw material needed for each part. The same code should be used by purchasing, by inventory control, by production planning, and by costing. In the event that one of them is using a different code, raw material might not be available for production or unnecessary stock might accumulate.

Each of these functions strives to solve its own problems and to gather data for its own use. In order to do so, each will issue instructions to the shop personnel, and most of the time these instructions will be contradictory. Emphasis will be placed on different forms or duplicate copies of the same form will be filled in and distributed to the many functions in the company. In such cases, the shop personnel will usually come to the conclusion that management does not know what it wants. Thus they may either use their own initiative or do whatever the functioner closest to them—or in the stronger position in the company—has asked them to do.

This problem extends even further. Thus whereas inventory control is concerned with the quantity and code number of transactions, the order number has no real meaning for them. Analogously, the cost of material has no real meaning for the production planning and control people, while for the costing department it is indispensable. This dichotomy of interests can lead to negligence in completing all the details in the form, thus rendering them valueless to some of the functions. Despite the use of the same data and forms, and each functioner being in possession of a copy, the quality of the information given to all the functions would not be the same. Integrated systems can solve this problem by elimination. There are no separate functions from the data processing point of view; there is only one data base and one single system that serves the overall needs of the company. Hence, the *eighth concept* is: *Problems should be looked at from a systems point of view and not in isolation.*

Solutions should come from the system. Improvising and solving a problem on the spur of the moment to satisfy the immediate problem will, in the long run, only create a larger problem.

2.3 Computerized Manufacturing System Design Concepts   31

We once had the problem of assigning a code number to forgings and castings. It was decided to regard it as the initial operation in the routing of the parts. Although it was a very good solution for forgings performed at the shop, the purchased forgings and castings remained troublesome, and problems arose in requirement planning and the assignment of inventory code numbers. The perfect solution for an immediate problem turned out to be a very bad decision from a systems point of view.

9. The manufacturing process is basically an engineering and production process (i.e., a creation process), while management in industry concerns itself with the financial aspects of the business. There are many aspects to the business side, some of which are common to all enterprises. Examples of these common aspects include cash flow, credit to buyers and from suppliers, investment, and public relations. However, such aspects as costing, budget, capital tied down in production, and return on investment are unique to the manufacturing industry.

The financial and production aspects utilize the same data and information. Capacity planning is based on the routing and time required for each operation, while the standard cost for each part is the translation of time into money, determined by multiplying the operation time by the work center rate and then multiplying the result by a constant.

The actual cost is the translation of the actual time spent performing each operation. The same data are used as feedback to the capacity planning phase. Production planning, in order to plan its activities, must possess information on the quantity of raw material in inventory and the quantity required for each part.

Translating these quantities into values by multiplying them by the unit price provides the data required for finance planning and control, customer orders, bill of material, routing, inventory, open orders, and all of the basic elements for production planning and control. The same elements may be used to prepare a dynamic budget for the plant, to plan manpower, and to plan purchasing of new equipment.

The data required by the finance functions and management are the same as those required by the production functions, the data being generated and gathered from the shop floor upward. The foreman must be informed on the minutest detail concerning the shop-floor level, from operations being performed to the name of the workers employed; at the capacity planning level, the name of the worker is immaterial, but the part number and operation are important; at the requirement planning level, the product and parts are important, while the operations are immaterial; at the customer orders level, the product is important and the parts immaterial; finally, at the top management level, the overall customer orders or group products are important. Thus the higher you go, the more

32   The Integrated Manufacturing System

concentrated the information becomes. Figure 2.4 depicts the information at different management levels. (This kind of information can be concentrated, but cannot be expanded.) Hence, the *ninth and tenth concepts* are: *The integrated data base system should capture data and information from the lowest source level available,* while *The management and finance systems should be extensions of the engineering and production systems.*

2.4   The Integrated Manufacturing System

The manufacturing integrated system (IMS) is illustrated in Figure 2.5. It can be seen that although separate phases exist, none is independent. Each phase is connected to the other phases by common data, and these data constitute the lowest common denominator. This system will function with respect to both built to order and built to inventory manufacturing; in fact, it represents a compromise between the two philosophies.

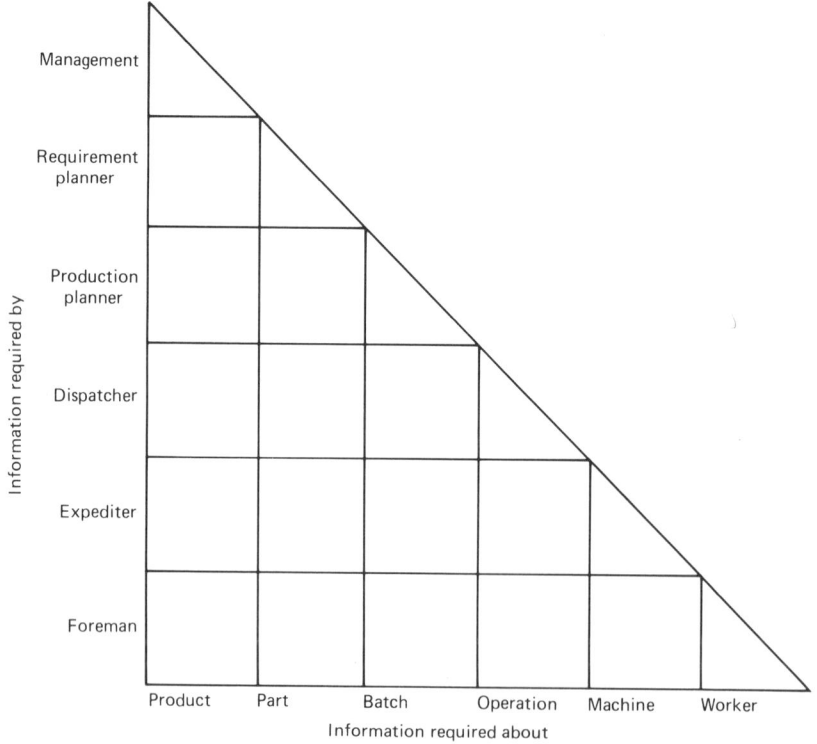

Figure 2.4   Information at different management levels.

2.4 The Integrated Manufacturing System

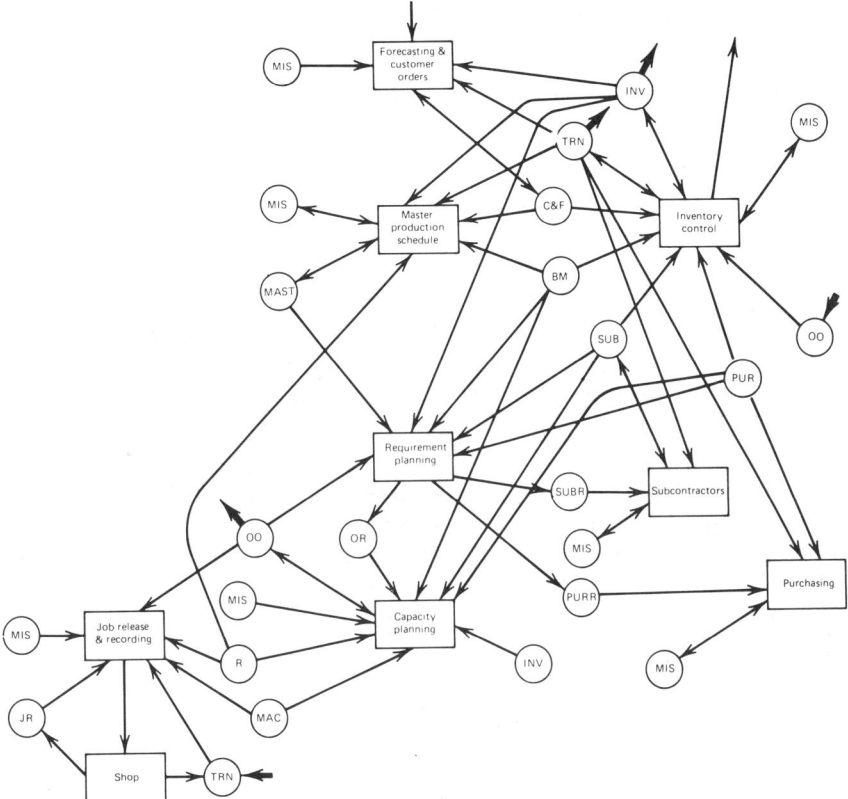

Figure 2.5 The integrated manufacturing system. Notation: MIS, miscellaneous files; INV, inventory status files; TRN, inventory transactions files; C&F, customer orders and forecasting files; BM, bill of material files; SUB, subcontractors files; PUR, purchasing files; OO, open order files; MAST, master production schedule files; SUBR, subcontractors requirements files; PURR, purchasing requirements files; OR, open order requirements files; R, routing files; MAC, machine files; JR, job recording files; □, application; ○, file; →, connection; ➤, same file.

Forecasting can make use of the inventory status and inventory transactions files, earlier forecasting files, and execution; it can also employ such miscellaneous data as sales history, market research, and management policy. This new forecast, together with confirmed customer orders, will constitute the company sales plan for the future.

The master production schedule will use the new forecasting and customer orders file, the bill of material file, the inventory and inventory transactions files, and the previous master plan in order to create a new master plan. It will also utilize such miscellaneous data as department and plant capacity, suppliers' capacity, and cash limitations. In this new master plan the identity of the source of the customer order begins to disappear. It does not schedule to satisfy a particular customer or a specific order; instead, all orders and forecasts for the same product are combined to give the quantity required per period.

Requirement planning is directed by the master production schedule file. It uses the inventory status, purchasing, subcontractors, open order, and bill of material files to compute the net requirement for each component and subassembly needed to satisfy the master plan.

Requirement planning combines identical parts of different products and spare part orders into one manufacturing lot. At this point the identity of the source of the customer order will completely vanish unless special precautions are taken to preserve it. The output of this computation is a list of parts and raw materials to be purchased, to be subcontracted, and to be manufactured in the shop. The list includes quantities, due dates, and connection links between parts, thus forming a product network.

Capacity planning updates the open order file together with the new requirements as computed by requirement planning. Using the data stored in the routing, machine, bill of material, inventory status, subcontractors, purchasing, and promised delivery date files, as well as such miscellaneous data as machine maintenance and breakage, tooling, and employee absenteeism, capacity planning computes the early start, late start, early finish, and late finish dates for all the operations to be performed. All of the above take into account load balancing and available capacity. The output is stored in the open order file.

Job release uses the open order file to select orders to be released for execution. Modifications on a day-to-day basis can be made with the aid of data from the routing, machine, and miscellaneous files. This phase also employs the job recording and inventory transactions files to update the open order file with respect to job status and completed jobs.

Purchasing has its own open order file and miscellaneous files specific to its task (e.g., lists of vendors that give their rating and credit system). New orders come through requirement planning and are received through inventory transactions.

Subcontractor data and orders are similar in concept to purchasing. In many places the purchasing department handles the subcontractors as well as the vendors. The only difference in the system is that it is possible to subcontract a single operation for a self-produced part. The request will come through capacity planning.

Inventory control is the juncture point of all manufacturing activities. It issues and receives components and is therefore closely connected to the open order files of production, purchasing, subcontractors, and customer

2.4 The Integrated Manufacturing System

orders; furthermore, it has its own basic inventory status and inventory transactions files. Inventory control utilizes the bill of material file as well as such specific miscellaneous files as storage location, goods in transit, and inspection.

Thus Figure 2.5 demonstrates that the IMS is basic and capable of solving most manufacturing problems. Moreover, it is also the key to the other systems (e.g., management, financial, and personnel) that operate in an industrial enterprise.

Figure 2.6 illustrates the application of the IMS to the management,

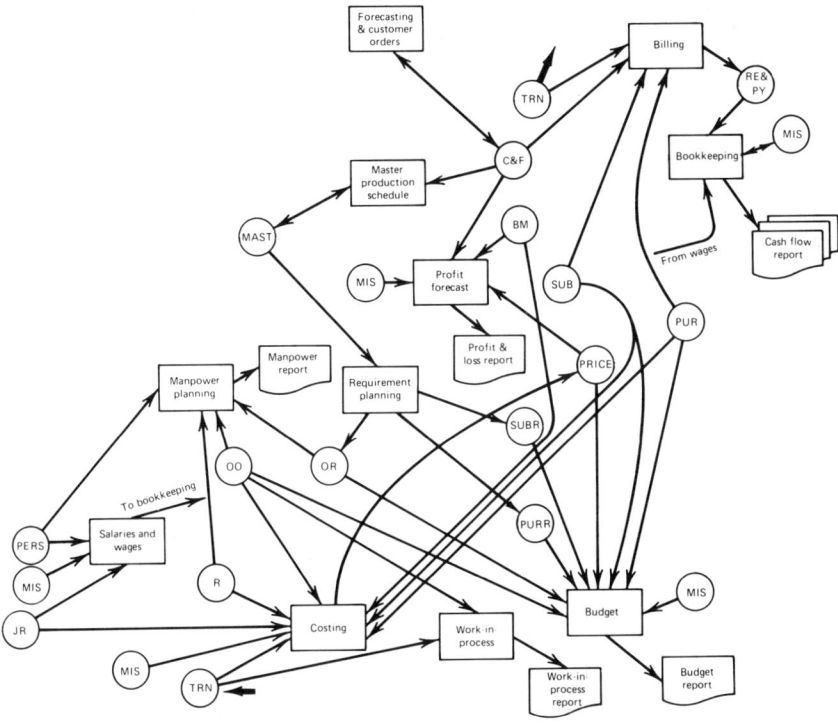

Figure 2.6 Application of the IMS to the management, financial, and personnel systems of an industrial enterprise. Notation: MIS, miscellaneous files; TRN, inventory transactions files; C&F, customer orders and forecasting files; BM, bill of material files; SUB, subcontractors files; PUR, purchasing files; OO, open order files; MAST, master production schedule files; SUBR, subcontractors requirements file; PURR, purchasing requirements files; OR, open order requirements files; R, routing files; MAC, machine files; JR, job recording files; RE&PY, receivable and payable bills files; PRICE, price files; PERS, personnel files; □, application; ○, file; →, connection; ➥, same file, ▱, report.

financial, and personnel systems of an industrial enterprise. Costing uses the job recording files to obtain information on the actual performance time of each operation in the shop, while the open order file supplies the standard time. The variance in time can be computed and then used by the manufacturing system to update the actual time in order to improve standards and obtain a more realistic capacity planning program. The miscellaneous files will contain data on the hourly rate of each department and cost center, and these data will enable both the translation of time into cost and also the computation of the actual cost and cost variance. The inventory transactions file is the source of information on the amount of raw material and number of components required for each part, subassembly, and assembly used in the manufacturing process. It also provides the material element for costing. Furthermore, the standards from the open order file and actual inventory transactions file will again provide the variance of the material element in quantity and cost. The purchasing, subcontractors, or bookkeeping file will provide information on the actual cost of the other costing elements.

Profit and Loss Reports

The customer orders and delivery files contain all the information on quantities delivered and the price of each. The bill of material and price files contain the information on the actual costs of each product. A combination of these data can be used to prepare an early estimate of profit and loss. (In cases where the hourly rate of direct labor does not include overheads, further data are required in addition to the above basic data.)

Cash Flow

Accounts receivable information by amount and date is available in the customer order and inventory transactions files. (This information will be used for billing and bookkeeping as well as in the payment collection system.) Accounts payable information by amount and date is present in the purchasing, subcontractors, and inventory transactions files. Wage and salary information is provided by a separate application of the manufacturing data. Fixed charges can be introduced by additional files.

Wages and Salaries

The job recording subsystem and file contain the data on the overtime, earned incentive hours, vacation days, and so on due to each company employee. The personnel file contains the rate of pay, income tax, special tax, and so on of each employee. Additional data in the special bonus file are used to compute wages and salaries. The job recording file is used to report the hours of direct and indirect labor in each department and cost center and for computing hourly rates of direct work; it also furnishes general management control data.

Dynamic Production Budget

The output of requirement planning is an outline of plant activities expressed in terms of quantities. The price file contains the data required to transform quantities into dollar values. The combination of these files and any additional data provides a dynamic production budget. In addition, an indication can be obtained on the basis of past data concerning anticipated profits.

Personnel

The open order shop file contains all the work scheduled for the departments in the shop. It covers all customer orders and the forecast or management production program for the future. All the work is specified by type of machine and date. This capacity plan may be translated into a manpower requirement by specifying due dates and required skill. The personnel file contains the present manpower status. The combination of these files will generate a variance manpower requirement to be used for recruiting, training, or layoffs.

The ratio of direct labor to indirect labor in each department and in overheads can be used to plan overall company manpower.

The manufacturing system is the basic system upon which the other systems of industry will be based. All of the systems—manufacturing, personnel, finance, and management—require the same reliable data. The activities performed are those actually required to meet the management production plan.

2.5 Design for Reliability of Computer Systems—General

Data processing handles a huge amount of data. A moderate-sized company processes about 750,000 input transactions every month. With this amount of manual input, data errors are inevitable. An error rate of 1% will result in 7,500 faulty input data items; these errors will be compounded to an extent that depends on the number of times these faulty data items are used (some input data items might be used as often as 30 times in different applications and reports). Consequently, an input data error rate of 1% might result in an average total of 75,000 errors in data processing files and reports.

Errors can cause considerable damage; for example, possible results could include purchase of material already in stock or failure to buy or produce an item required for assembly. Some errors produce less serious results, such as paying a debt to the wrong vendor, while others may be simply unpleasant and add to the large stock of jokes told about the stupidity of computers.

Special care is usually taken to reduce the number of errors that originate at the manual input end of data processing. However, the measures employed are often not sufficient, since even an error rate of 0.1%—about the lowest limit that can be expected—is intolerable. People make mistakes and there is nothing that can be done about it. Thus additional techniques must be employed to detect and correct these errors. A good data processing system will have about 70% of its software dedicated to this purpose and only 30% to handling the application as specified by the users. The famous "GIGO" (garbage in garbage out) used by speakers who pretend to understand data processing is not mandated. A good data processing system is cognizant of the error problem and can thus lend a different meaning to "GIGO"—garbage in gold out. This system will be able to discover and correct almost all errors and then proceed with the data processing.

The following rule should be kept in mind when designing a data processing system: *Any bit of information is guilty (of error) unless proven otherwise.*

Errors can originate from three sources:

1. *Data preparation.* Due to carelessly designed forms, there may not be enough room to record the information required. Instructions may not be clearly stated, the forms may be illegible, and errors in filling in data may occur.
2. *Data transfer (form shipment).* Some of the forms that are sent to data processing centers may be lost in the shipping, at the receiving end, or in the Data Processing Department.
3. *Data entry.* Keypunching, being a manual operation, is not error-free. Once the data exist in computer-readable form, it is extremely unlikely that errors will be introduced into the system. It is possible that the computer operator may use a wrong tape. This type of mistake, however, will soon be discovered and is correctable.

2.5 Design for Reliability of Computer Systems—General

Data validation programs should be designed in such a way that they not only detect errors, but also point to the source of the error. It is not of much use to know that out of 1,000 documents sent to data processing, only 999 have been processed. It is important to know which document is missing. Even if 1,000 documents are processed, it is still possible that one may be missing, since another may have been keypunched twice. It is not of great value to know that the balance sheet on bookkeeping does not balance—it is important to know where the fault lies. Finally, it is more important to know which transaction is at fault and what the correct code number is than to know that one of the inventory transactions carries a wrong code number.

In the following sections some of the suggested standard procedures of error detection are discussed.

Numeric Codes

Numeric codes should be used whenever possible. It is much easier to make an error in the alphabetic field and much easier to detect an error in the numeric field; "Bob Smith," "Bob Smith," and "Smith Bob" are three different names as far as the computer is concerned. Handwriting can cause confusion, since some letters may look alike:

	A B C D E F G H I J K L M N O P Q R S T U V W X Y Z
Capital letters	Q O V U
Small letters	E C J l N M
Numbers	6 1

If the use of the alphabetic field is unavoidable, special care should be taken to avoid the use of ambiguous letters for coding.

Keypunch and Verify

Each document is transferred into a computer-readable form by keypunching, which being a manual operation is not error-free. A verifier is provided in order to check punching accuracy. The verifier keys are depressed in the same order as in keypunching, that is, as if the same information was being rerecorded. If the punch position in the record does not correspond to the key depressed by the verifier, an error signal appears. It can then be determined whether the error lies in the punching or in the verification. The verification should not be performed by the same person who did the keypunching.

Check Digit

A numeric field may become a self-checking number if a check digit is added to the right of the basic number (unit position). The two common methods of obtaining the check digit are modulus 10 and modulus 11. In modulus 10, the unit position and each alternate position of the basic number are multiplied by 2, while the other positions are multiplied by 1. Add the digits of the products (not the products themselves) and divide by 10. Subtract the remainder from 10. The result is the check digit.

Example

Basic number	9	5	4	2	4	5	8
Multiply by	2	1	2	1	2	1	2
Product equals	18	5	8	2	8	5	16

Add digits   1 + 8 + 5 + 8 + 2 + 8 + 5 + 1 + 6 = 44
Divide   44 divided by 10 = 4 plus a remainder of 4
Subtract   10 − 4 = 6
The check
digit is   6
The self-checking number is   9 5 4 2 4 5 8 6

In modulus 11, the digits of the basic number are multiplied by factors of 2, 3, 4, 5, 6, 7, 2, 3, 4, ... sequentially (starting from the unit position and going toward the high-order digit). The sum of the products is divided by 11 and the remainder subtracted from 11. The result is the check digit.

Example

| Basic number | 9 | 5 | 4 | 2 | 4 | 5 | 8 |
| Multiply by | 2 | 7 | 6 | 5 | 4 | 3 | 2 |

Add the prod-
ucts   18 + 35 + 24 + 10 + 16 + 15 + 16 = 134
Divide   134 divided by 11 = 12 plus a remainder of 2
Subtract   11 − 2 = 9
The check
digit is   9
The self-checking number is   9 5 4 2 4 5 8 9

In modulus 11, numbers or codes that have a remainder of 1 and, therefore, a check digit of 10 cannot be used. A check digit of 11 (zero remainder) is replaced by 0.

Naturally, everyone is free to invent their own system of computing the

2.5 Design for Reliability of Computer Systems—General

check digit, but modulus 10 and modulus 11 have an advantage in that they can be checked in the keypunch and verifying machines. This is one good feature of the keying equipment. However, if keypunching is regarded as a mass-production operation, the checking of errors should be transferred to the computer, which can detect more errors and furnish a complete error report.

Absolute Check

Each record is checked independently for errors. The checks are made as follows: field—according to the definition of the field in the record, numeric or alphabetic checks are made; the value of the number—for example, in a date field the month is numeric, but not 0 nor greater than 12, and the codes are checked against the permissible-code table; the interrelationship between fields in the record is checked for errors—for example, the inventory transaction of an item with a reject status code cannot have an issue for assembly code; a check to verify if all the fields are present according to codes on the record—for example, a "start of job" code on the record must be followed by the identity of the job.

The absolute check will detect errors in the early stage of processing, thereby saving computer time. Moreover, it will eliminate program checks in the execution of the programs. (A program check is caused by a command to the computer that cannot be processed, e.g., add 20 + Bob.)

Totals and Hash Totals

The forms of critical jobs (such as salary or bookkeeping) sent to data processing should contain a total sum that will be used to compare the sum of the individual records as read by the computer with the given total. If the sums do not match, either an error is present in one of the records, a record is missing, a record has been added, or all three are true. It is recommended that, in addition to the total, a subtotal be given after 20 to 25 records (multiline forms). Since an error might exist in the account number, employee number, or any other leading number, a total and subtotal of these numbers should also be used. These totals and subtotals are called hash totals and are of no particular significance.

Sequence Number

The documents and forms used should carry a sequence number. The numbers may be printed on the forms or a numerator can be used if printing is impractical. The documents sent to data processing should be accompanied by a mailing list stating the first, last, and any missing sequence numbers in the shipment. The sequence number should be keypunched as part of the record. A computer program will check the completeness of documents before processing the data, and an error report will specify what documents are missing or whether there is duplication.

When files are being updated, additional errors may be detected. Some of these errors will be self-detecting, as shown by the following two examples:

1. *Changing or deleting a record that does not exist or adding a record that already exists.* It is obvious here that an error has been made in the key of the data or that the operation code is wrong. Some errors of this type will not be self-detected.

2. *An inventory issue transaction in which the material code is in error, but the wrong material code exists.* In this case two errors are introduced into the file. The quantity balance of the original item will show a greater quantity than is actually present on the shelf, while a lower quantity of the second item will be shown by the file. The following techniques detect this type of error:

- *Zero check.* At some point in time, if not in all transactions, the correct balance is given and keypunched on the transaction. The computer updating program will compare the given balance with the computed one. If an inequality exists, an error message will be displayed.
- *Auxiliary data and files.* It is sometimes a good idea for error-detection purposes to obtain the same data in two forms. For example, with respect to salary, it is useful to state both the rank and gross basic salary. An auxiliary table that includes the correct salary for each rank will be used to check the record. Another example is provided by inventory, where it may be of value to state the inventory item code number as well as the unit of measure.
- *Logical limits.* People, consciously or unconsciously, use their better judgment to evaluate the information they receive. Thus they will automatically detect an error if told that someone has paid $1,000 for a pencil. The computer does not have this quality. For the computer $1,000 is a number and therefore the information is acceptable. Simulation of the human mind should be incorporated into computer programs whenever possible. Thus, for example, with respect to the

performance of a job, an efficiency of 150% and above or of 80% and below should alert the computer to possible errors in the job code, in the time standard used, in the quantity reported, or in the elapsed time reported. If the price of an inventory item recently received is more than 20% of the price previously paid, the computer should again be alerted to the possibility of errors.

Figure 2.7 shows the typical flow of a batch processing job in the computer.

In interactive on-line systems, some of the data validation checks cannot be performed, since each record is treated independently. In these cases we must hope that whoever sits at the terminal will have good enough judgment to detect errors.

2.6 Design for Reliability—The Integrated Manufacturing System

Manufacturing systems must be reliable and error-free. Errors are a crucial factor; they can cause parts to be produced in the wrong quantities, overloading machines that produce parts which will become dead stock, while the required part is lacking. Errors cause irrecoverable waste of manpower, machines, and money. The standard error-detection techniques discussed in the previous section should be employed; however, they are not sufficient. Additional, meaningful checks must be made. The integrated manufacturing system (IMS), whose concepts were discussed in Section 2.3, is a chain of activities in which each stage is dependent on the previous one and connected to the following one. Data are transferred throughout the system (see Figure 2.5) from one application to another. From the beginning point of the manufacturing cycle—customer orders or forecasting—and down the line, activities are planned, dependent, and predictable. The requirement planning output is a list of all the items and raw material that should be subcontracted or purchased; it includes quantitites and due dates. The capacity planning output is a list of all operations that must be performed in the shop; it includes quantities, dates, departments, and machine numbers. If manufacturing proceeds according to the master production schedule and we have faith in the requirement and capacity planning programs, then we expect specific activities to be performed at specific times. Moreover, if any of these activities does not take place, it is a good indication that something has gone wrong in the process, which in itself represents a very valuable piece of information. All of this information is present in the files of the inte-

The Integrated Manufacturing System

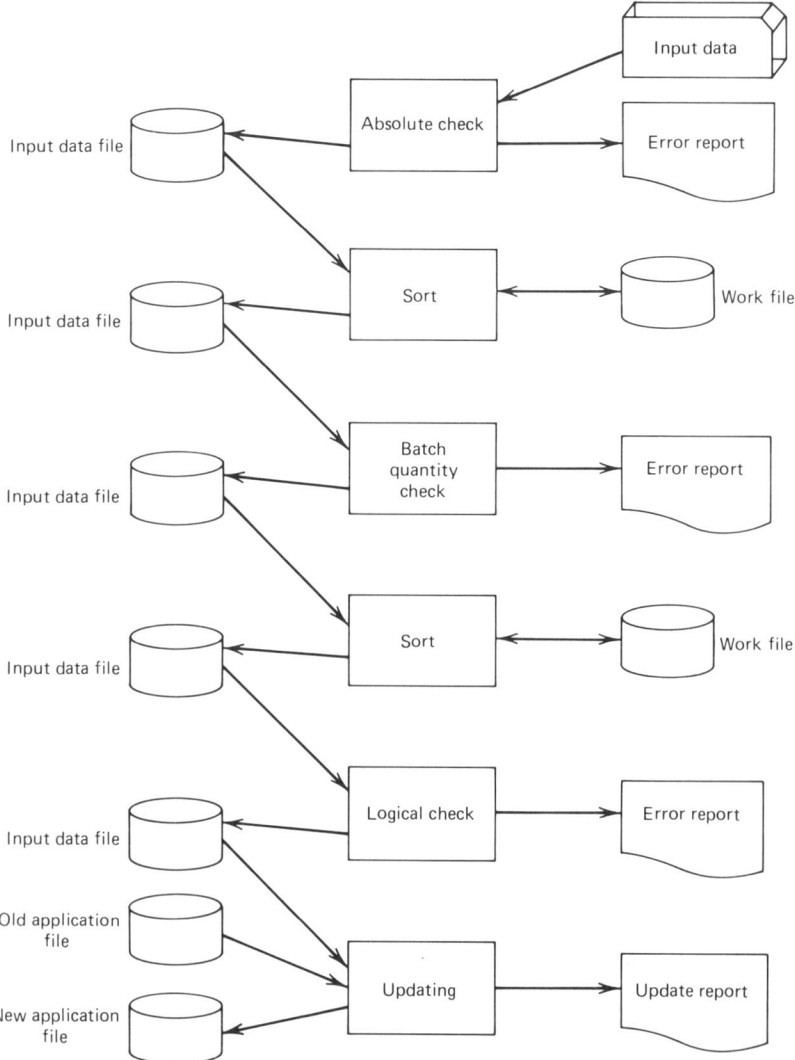

Figure 2.7 Typical batch processing flow.

grated system and can be used to increase reliability, if the following concepts are applied:

- Manufacturing activities are predictable.
- Use "two-way" data processing.
- Eliminate reporting of any input data that can be generated internally in the system. (One does not make mistakes if one does not do anything.)

2.6 Design for Reliability—The Integrated Manufacturing System

The following examples will demonstrate the application of these concepts.

A purchasing transaction with a new order code is compared to the purchasing requirements file of requirement planning. The item number, quantity, and required delivery date of the order must correspond to one of the records in the requirements file; the price stated in the order must be reasonably close to the limits set by the pricing file; and the vendor code or name must appear in the vendor file. Only if all of these checks are satisfied will the transaction be accepted for updating. The new order will be added to the purchasing open order file. At the same time, the requirements file will be updated and the quantity added to the "quantity purchased" field; the pricing file will be updated with respect to the new unit price; and the vendor file will be updated with respect to the order given to the vendor.

An inventory issue transaction is checked against inventory level and with respect to the "action code." The issue quantity cannot be greater than the inventory balance of this item. If it is a shipment to a customer, then it will be comapred with the customer open order file. The customer code or name and the item number must correspond to those in the file, while the quantity must be equal to or less than that ordered. In addition, the date might be added (within reasonable delay or early shipment) to the validation test. Only if all of these checks are satisfied will the transaction be accepted for updating. The inventory file will update a new balance for this item; the customer open order file will be updated by the quantity shipped; and the billing file (and, likewise, bookkeeping records) will also be updated.

If the "action code" is an issue for assembly in shop, the item number will be compared to the bill of material in order to ensure that this item is required for the proposed assembly. The shop open order file will be used to validate the fact that such an assembly has been ordered in the right quantity and for the right date. Extra checks can be made to verify that all the other parts required for the assembly are available. Only if all of these checks are satisfied will the transaction be accepted for updating. The inventory file will update a new balance for this item; the costing file will accumulate the cost of these items in the assembly item record; and the item requirements file will be updated by the quantity issue.

In a job recording transaction, the sequence of operations will be checked and the quantities reported in each operation. The first machining operation cannot start if raw material has not been issued.

The approach described in the preceding examples will, with minimum manual input reporting, provide a reliable and error-free system. However, there are some unpredictable activities in manufacturing, and the system must not ignore them. (These unpredictable activities will be treated in spearate chapters of this book according to the nature of the activity.) A special "action code" will be used to inform the system that a transaction should not be checked thoroughly and that it should not update any other file besides the one directly addressed.

The use of this special "action code" should be limited to genuinely unplanned activities. It can severely damage system reliability and manufacturing activities if used to overcome planned activities that have not passed validation tests, since the interdependent file will not be updated and wrong requirement planning will result. It is recommended that the use of this special "action code" be limited to authorized personnel, since they must approve any unplanned activity that is going to take place.

"Two-way" data processing poses a difficult problem for data processing personnel with respect to the manner of file organization, since it calls for simultaneous access to records of different files, each with a different key. There is no sequence of the required records in the slave files (and sometimes not even in the master file), and updating of any file before all have been checked and input data validated is forbidden.

This requirement of file upkeep is a problem. It can be done by sequential file organization, but only with great difficulty. Sequential file organization is a time-consuming process and data have to go back and forth between files (i.e., updating "on probation"). From the manufacturing point of view, it is a "data base," all the required data being captured and ready for use; however, from the data processing point of view, the sequential file organization may become a source of errors.

A better way to maintain the files is by a special file-organization technique called "Data Base." This type of file organization is of service to data processing personnel only and permits more efficient use of the computer. It has nothing to do with the manufacturing system.

Chapter Three
Engineering in the Manufacturing Process

Engineering data, that is, bill of material and routing, are essential to the integrated manufacturing data base control system. These data are subject to a great number of changes, since for various reasons companies are constantly altering the product design. The addition of new products, facilities, and toolings can lead to changes in the routings or even to the introduction of new ones. This information must be communicated to the manufacturing system and computer programs so that all engineering data are maintained and updated. Since these data are basic, no truly meaningful validation check as to optimality can be made or computed; the system can only check for administrative errors. Optimum product design and optimum process planning are not of concern to the IMS. The field of computer-aided deisgn (CAD) is designed to assist the product designer. It can, without doubt, improve the quality of design and speed up performance; however, up until the present, it was not in any way connected to the integrated system. The IMS, despite its sophisticated mathematical approach and algorithms, can only provide results whose quality is limited by the quality of the basic engineering data on which it operates. While certain design features and certain item dimensions are critical to the operation, others are immaterial and are specified only because the engineer must turn out a unique set of drawings and specifications. It is estimated that about 70% of the dimensions specified, which bind the production engineer, fall into the second category. Under the present-day systems, this situation is not utilized by the production engineer. Hal-Technology, introduced in Part III of this book, is a system that takes advantage of this situation for the benefit of the overall integrated manufacturing process. Experience has made it apparent that these two phases—engineering and production—are carried out by two distinct types of engineers who seldom understand the problems of their counterparts. In this chapter the engineering function will be outlined for the benefit of the production engineer.

3.1 Product Design—Engineering Design

Engineering is the application of available scientific and empirical knowledge in the creation of an appliance or machine meant to perform a given task. This given task is a need defined by either a customer or management, usually in a short qualitative statement. The designer is a problem-solver who applies such fields as physics, mathematics, hydraulics, pneumatics, electronics, metallurgy, strength of materials, dynamics, magnetics, and acoustics in order to find the solution, namely, the new product.

There is no single solution to a design problem, but rather a variety of possible solutions which surround a broad optimum. The solutions can come from different fields of engineering and apply different concepts. The driving power in a machine, for example, can be electric, hydraulic, pneumatic, mechanical transmission, or an internal combustion engine. Among all of these possible solutions, there is an optimum one that is determined by the criteria of the designer.

The designer is bound by constraint conditions that arise from physical laws, the limits of available resources, the time factor, company procedures, government regulations, and morality.

The design is iterative. The designer continually reexamines his previous decisions in the light of new information gleaned as the design progresses. New, random ideas and concepts are applied until satisfactory results are obtained.

The designer faces the problem of predicting the performance of the design. There are many uncertainties, since not every characteristic can be computed on the basis of theoretical scientific knowledge or backed by practical experience. Nevertheless, the designer must make decisions, take responsibility, and hope to achieve an acceptable solution.

Successful designers must be knowledgeable in the following areas:

- Engineering science.
- Engineering processes.
- Engineering materials.
- Engineering costs.
- Standard components on the market.
- Company standards and design policy.
- Competitors' product design.

3.1 Product Design—Engineering Design

They must also be skillful at:

- Problem and task definition.
- Data collection, literature survey, and application.
- Mathematical modeling and manipulation.
- Combining previously unrelated bodies of knowledge to create new combinations.
- Thinking in three dimensions.
- Developing ideas by means of sketches and drawings.
- Anticipating problems in a new design concept, and estimating their ease of solution.
- Making decisions under the constraints of incomplete information and conflicting requirements.
- Communicating to all levels.

Thus a successful designer must have the following attributes:

- Curiosity.
- Creativity.
- Flexibility.
- Openmindedness.
- Patience.
- Judgment.

There are two distinct phases in design work:

1. *Design of the basic concept of the solution.* In this phase designers employ their creativity. They are able to let their imaginations run free and come up with any wild idea. The more extravagant the ideas, the better the designer.
2. *Decision and solution specification.* In this phase designers employ patience and technical know-how. They are constrained by rules, procedures, mathematical equations, and standard communication techniques.

It is not easy to switch one's state of mind and work on both phases; however, it is essential. Frequently, the difference between good and bad design resides in a lack of attention to details rather than in the basic concept. Details are frequently left to junior designers or draftsmen who are not fully aware of the problem and its solution. However, the process planner and the production engineers are obliged to accept these drawings without question. The general view is that designers (of the basic concepts) are born rather than made. Design is a highly individualistic, intuitive process. It is very rare that designers are able to describe how and why they have chosen a particular solution. Nevertheless, good designers try to convey this subjective process objectively by following a certain pattern. This pattern represents a general problem-solving technique rather than a solution to a particular problem.

The design process cycle is shown in Figure 3.1.

Goal

The designer's work must always be directed toward a goal. This goal is usually stated in general nonengineering terms without any implication as to the means to be adopted to achieve it.

Task Specifications

It is important that the designer does not rush into solving the problem as stated in the goal. The purpose of the task must first be understood, and then must be converted into a set of quantitative engineering specifications. For example, if the goal is to design a conveyor belt, it should be realized that the purpose of the endeavor is to move items from one place to another. The conveyor belt is only one possible solution. Another possibility to be considered is rearrangement of the shop-floor layout in order to eliminate the need to move items.

The goal in designing an air-conditioning unit is to create comfortable conditions of temperature and humidity. If the problem is initially stated in broad, general terms, more possible solutions will be considered, thus enabling better solutions to be found.

The second stage is to transform the general terms used in the task

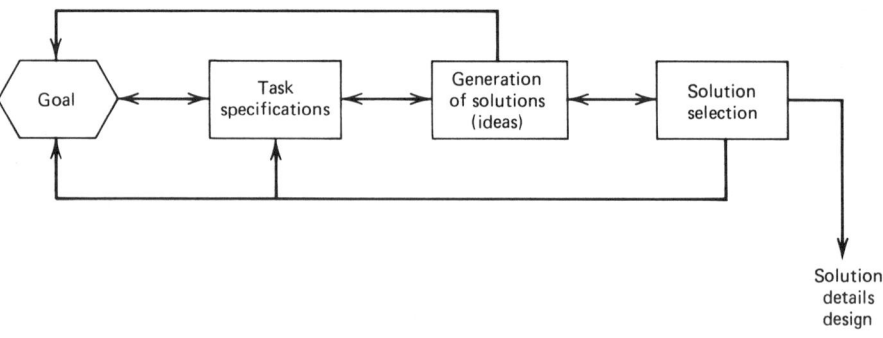

Figure 3.1 The design process cycle.

3.1 Product Design—Engineering Design

specifications into values. This will be done by collecting information and by computations. The term "comfortable conditions" used in the air-conditioning example must be converted into a statement of the form "room temperature of 22°C and relative humidity of 50%." Such factors as room size, the normal temperature in the area, time required to reach the desired conditions, the wall sizes and locations, and the number of people in the room must be specified. In addition, the amount of heat transfer and the air flow must be computed in order to reach a good engineering task specification. The engineering task specification does not worry about air-conditioning; it concerns itself with specified values of heat transfer.

The desired conditions stated above represent the primary objective to be met. However, there are many desirable but not mandatory secondary objectives for the designer to consider. In the above example, these would include minimizing the size, noise level, and cost of the air-conditioning system.

The designer must specify all of the secondary objectives before embarking on the evaluation and selection of the solution. However, whereas all solutions must fulfill the primary objective, they need not achieve all of the secondary ones. In fact, they are usually unable to do so, since many of the secondary objectives conflict with each other. Thus the extent to which the secondary objectives are fulfilled will provide the scale by which the solutions are evaluated. Consequently, it constitutes the difference between good and bad design. In order to be as objective as possible, the designer must rate the relative importance of these secondary objectives before selecting a particular solution. These ratings will later be used to evaluate ideas and decide on the best solution.

In general, secondary objectives might cover the following topics:

- Ease of operation.
- Durability (product lifetime).
- Reliability (low maintenance).
- Efficiency (low operating cost).
- Safety.
- Ease of maintenance.
- Noise level.
- Weight.
- Floor space occupied.
- Aesthetics.
- Cost.
- Ease of installation.

- Ease of storage.
- Ease of transportation.
- Ease of production.

Generation of Solutions (Ideas)

This is a pure innovation process depending on the creativity of the individual. However, few people actually make use of their full imaginative potential. The use of the following techniques can aid in the realization of potential creativity:

- *Awareness that there are many approaches to solving a problem.* Care should be taken to avoid a predisposition to particular methods or ways of thinking. (The literature refers to this as a "set.")
- *Brainstorming.* Criticism and the fear of criticism are known to inhibit creative thinking. The idea-generating stage should be separated from the analysis stage. Creative thinking, together with the multiplicity and free flow of ideas, should be encouraged; in other words, quantity is important. Evaluation is not permitted at this stage.
- *Talking to people about problems.* The simple act of explaining a problem and its difficulties, even to an inattentive listener, can provide a fresh stimulus that allows the problem to be viewed in a new light.
- *Inversion.* Invert the problem. Let moving parts become stationary and stationary parts be put in motion. Turn inputs into outputs. Lay it sideways, turn it upside down.
- *Analogy.* Think of solutions to similar problems in nature, literature, science fiction, and so on.
- *Hypothesis.* Overcome difficulties by supposing that they have been solved and carry on the ideas from this starting point. Imagine situations, hoping that if they generate new ideas, a practical solution will follow.
- *Systematic search for ideas.* Break the problem into subproblems and concentrate on ideas for each stage separately (from the top downward).

Solution Selection

This is a pure analytic process. The designer must apply one solution to the problem at hand. A good designer generates many solutions, but is

3.1 Product Design—Engineering Design

aware that the best solution possible has not yet been found. However, there are time limits, and at this point the designer must select the best available solution and continue to develop it. Thus solution selection is a decision process.

All alternative solutions must achieve the primary objective. Therefore, the overall performance of the alternative solutions is evaluated according to the extent to which the secondary objectives are fulfilled. This creates a problem, since each secondary objective is measured with respect to a different scale and in terms of different dimensions.

The dimensionless decision matrix is one useful tool that the designer can employ to assist in decision-making. Secondary objectives that have similar dimensions (e.g., minimum cost, operating expenses, and maintenance expenses) must be treated by a conventional mathematical technique. The use of the dimensionless decision matrix in such cases will result in biased, illogical decisions.

Figure 3.2 shows the dimensionless decision matrix; it consists of the following:

- *Secondary objective column.* This column lists the secondary objectives that the designer wishes to use in the evaluation.
- *Weight column.* This column lists the relative importance of the secondary objectives. The least important secondary objective is given the value of 1, while the others will have values greater than 1, the specific

Secondary objective (i)	Weight	Alternative solution (j)							
		$j = 1$		$j = 2$		$j = 3$		$j = 4$	
		Rating	Value	Rating	Value	Rating	Value	Rating	Va
$i = 1$	W_1	R_{11}	$W_1 \times R_{11}$	R_{21}	$W_1 \times R_{21}$	R_{31}	$W_1 \times R_{31}$	R_{41}	
$i = 2$	W_2	R_{12}	$W_2 \times R_{12}$	R_{22}	$W_2 \times R_{22}$	R_{32}	$W_2 \times R_{32}$	R_{42}	
$i = 3$	W_3	R_{13}	$W_3 \times R_{13}$	R_{23}	$W_3 \times R_{23}$	R_{33}	$W_3 \times R_{33}$	R_{43}	
$i = 4$	W_4	R_{14}	$W_4 \times R_{14}$	R_{24}	$W_4 \times R_{24}$	R_{34}	$W_4 \times R_{34}$	R_{44}	
$i = n$	W_n	R_{1n}		R_{2n}		R_{3n}		R_{4n}	
Subtotal $ST_j = \sum_{i=1}^{n}(W_i \times R_{ji})$									
Certainty		C_1		C_2		C_3		C_4	
Total $P_j = C_j \times ST_j$		P_1		P_2		P_3		P_4	

Figure 3.2 The dimensionless decision matrix.

value depending on their importance. The number of secondary objectives should be considered when assigning weights.

The assignment of weights is a subjective task. However, designers who have to reach a decision must bear in mind that their decisions and designs will be judged by management and, in the end, by product marketability and customer satisfaction.

• *Alternative solution columns.* These columns list the alternatives remaining after the rough evaluation. Some alternatives, especially if the brainstorming technique has been used, can be deleted simply by making a rough comparison with the other ideas.

• *Certainty row.* The entries in this row are expressed as a coefficient that is less than one. The certainty coefficient permits the inclusion of solutions that the designer is not familiar with, such as those that have appeared in the literature, those that have been used in other fields and problems, or those that appear theoretically feasible. A good designer should be openminded and consider unconventional solutions that have never been tried in practice. However, designers are not scientists, but engineers, and as such they will be judged not by their ideas, but by the results they achieve. Thus to be on the safe side, designers will stick to sure and familiar solutions. The certainty coefficient represents the amount of chance they are willing to take. Its value (expressed in percent) defines the advantage of the uncertain solution over the best solution in the dimensionless scale; hence, it is used to lower the overall rating received by the uncertain solution. For example, a laser beam provides one potential alternative metal-removing technique. The designer is familiar with the literature on lasers, but has never used it in practice. This is no reason to disqualify it as an alternative; however, since it is the designer's responsibility to develop a successful design, unfamiliar alternatives must be regarded as somewhat risky. Thus the designer might decide that only if the laser alternative is superior to the other alternatives by more than 30% will taking the responsibility of using this idea be worth the risk. In this case, the designer will be using a certainty coefficient of 100% − 30% = 70% (or 0.7).

In order to maintain objectivity, the weight and certainty coefficients must be assigned as the first steps in solving the dimensionless decision matrix.

• *Rating column.* The entries in this column will be graded on a dimensionless scale of 1 to 10, where 10 is assigned to the alternative that best satisfies the secondary objective under consideration; the other alternatives will be rated relative to this best alternative in a decreasing scale. More than one alternative can be assigned the same value (including the

3.1 Product Design—Engineering Design

value 10); however, at least one alternative for each secondary objective considered must have the rating 10. Designers must depend on their judgment and estimates, since many details are as yet unknown. It is their job, and they will be evaluated on the basis of their achievements and decisions. They might choose the alternative they prefer or "feel" to be the best. The dimensionless decision matrix is their tool, and they can use it any way they like; it might assist them in selling their ideas to management, but above all it helps them in their quest for a good and objective decision. Designers can also use the advice and judgment of others in the rating of the alternatives if they so desire. Dimensionless secondary objectives, such as aesthetics, will be rated by intuition. The secondary objectives that have dimensions, such as minimum weight and cost, can be rated by intuition or by estimated values. If the secondary objective is minimum weight, the estimated weight can be used in determining the rating, as shown in the following table:

	Alternative				
	1	2	3	4	5
Estimated weight (kg)	10	5	7	5	8
Minimum value = 10 points			5 kg = 10 points		
Compute	$10 \times 5 \times \frac{1}{10}$	$\frac{1}{5}$	$\frac{1}{7}$	$\frac{1}{5}$	$\frac{1}{8}$
Exact rating	5	10	7.14	10	6.25
Rounded rating	5	10	7	10	6

Since the above calculation is only an estimate, whole numbers may be used and the ratings computed and rounded as desired.

The selected alternative is the one for which the following expression has a minimum value:

$$P_j = \left(\sum_{i=1}^{n} (W_i \times R_{ji}) \right) \times C_j$$

where W_i = weight of secondary objective i
R_{ji} = rating of alternative j for secondary objective i
P_j = overall value of alternative j
C_j = certainty coefficient of alternative j.

The solution can be carried out as shown in Figure 3.2. It should be borne in mind that this is not a scientific decision; the weights, rating, and certainty have been assigned values by intuition or estimation. If the differences among some of the alternatives are small, the designer is entitled to choose any one of them, regardless of the overall algebraic value.

Solution Details Design

The concept and general idea of the solutions have been established in the previous stages. This step enters into the details of design.

At the lower level (detail design), new problems will usually arise, new goals will be set, and the design process will repeat itself until all details have been decided upon. The lower the level, the more technical the problems become, for example, how to fasten machine numbers (by bolts and nuts, rivets, welding, gluing, etc.) or how to support a rotating shaft (by sliding bearing, ball bearing, roller bearing, magnetic bearing, etc.).

3.2 Performance Evaluation and Computation

The quality and reliability of the designed product are determined and controlled by the designer. In the dimensionless decision matrix, all-embracing secondary objectives were considered; as a result, the best alternative and its objective rating were obtained by the designer. The quality is a measure of how closely the product conforms to the objectives initially set. Cadillac and Ford, for example, both meet the primary objective of providing a convenient means of transport. However, the two companies intentionally choose different compromises among the conflicting secondary objectives.

Reliability is defined as the probability that the product will perform a required function under given environmental conditions for a specific period of time. The reliability is measured mainly in terms of the failure rate. To avoid failure, the designer applies mathematical procedures. A good designer will distinguish between the "mode of failure" and the "failure mechanism."

The mode of failure is defined by that specific operational aspect of a part or of the whole product which fails to perform or performs improperly. For example, a hydraulic cylinder, has the following mode of failure: It might remain in position when intended to be moving; it might move when meant to be stationary; it might be unable to shift a load; or it might have uneven motion.

The failure mechanism gives the specific causes that produce the mode of failure. In the above example, leakage through the seals, excessive friction, inadequate support to the piston rod, excessive deflection of the piston rod, excessive elongation of tension rod, and so on could be responsible for the failure. The designer can apply the following procedure analysis:

1. Determine the mode of failure.

3.2 Performance Evaluation and Computation

2. Define the failure mechanism.
3. Select the theory of failure.
4. Set up a mathematical model to obtain the relations between variables.
5. Solve the mathematical expression and assign dimensions to the variables.

Simple assumptions, for example, that the materials are homogeneous and ductile, must be made in order to construct the mathematical model describing the physical situation and predicting the behavior of the element being designed. The designer must be aware of these assumptions and decide if they are applicable in the particular case.

In practice, the product may fail for one of the following reasons:

- Human error in the use of the product.
- Human error in design.
- Human error in manufacturing.

Designers will be judged by the final performance of the product and thus will take measures to ensure its proper performance despite the existence of these errors.

Potential errors in design and use of the product can result from the fact that:

- The designer has not foreseen all possible modes of failure; this can be due to inexperience, lack of imagination, inability to visualize how and under what environment the product will be used in practice, lack of time for careful thought, or even plain carelessness on the part of the designer.
- The designer has foreseen the mode of failure, but has been unable to select and set up a mathematical model.
- The designer was successful in the above steps, but has made a mistake in the calculations or in the manipulation of the model.
- The designer was willing to accept a small risk.

Factors of Safety

To ensure against failure resulting from their willingness to accept a small risk, designers provide a margin of safety—the "safety factor." The safety factor is determined as a ratio of the design strength to the applied load and is always greater than 1. For mechanical components it is customary to use a factor of from 4 to 40. The safety factor can be obtained by considering each determining factor separately:

SF	$= FA \, (L_1 * L_2 * L_3 * S_1 * S_2 * S_3 * S_4 * S_5)$
where SF	= safety factor
FA	= working safety factor
	= 1.0 if the failure is not serious
	= 1.2 if the failure is serious
	= 1.4 if the failure is very serious
L_1	= uncertainty in magnitude of load applied
L_2	= nature of applied load
L_3	= uncertainty in load distribution
S_1	= variation in material properties
S_2	= manufacturing effects
S_3	= environmental effects
S_4	= stress concentration
S_5	= design assumptions.

The following values are recommended for $L_1, L_3, S_1, S_2, S_3,$ and S_5 in the literature:

Very good	1.1
Good	1.3
Fair	1.5
Poor	1.6

The following values are recommended for L_2:

Light shock load	1.2
Medium shock	1.5
Heavy shock	2–3
Impact	>3

Another approach to the selection of the safety factor is illustrated in Figure 3.3. Both the load capacity (strength) and the actual load (stress) are not fixed values, but due to the nature of the design have a certain distribution around a mean value. The specific shape of the distribution curve depends on the particular problem. The safety factor is defined as the ratio of the mean load capacity to the mean actual load. The overlapping area of the distribution curves indicates the probability of failure. The designer can choose any desired reliability value by using statistical theory to compute the corresponding safety factor.

An excessive safety factor gives the designer peace of mind and security; however, it can also give rise to severe penalties in weight, size, and cost.

3.2 Performance Evaluation and Computation

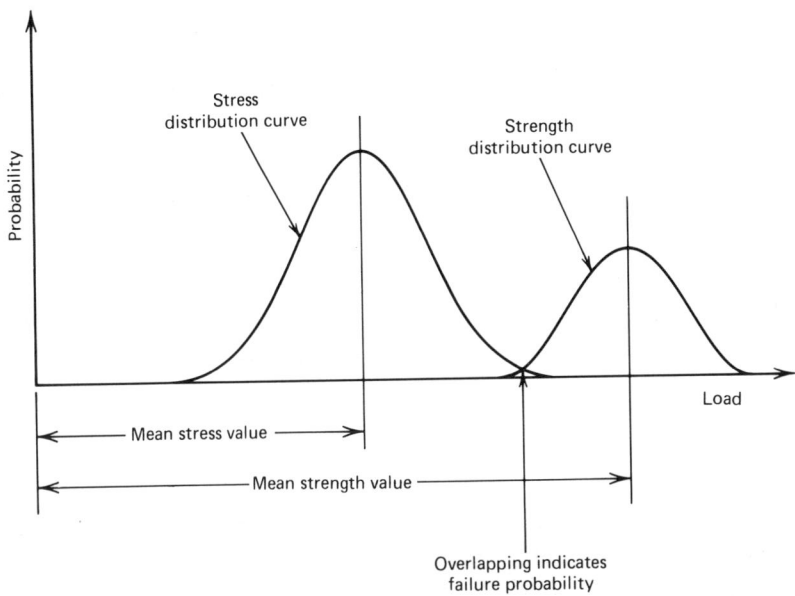

Figure 3.3 The relationship between the distributions of strength and stress.

Tolerances

Potential errors in manufacturing can either affect production performance, product life, and product assembly or have no significant effect on the product at all. In manufacturing, it is impossible to make each dimension and characteristic agree exactly with one specific value. Every element will deviate from the theoretical dimension to some extent. In many cases even gross deviation from component geometry and characteristic can exist with no significant effect on product performance. On the other hand, in some cases a microscopic deviation can have a catastrophic effect.

To ensure against failure due to human error in manufacturing, the designer specifies the permissible deviation, that is, the acceptable range of values. In other words, the designer specifies a tolerance.

In mechanical parts there are three types of dimensional characteristics which need to be controlled by tolerances: (1) size, (2) shape, and (3) location. There are three classes of fit between mating parts (e.g., shaft and holes):

1. *Loose fit.* Used for dynamic fit.

2. *Neutral fit.* Used for static fit with no load.
3. *Tight fit.* Used for static-fit loaded parts.

There are two methods of applying the tolerances:

1. Basic hole system.
2. Basic shaft system.

Which system is adopted depends on the method(s) of processing and the state of the raw material prior to processing. A hole is usually made by fixed-size drills, reamers, broachers, and so on, the size being controlled by the tool and not by the operator. A shaft is made by such adjustable tools as cutting devices and grinding wheels. Since in this case the size is controlled by the operator, the basic hole system is most commonly used. The basic shaft system is preferred when, for example, cold-drawn finished shafts are used as raw material and no machining is carried out prior to use.

There are two systems of specifying tolerances:

1. *The Unilateral System.* Here the tolerance is applied only to one side of the basic dimension—plus for the hole and minus for the shaft. The basic dimension is the minimum size of the hole or the maximum size of the shaft.
2. *The Bilateral System.* In this system the basic dimension is the designed size of the shaft or hole, the tolerance being equally applied to both sides.

The Unilateral System allows tolerances to be increased or decreased without the basic dimensions of the drawing being affected; therefore, it is the most commonly used system.

The grade of fit defines the relative looseness or tightness. The tolerance is simply a tool employed by designers to ensure that the actual part corresponds to the design. Their interest lies in the "allowance," that is, the allowed difference in dimensions between the mating parts. Designers may specify and divide this allowance between the shaft and the hole in any way they choose, as long as the total allowance is preserved. Figure 3.4 demonstrates this freedom. In making their choice, designers will usually refer to the standard systems of tolerance and charts that are so well represented in the literature.

Tolerances are applied not only to diametric dimensions, but also to longitudinal dimensions and assemblies. The calculation of tolerances is always based on the allowances. For example, an assembly of five components in a row may have an allowance of 0.030 in. in the assembly. In

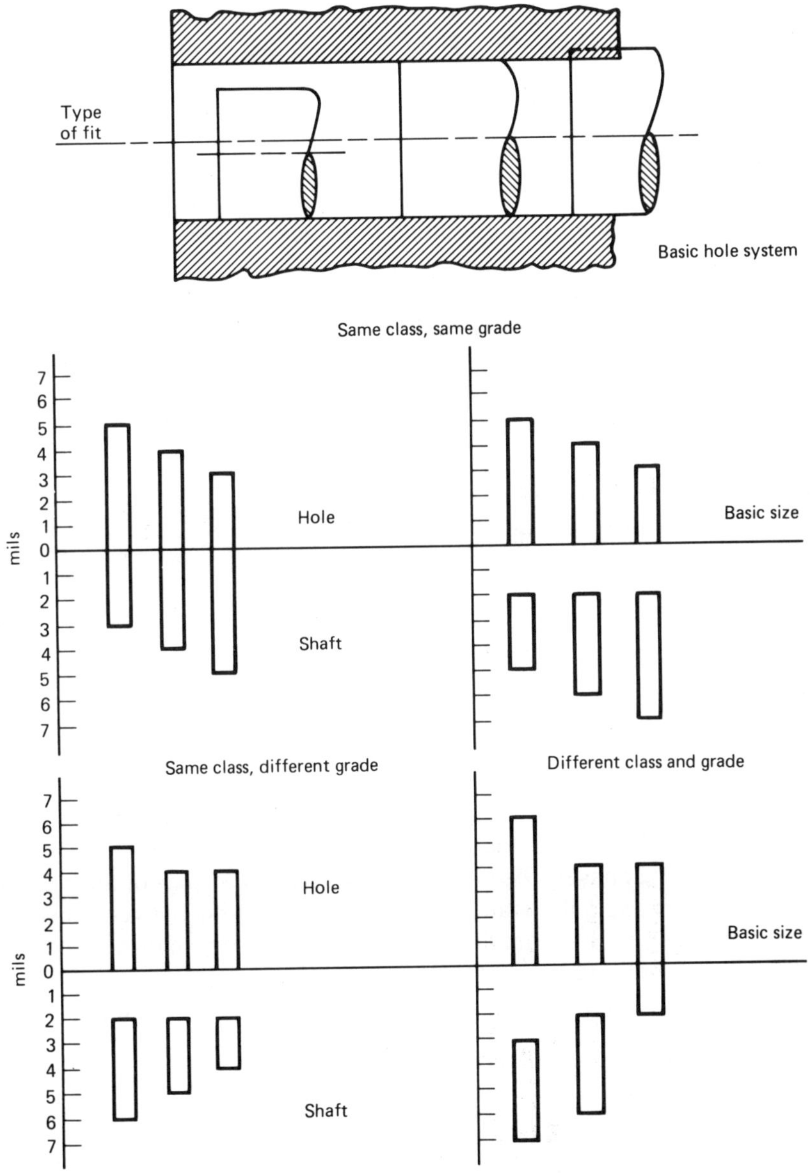

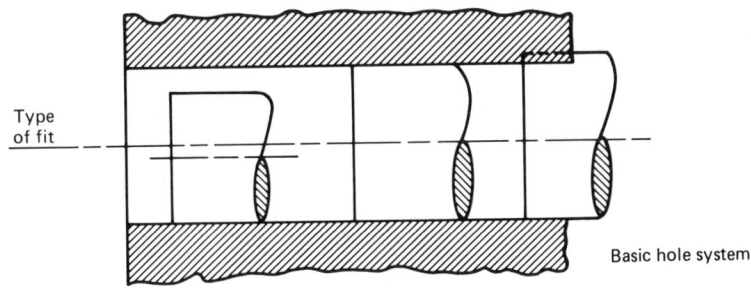

Figure 3.4 Tolerance system.

order to assign a tolerance to each component, an even arithmetic distribution can be used and each component assigned a tolerance of 0.030/5 = 0.006 in. If an uneven distribution is used, the arithmetic sum of the component tolerances will equal the assembly tolerance. This type of distribution can result in tight tolerances.

If the components are manufactured individually, and each one deviates along the full range of the tolerance, statistically controlled conditions prevail. In such cases statistical tolerances can be used. The principle involved is that the standard deviation of an assembly is equal to the square root of the sum of the squares of the standard deviations of the dimensions involved. In the tolerance field this principle is expressed as follows: *The assembly tolerance is equal to the square root of the sum of the squares of the component tolerances.*

Thus, in the example considered above, the assembly tolerance—which consists of five component tolerances—is 0.030 in. If evenly distributed, each component tolerance (TL) will be as follows:

$$0.030 = \sqrt{5 * (TL)^2}$$
$$TL = \frac{0.030}{\sqrt{5}} = 0.0134 \text{ in.}$$

This is considerably different from the component tolerance of 0.006 in. obtained with the arithmetic distribution.

Tight tolerances afford the designer piece of mind and security; however, they also raise the cost of the product. This is an area of constant conflict between the manufacturing department and the designer, the former preferring to work with loose tolerances. However, the question of tight versus loose tolerances is of no interest to the production and process planning phases; their concern is with the drawings of the components only. Figure 3.5 illustrates the general relationship between cost and tolerance, using as an example the percent increase in cutting cost as a function of improving the surface finish.

The increase in cost with tigher tolerances is due to the following:

- More machining operations.
- More rejected parts.
- More expensive gauges.
- More expensive quality control.
- More expensive machining tools and fixtures.
- More set-up time for machining.
- Higher wages (skilled labor is required).

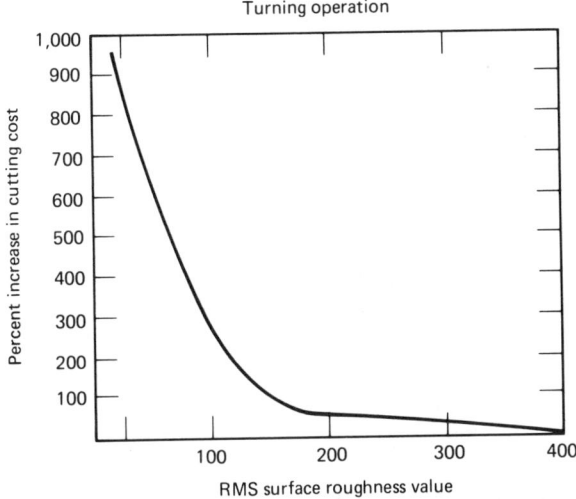

Figure 3.5 Cutting cost versus surface finish.

Basic Dimensions

Mathematical models and computations are used to assign basic dimensions to components. It is essential that the component be uniquely defined both geometrically as well as with respect to material; nothing should be left to the decision of the operator in the shop. However, many component dimensions are completely unimportant; large variations can exist without affecting product performance. When no parameters exist (as is true, e.g., with respect to the size of clearance holes, air vent holes, peepholes, cover plates, and handles), a mathematical expression cannot be selected. In other cases, a mathematical module can be selected, but the number of variables exceeds the number of equations and hence the mathematical equations cannot be solved unless the designer arbitrarily selects values for some of the parameters (variables).

For example, the tension σ that a simple mechanical member is subject to can be computed by

$$\sigma = \frac{P}{A}$$

Where P is the applied load and A is the cross-sectional area. Although the area is composed of two dimensions, only their product is important in determining strength.

Engineering in the Manufacturing Process

In helical-spring design, the equation for the force gradient κ is

$$\kappa = \frac{Gd^4}{8D^3N}$$

where G = modulus of rigidity of the material
d = wire diameter
D = mean coil diameter
N = number of active coils.

Any combination of these four variables which produces the desired value of the force gradient will be satisfactory. We could try to optimize and state that the weight of the spring W should be the minimum, the weight being given by

$$W = \rho\,(\pi\,DN)\left(\frac{\pi\,d^2}{4}\right) = \left(\frac{\pi^2}{4}\right)\rho\,Dd^2N$$

where ρ is the weight density of the material. We could also try to add the constraint of natural frequency; however, there are still too many degrees of freedom. The designer would discover that in order to solve this problem, the largest possible arbitrary value would have to be assigned to D—the mean coil diameter—so that the other variables could be solved for optimally.

The term ampere-turns (NI) often appears in equations relating to electromagnetic devices. The product of the current and number of turns in the coil is important, and the designer must specify the value of each one separately. Even optimization does not assist much in this case; the designer must arbitrarily define one variable in order to be able to proceed.

In the sense of the preceding examples, constraints are very useful to designers, since they allow them to formulate additional equations and thereby reduce the degree of freedom they have. Standards are regarded as constraints; the drawing precisely specifies the components and is binding on manufacturing and production. However, the manner in which each dimension, characteristic, and tolerance has been assigned and the reason for it should not be forgotten.

3.3 Standardization

There are several types of standards which the designer must consider:

- National and international standards.
- Market standards.
- Company standards.

3.3 Standardization

Standards specify the current engineering practice in common design situations. In effect, they save designers time and effort by making routine decisions for them; they reduce cost and the unnecessary variety of common components, thus allowing special mass-production machinery to be used; they resolve possible legal complications in cases where safety is an important factor; finally, standardization also increases the marketability of a product by making it possible to interchange it with other available products. Approval by official bureaus is an additional advantage. However, in advanced, sophisticated designs, the designer must be careful not to permit excessive use of standards to hinder improvements in engineering. Standardization restricts progress by tending to stabilize conditions at an existing level.

National Standards

There are two types of national standards. The first is regulated and enforced by the Government to protect public safety (e.g., building codes, boiler codes, and safety devices in automobiles). These standards are mandatory, and the designer must comply with them.

The second type guarantees quality to the purchaser. It is a government service to the public executed by such agencies as the National Bureau of Standards. Although these are voluntary standards, the stamp "Approved by the National Bureau of Standards" carries considerable weight in marketing.

The National Bureau of Standards will test and establish working standards at the request of private industry, other government agencies, or consumer organizations. It will, as in the case of dry cells and batteries, establish standard values, interchangeable sizes, and minimum battery life. The standards are developed by business and consumer groups and usually represent a compromise between the demands of consumers and the restrictions of manufacturing. Product standardization frequently protects and benefits consumers by assuring them of products that are interchangeable, uniform in quality and performance, and often fairly priced.

Market Standards

There are many common problems in design, such as how to fasten components and how to transfer motion. A single solution can often be found to a set of problems; for example, a screw is a solution for fastening components, a gear is a solution to the problem of transfer of motion between shafts, a bearing is a solution for how to support a shaft, and a resistor or potentiometer is a solution to the control of voltage drop and current flow. Each of these components requires special know-how with respect to design and special, economical production machinery. There is not much sense in each plant and designer manufacturing their own items. Hence, such organizations as the American Standards Association (composed of industry representatives, engineering societies, and several government departments) coordinate and approve engineering and industrial standards. Plants have been specifically built to produce these standard items; they have the most qualified designers and the appropriate equipment to manufacture these items, while at the same time providing catalogs that describe their properties and explain how to select the correct item for a given job. Thus the designer's work is confined to selecting rather than designing standard items or standard materials. The idea of standard components and their use is of great benefit. Their application is spreading from components to assemblies and to units of a different nature, such as electric motors, hydraulic pumps, cylinders, valves, power supplies, amplifiers, and gearboxes.

The range of standards is very wide, and designers have a certain degree of freedom in selecting the standard items. If for any optimum consideration the design has to be modified, it is highly probable that it will be possible to use an alternative standard item.

Company Standards

There are three types of company standards. The first type is procedural, and it includes drawing sizes, coding systems, and tolerance policies.

The second type encompasses the component and dimension standards that the company has adopted; it is a subset of the general market set. The sizes of holes, bolts, screws, fittings, and so on are standardized, thereby reducing the inventory level of these items; tooling, inspection gauges, jigs, and fixtures are also standardized.

The third type of company standard relates to manufactured products. Its primary benefit consists in allowing the utilization of company know-how to reduce the number of items produced, while simultaneously maintaining product diversity and design flexibility and reducing the time required to develop a new product.

To give a few examples, a company that produces the linear variable differential transformer (LVDT), or is familiar with its technique, will base a complete line of measuring instruments on this item; however, a second company that produces strain gauges, or is familiar with their technique, will base the above line of measuring instruments on strain gauges. A company that produces home appliances will standardize electric connectors, device attachments, mechanisms, and so on. Finally, a company that produces cigarette lighters will utilize the same ignition mechanism in several series of lighters.

The establishment of company standards is a good procedure for ensuring optimum efficiency. However, individual designers may find that their particular designs are constrained by these standards or that they do not have sufficient freedom in design. It is up to them to convince management that the standards are too restrictive and costly, and thus should be modified.

3.4 Design for Production

The product designer should bear in mind the manufacturing process that will be used to produce the designed part. Each manufacturing process has its advantages, capabilities, and limitations. The cost of a part can be kept to a minimum if its features and dimensions correspond with the capabilities of one of the available processes. Otherwise, the cost might be excessively high or production impossible. Designers do not define the process plan, but rather steer toward utilization of existing processes, preferably one available in their own plants.

The manufacturing process can be divided into the following categories.

Forming from Liquid-Casting

To form a part by liquid casting, the raw material is heated to its liquid state and then poured into a mold of the desired form.

Casting is the most economical method for producing complicated shapes in quantity; however, not all materials are castable. Casting is susceptible to internal porosity resulting from shrinkage and the presence of gas. The flow of material in the mold in thin channels (wall thickness) is a problem that should be considered. The cost of the mold is high, and the dimensions and surface finish often cannot be kept to tight tolerances unless die casting or precision casting is used.

Forming from Solid

This type of forming can be divided into three subgroups:

- Hot working, including hot rolling, forging, and extrusion.
- Cold working, including cold rolling, cold extrusion, stamping, bending, spinning, stretch forming, and shearing.

- Forming from powder, including powder metallurgy and plastic moulding.

Rolling is the cheapest method for shaping metals. Rolling into bars, plates, or sheets is executed by passing the ingot between rolls that grip the metal and draw it through, compressing it and reducing its cross section while increasing its length.

Not all materials are formable. The process is limited to simple shapes; it is susceptible to changes in such metal properties as tensile strength, hardness, and ductility; and crack development during the process can be a problem. Dimensions can be kept to tight tolerances.

Forging is defined as the working of a piece of metal into a desired shape by hammering or pressing, usually after heating to improve its plasticity. The metal may be shaped by drawing out, which decreases the cross-sectional area and increases the length; by upsetting, which increases the cross-sectional area and decreases the length; or by squeezing in closed impression dies.

Forging has better mechanical properties than casting; therefore, parts that must withstand severe stresses are preferably made by forging. Some metals are more difficult to forge than others. Die costs for forgings are generally higher than the cost of the casting molds, but forgings usually sell at a higher price than castings. However, many intricate and cored shapes possible in casting cannot be forged. Closed impression die forgings are limited with respect to size. Defects in forgings can result from faults in the original metal, die design, heating, or forging operation. Dimension control is difficult due to shrinkage, die wear, or die strikes out of alignment.

Spinning and stretch forming are better suited for small-quantity production. Some of the work executed cannot be duplicated by other processes. Thin wall dimensions can be controlled, and if done properly no change in material properties will occur.

Powder metallurgy is a technique by which metals or alloys are first produced in powder form and subsequently made into metal parts by die and press. It is possible to combine metallic powder with nonmetallic powder to produce useful material combinations, such as cutting tools, electric motor brushes, and self-lubricating bearings. The chemical analysis of the parts is closely controlled, although not all materials can be processed.

A high production rate can be obtained with the powder metallurgy technique, although the initial cost of the dies is high. Tight dimensional tolerances and good surface finish can be maintained, while highly skilled labor is not required. There is a size limitation due to press and die requirements, and the raw material is expensive. Intricate shapes cannot be produced, and complicated dies may be required for some parts that initially appear suitable for this technique.

Forming by Metal Removing

This technique encompasses all types of metal-cutting operations such as turning, drilling, boring, shaping, slotting, milling, grinding, broaching, sawing, and reaming. The machines employed are usually universal machines that utilize universal and commercial tools. Tight dimensional tolerances and good surface finish can be maintained, while no change in metal properties occurs during manufacturing. The metal-removing process is best suited for the production of small quantities. However, with special tooling, jigs and fixtures, automated versions of the machinery, machines of the numerical control type, and transfer lines, the process may be adapted to a large-quantity, mass-production technique.

Fabrication by Joining Parts

This technique involves the use of welding, brazing, soldering, and adhesives to join several components into one complicated part. It is often used in conjunction with a metal-removing process to eliminate excessive waste of raw material. Welding results in a strong bond between components; however, since it is a high-temperature process, the metal being heated to the melting point, distortion, shrinkage, and changes in chemical properties can occur. Not all materials are suitable for welding; usually, if tight dimensional tolerances and good surface finish are required, extra operations should be applied. Good welding joints will have at least the same strength as the components themselves.

Soldering and brazing result in weaker joints than welding. They use filler materials whose melting points are lower than those of the component materials, and both can join a wide range of metal combinations.

Adhesives can be used to join different types of materials, for example, metal to wood, plastic to plastic, and metal to metal. A strong joint, with no distortion or shrinkage, can be obtained. Rubber-base adhesives produce elastic and vibration-resistant joint parts. Special surface treatment is required, and the designer must realize that the strength of the joint is not uniform in all directions (usually, the joint is strong in shear and weak in peel).

Heat Treatment

This technique includes such treatments as annealing, normalizing, quenching, tempering, carburizing, induction, and hardening.

Heat treatment involves the heating and cooling of a metal alloy in order to increase the hardness of the material, soften the material (annealing), relieve various stresses induced in processes, alter the structure of the material (thereby improving machinability), and improve the physical properties of the material.

Assembly

This technique involves the mechanical joining of parts by threads, bolts, rivets, and so on.

Selection of the Process To Be Used

The following design factors have a bearing on the selection of an appropriate manufacturing process:

- Selection of material.
- Size of part.
- Complexity of form.
- Section thickness.
- Dimensional accuracy.
- Appearance and surface finish.
- Quantity required.
- Cost of raw material, possibility of defects, and scrap rate.
- Subsequent processes.

The selection of a manufacturing process is not always straightforward. The same part can often be manufactured by using any of the available processes. For example, a machine frame can be cast or constructed from profile bars that are welded, riveted, or bolted, the stress and strain computations being different in each case. However, the strength constraint is not always the controlling one; it is possible that the construction constraints will control the design. Figure 3.6 shows how a simple frame segment might be manufactured using different processes:

1. *Casting.* The thickness of the fin and the adjoining fillets has to be large in order to allow metal flow in the die. As far as strength consid-

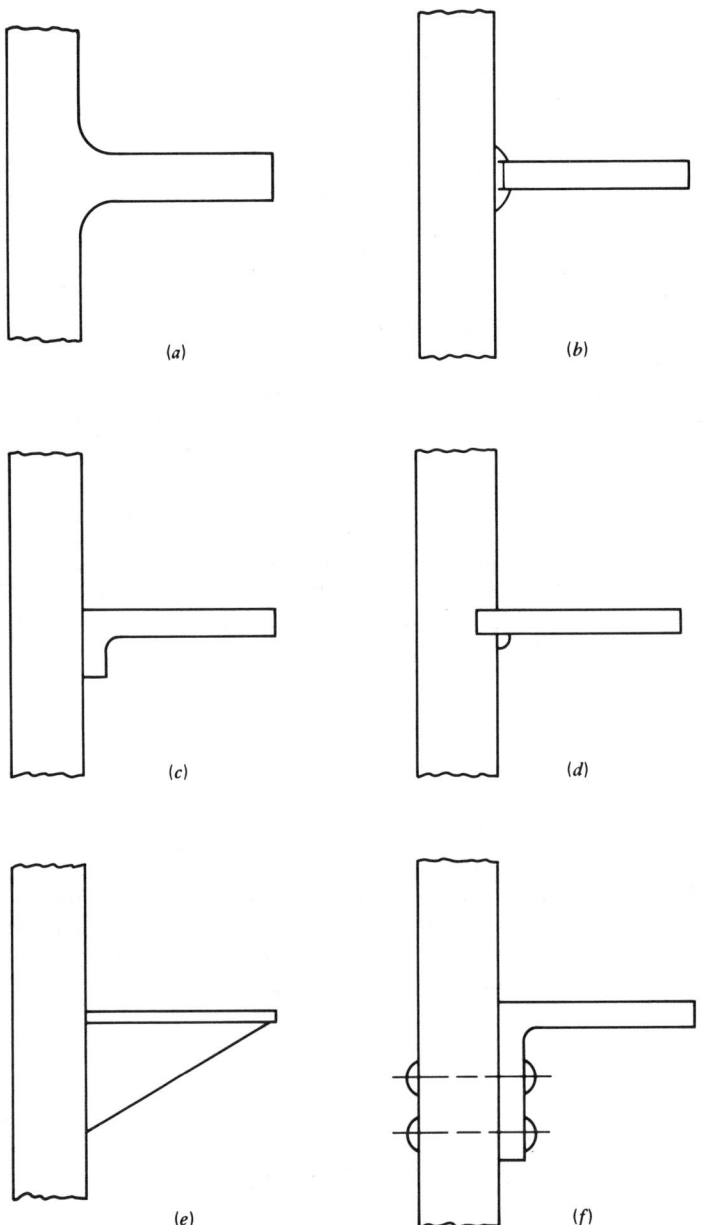

Figure 3.6 A simple frame segment manufactured by different processes. (a) Casting. (b) Welded joint. (c) Welded joint with bend of fin. (d) Welded joint with slot in body. (e) Welded joint with support plate. (f) Riveted or bolted joint.

erations are concerned, it may be possible to reduce thickness significantly.

2. *Welded joint.* The thickness may be controlled by strength constraints and be significantly smaller than that in casting. There is a minimum thickness required for welding, but it is much less than in casting. The problem lies in the joint itself. Butt welding is weak in this case, and in order to increase welding strength and ease part positioning, a bend of the fin or slot in the body is recommended.

3. *Welded joint with support plate.* For added strength and reducing fin thickness, support plates can be used.

4. *Riveted joint.* The fin must have an "L" or "T" shape in order to leave room for the rivets. The thickness can be controlled by the strength constraint. A sharp corner at the top of the fin may result. The length of the bend depends on the number of rivets required to resist the load.

5. *Bolted joint.* The consideration used for riveting holds in this case also, with the provision that sufficient space be available for a thread in the body. If the body is thin, either a screw and nut or self-taping screws can be used, depending on body thickness and accessibility for a wrench or screwdriver.

Product designers are not process planners. However, what they had in mind during the design stage significantly affects the manufacturing process and the process planning. They do not go into details of the manufacturing process, but usually work by intuition. Parts that were designed with a specific manufacturing process in mind might turn out to be very difficult to process. The process may be changed. In such cases, it should be remembered that parts are designed subject to functional, strength, or manufacturing constraints. Part drawing must not always be looked on as a constraint; it might be an artificial constraint if the manufacturing process is the controlling factor in part design.

3.5 Process Planning

The purpose of process planning is to define in detail the process that will transform raw material into the desired form. It handles each part of the product as a separate task and can be looked on as an engineering design function for which the primary objective is to define a process, cost and production lead time being secondary objectives and the designed part, quantity, facilities available, tooling and labor serving as constraints. The

3.5 Process Planning

technique recommended for application to the engineering design phase is also valid in process planning; however, there are less alternatives available. In Section 3.4, which presented a review of the currently available manufacturing processes, it was shown that each such process is unique and applicable to a specific need. The product designer had a specific process in mind, and this affects the geometry of the part. Process planners are bound by the part drawing and, therefore, are forced to consider a particular process. However, they are still free to define any process that they consider appropriate. The available processes usually overlap to some extent, and thus, for example, it is almost always possible to replace a forming process by a metal-cutting and fabrication process.

The quantity required is also a major parameter with respect to process selection. A general equation for the direct part cost is as follows:

$$C = \frac{C_T}{Q_1} + \frac{C_s t_s}{Q_2} + C_L t$$

where C = overall cost per part
C_T = overall tooling cost
C_L = hourly rate for direct labor
C_s = hourly rate for set-up labor
Q_1 = total quantity to be produced by the tool
Q_2 = batch-size quantity
t_s = set-up time
t = direct manufacturing time.

Equations of this nature express the minimum cost as a function of quantity, and the optimum quantity is dependent on the direct manufacturing time. In practice, these equations become even more complicated, since the direct manufacturing time is a function of the tooling cost; a complicated and expensive tool will usually reduce the time. If we assume a continuous process and include the set-up cost as part of the tooling cost, the previous cost equation becomes

$$C = \frac{C_T}{Q} + C_L t = \frac{C_T}{Q} + C_L \cdot f(C_T)$$

The optimum tooling cost can be derived from this equation. Even in a predefined process, the process planner has the freedom to specify tool design and thereby affect the cost of the parts.

Process planning, like engineering design, is an innovation process. There are merely general physical rules to be considered, and alternatives cannot be generated by mechanical means or formulas. It is the responsibility of process planners to use their creativity and come up with alternatives. Quantity, which is a dominant factor in process planning, is not engineering data. During product design and process planning, one might work with rough estimates of the quantity required; however, once the product is designed and the process planning defined, they serve as constraints to such economic quantity computations as master plan preparation and economic lot size. In mass production, it is almost impossible to adjust the manufacturing process to changes in quantity; however, it is possible in batch-type production. A dynamic process plan that will comply with economic lot size, but not dictate it, is required. This need is seldom realized.

In batch-type production the constraint of available facilities has a similar effect. Process planners are aware of the facilities in the shop. However, they do not know the load on the machines. There are alternatives in machining, but there is no way to ascertain which machine will be overloaded and which underloaded other than by rough estimates and assumptions. Process planners take these considerations into account. Usually (and rightly so), they aim at an economical—from their point of view—process plan. Balanced machine loading is neither their responsibility nor even a topic requiring their consideration. In fact, it is impossible to know the load if one does not know the process. It is a loop problem, and in normal situations the process planner will simply select a "better" machine. Due to this phenomenon, about 30% of the machines will be overloaded and the rest underloaded, which again indicates the need for a dynamic process plan.

Designing an economic process calls for thousands of computations. One has to examine all possible combinations of operations, machines, tools, and so on. It is a huge job, almost impossible to perform without the aid of a computer. Since no computer programs are available today process planning is unfortunately human-oriented activity. It is highly dependent on individual skill, human memory, reference manuals, and, above all, experience.

A study of this situation was conducted. Drawings of parts having different complexities were given to four process planners of different backgrounds, and they were asked to define the process. It was amazing to note that the number of different processes defined per part coincided with the number of process planners participating. Figure 3.7 presents the results of this study for the case of the simplest part used. In order to produce the 40-mm hole, the four process planners recommended the following four sets of operations:

1. Drill 10 mm, drill 38 mm, and bore to 40 mm.
2. Drill 20 mm, drill 38 mm, and bore to 40 mm.

Operation number	Process planner			
	One	Two	Three	Four
1	Machine first face	Hole drilled in two steps: a. 20 mm dia. b. 38 mm dia.	Outside surface—50 mm dia.—turned	Hole drilled to finish in two steps: a. 30 mm dia. b. 40 mm dia.
2	Hole finished in three steps: a. Drill 10 mm b. Drill 38 mm c. Bore 40 mm	Machine first face	Hole drilled to finish in one step with drill of 40 mm dia.	Outside surface—50 mm dia.—turned
3	Outside surface—50 mm dia.—turned	Cutoff	Machine first face	Machine first face
4	Cutoff	Machine second face	Cutoff	Cutoff
5	Machine second face	Outside surface—50 mm dia.—turned	Machine second face	Machine second face
6		Hole finished to 40 mm dia. by boring		

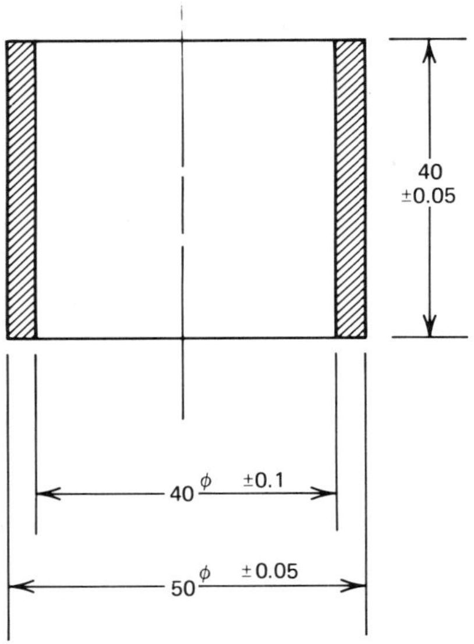

Figure 3.7 Process plan of the same part by different planners.

3. Drill 40 mm.
4. Drill 30 mm and drill to 40 mm.

A simple glance at the processes defined reveals the past experience of each planner. The first was previously a foreman of a precision parts department, the third did machining of heavy equipment parts, the second was an instructor in a technical school, and the fourth worked in a job-shop.

The same part drawing was given to the same process planner a few months later, and the specified process differed somewhat from the previously defined original. Thus mood should probably be added to the list of controlling parameters.

This situation is neither unique nor surprising. In one company a sample of 425 gears, relatively simple in nature, was studied. In all, 377 different process plans (operation sequence and machine groups) and 57 different types of machine requirement were recorded. A survey conducted in any plant would probably result in similar findings. In effect, at present there is no such thing as a unique, most economical process.

The process planning becomes the routing, which is basic data that are adopted unquestioningly by production and finance personnel. The quality of their work, no matter how sophisticated, is dependent on the quality of the process planning and product design. Process planning is a key function, dictating the cost of the product and the efficiency of the manufacturing system; however, it is a human-oriented activity. At present, extensive effort is being invested in making process planning a systematic and consistent activity. One approach is a retrieval system; this system will be briefly discussed in Section 3.7. Another approach is a systematic generative-type system; Part II of this book is devoted to this system.

3.6 Group-Technology

Group-Technology (GT) is a manufacturing philosophy aimed at increasing productivity in manufacturing of the job-shop type. The U.S.S.R. and Germany invested the main effort in developing GT in the early 1950s. There are many definitions of GT, and they are continuously changing as the scope of GT changes and as it becomes apparent that some planned activities cannot be accomplished by GT. On the other hand, it is realized that this technology can serve as a solution to additional activities. One of the first definitions was given by E. K. Ivanov (10), who stated that *"the main goal of GT is to produce single or small-quantity items using mass-production techniques."* To achieve it, we must use semiautomatic

3.6 Group-Technology

machines (such as a capstan lathe) equipped with tools that are suitable for a certain family of parts. Nevertheless, each part will not necessarily require all of the tools. It is also necessary to design and build special chucks and conveyors, which will help in handling and transferring parts between machines. Ivanov claims the achievement of a 270% rise in labor productivity and a 240% rise in shop output.

From the work of V. A. Petrov (1) on GT, we see that the same trend exists in the U.S.S.R. In his work he emphasizes the importance of designing special machines that can be employed in the manufacture of a certain family of parts, while stressing the importance of designing special chucks, such as universal prisms, for use in a milling machine, which will save set-up time. He goes into great detail on the question of production planning and machine loading, suggesting that if the plant does not have adequate capacity for loading these special-purpose machines, it should be supplied with parts from neighboring plants.

If we follow the literature on GT from the U.S.S.R. on its way west, we can sense the realization that the first goal of GT is probably something that cannot be achieved. However, since the basic idea is a sound one, this technique should not be altogether abandoned, but, instead, put in the proper perspective. Thus we find V. B. Solaja's definition of GT:

> Group-Technology is the realization that many problems are similar and that, by grouping together similar problems, a single solution can be found to a set of problems, thus saving time and effort.

And in "Engineering," 1968, we find this definition:

> Group Technology is the technique of identifying and bringing together related or similar parts in a production process in order to utilize the inherent economy of flow production methods.

The general manufacturing philosophy of GT is now accepted, although it had been practiced under different names, or without any label whatever, even before receiving formal recognition.

In order to practice GT as a systematic scientific technology, tools for the identification of the groups must be prepared. In translated Soviet literature of recent years, there are no papers on the subject of GT. However, in Western literature we find many articles that actually deal with the "tools" for GT and with classification and coding systems. (Sometimes it seems that the tool becomes more important than the product.) Insofar as we are supposed to have classification and coding systems, the goals and applications are expanding beyond the original requirements, and the broad meaning of GT now covers all areas of the manufacturing process:

- *Design.* There have been improvements and savings in engineering design due to new methods of data retrieval, elimination of the duplication of drawings, and modification of older, similar drawings.
- *Material management and purchasing.* The use of groups of materials has led to greater purchasing efficiency, lower stock levels, and savings in procurement.
- *Process planning.* Savings in process planning have resulted from using the same process for a family of parts.
- *Production control.* Improved scheduling has resulted due to the similar setup of groups.
- *Manufacturing.* The use of flow-line systems, machine groups, and manufacturing cells has led to greater efficiency.
- *Management.* The grouping of parts for the evaluation of vendor quotations, cost, manufacturing load estimate and so on has led to improvements and savings.

The main differences between GT and the conventional manufacturing system (including the IMS) lie in two areas:

1. *The engineering phase.* This phase was not treated systematically in the IMS. Modern GT offers systematic treatment. Although it merely involves a retrieval of parts by similarity, the savings attributed to the use of GT are impressive. For example, a company reported that about 2,500 new parts were released annually, while about 30,000 active parts were in its design files.

It has also been reported that from 5 to 10% of the annual output of new parts could be avoided by the proper use of classification and coding systems. Thus a company can save from $237,500 to $475,000 just by reducing the duplicated design. This is without making anything; it simply involves preparing the part to go to shop floor and be made.

This subject will be elaborated on in Section 3.7.

2. *Target of attack.* The IMS tries to increase productivity by using capacity planning to attack the direct machining time. Group-Technology, on the other hand, is concerned with the lead time. It is claimed that only 5% of the lead time in producing a part is direct working time, whereas 95% of the time the part waits in the shop. Furthermore, the 5% can be divided into 30% actual machining time and 70% for positioning, chucking, gauging, and so on. Hence, only 1.5% of the lead time is actual machining time, and GT directs its effort toward reducing lead time by attacking the remaining 98.5%. One way to achieve this is by organizing the plant layout according to work cells rather than functions. A work cell is a unit that includes all of the machines required to produce a family of

parts. Raw material enters a cell, and a finished part emerges. The reported success in reducing lead time is very impressive.

In order to utilize the benefits of GT, a classification and coding system is required. Ham (2) defines the requirements of this type of system as follows:

> The system must be both design or form oriented and also production or process oriented.
>
> Care must be taken in formulating the process code so that the system can accommodate future changes in technology which may produce entirely new methods of processing the part family.

To make these demands of a classification and coding system is practically asking for the impossible. Many of the reports on successful GT applications have come from studies in which the main work on the manufacturing concept was done with families that had been organized by human effort. Engineers have tended to view each part produced in the company and make a human decision, relying on their memory and on the flexibility of the human mind. Perhaps they made mistakes, but the stress was on the effort toward goals and not toward tools required to achieve the goals. Any classification and coding system will contain errors, owing either to the rigidity of the system or to human error in the classification and coding processes.

A. Houtzeel (3), in his paper describing the Miclass (4) system and his experience of seven years in development and testing, writes:

> Errors are a serious problem for all conventional classification systems. Classification may start off accurately but within a few months errors may run as high as 30 to 40%. Miclass has been carefully designed to avoid such problems and, with computer classification, the normal error rate can be kept less than 5% and generally under 2%.

One version of a classification and coding system and its application is illustrated in Figure 3.8.

Classification and coding systems can be categorized as design-oriented, production-oriented, or resource-oriented.

Design-Oriented Classification and Coding Systems

The first attempts at creating classification and coding systems for components were specifically based on considerations of shape and size.

G. Halevi (5) has proven that the most economical process for components that have exactly the same shape and size will vary according to other considerations and requirements, such as quantity, raw material,

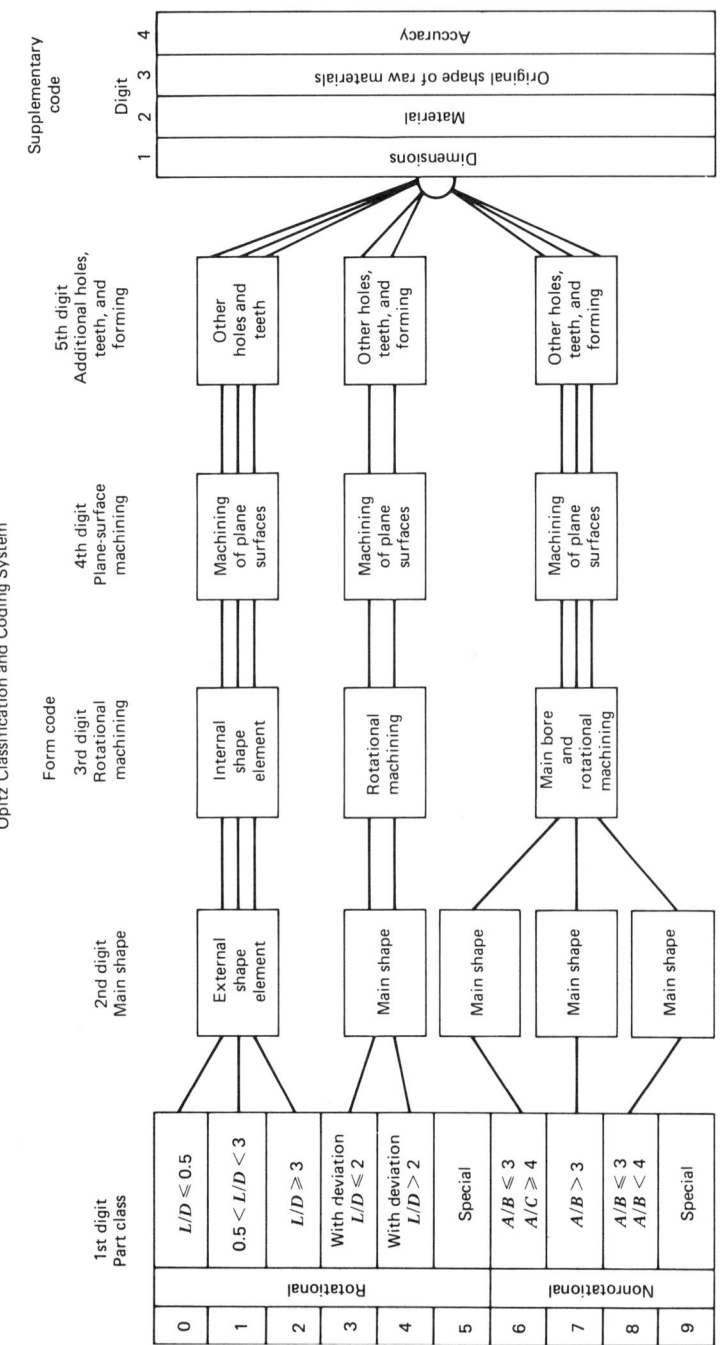

Figure 3.8 A classification and coding system.

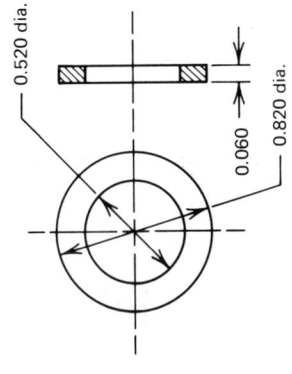

Opitz
Dr. H. Opitz
Aachen, West Germany

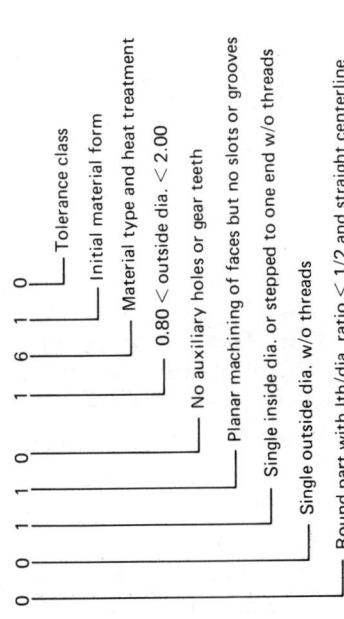

Figure 3.8 (*Continued*)

and optimization criterion. Similar conclusions are found in the paper of W. J. Hancock (6) from Group-Technology International, England, in which he writes:

> While they met with some measure of success, it was found that the systems available at that time (early 60's), were design-oriented, and, projected into production-oriented systems, were not giving the results desired. Today, it is increasingly recognized that classification and coding of components by shape, etc. will not give the results desired.

Hancock goes further in developing his ideas and comes to a clear conclusion:

> From my own experience in this field, and that of my colleagues, my conclusion is that no one can devise a coding and classification structure as a sieve or riddle by which, with the appropriate number of shakes, rattles and rolls, component families will emerge, around which we can design Group-Technology cells.

Production-oriented Classification and Coding Systems

The inadequacy of the design-oriented approach thus leads to the use of a production-oriented classification and coding system. Naturally, as a result, the design utilization of the GT technique is neglected. However, this is permissible, since it constituted a secondary rather than a primary objective.

This classification and coding system should be based on process planning; however, the present situation in process planning [excluding the work of Halevi (7)] is a strictly human-oriented activity, highly dependent on individual skills, human memory, reference manuals, and experience. Due to the fact that more than one engineer (or process planner) are involved in deciding and specifying how things should be made, in the current process planning file we find different plans, different machines, and different sequences for the same or similar components. Thus if we use the current process plans as a basis for a classification and coding system, we probably will not achieve the results desired. Furthermore, Hancock (6) has found and stated that:

> Many of the characteristics associated with production are either totally variable, or at best, semi-permanent. These include the production quantity, the manufacturing methods and the technological changes. To build these into a code structure would be very difficult.

The production-oriented approach to classification and coding has achieved some success in certain manufacturing industries, but has left many unsolved problems in others. It may be a good approach in certain industries, but even there, I am afraid, it will only achieve a short-term success, possibly proving disastrous in the long run. It is a rigid, unbased system that will not adapt itself to technological changes, but will, therefore, be a hindrance to technical development.

Resource-Oriented Classification and Coding Systems

The failings of the production-oriented approach have led us to a resource-oriented classification and coding system. This approach has had different names in different countries and periods. In the U.S.S.R., it was called "flow-line production" (before its disappearance from the literature there). In western countries it is referred to as "production cell," "automated job-shop," or "multistation manufacturing system."

The work-cell method calls for a machine layout according to a component flow analysis (CFA), in which a component will enter a work cell and be terminated there. Hence, one work cell or department might include sawing, milling, turning, drilling, slot milling, grinding, lapping, and deep drawing.

Technological improvements can be made in each work cell according to the specific components to be manufactured in it. The extreme is the automated job-shop, where a transfer line and automatic chucking are installed.

The work-cell method represents the latest approach to GT (the cell concept, under different names or without any name at all, was used in the 1950s, later, this concept was abandoned and functional machine layout was preferred), and the reports on its use inform us that it is "immensely successful," "there was an average saving of 70% of machine time," "work in progress reduced by a ratio of 8 to 1," and so on.

There is no reason to doubt the accuracy of these reports or that this is an excellent technique for certain types of industries.

A preliminary study of the components with the aid of a CFA or an operations matrix established the optimal arrangement of work cells. Their classification and coding will be done by analysis of workpiece specifications so as to obtain the potential manufacturing work cell. Naturally, if the workpiece mix changes, a new study of the CFA should be made. Rearrangement of the work cells and reclassification and recoding of the components should be carried out.

Before installing and using the work-cell concept, one should bear in mind the problems that will arise:

- *Reorganization.* The ability to reorganize the machine layout of the plant and, probably, to repeat the reorganization every so often.

- *Work-cell supervision.* In the functional layout, it is required that the foreman be an expert in one type of manufacturing. In the work-cell concept, in order to supervise and instruct the workers, the foreman must be an expert in several fields, such as milling, turning, and grinding.
- *Work distribution.* With the functional layout there was no problem in transferring an operator from one machine to another, since they were of the same family and the operator had the training and skill to operate them. In the work-cell environment, however, this cannot be done.
- *Shop-floor control.* Since the work cell is built around the most complex workpiece-processing route within the group, it is possible that a condition of a single machine overload in one work cell and underload in another will occur for a certain type of machine. Hence, machine utilization and work balance will be low.
- *Production planning.* A condition in which one work cell is overloaded and another underloaded can occur.

Trying to solve the last two problems by transferring work for a specific machine in a cell from another cell will soon lead to a "new concept" of rearrangement of the machines according to a functional layout (which is what happened in the past).

Actually, the work cell comes close to the initial goal of GT: Each cell is regarded as a manufacturing unit, that is, a machine. Present-day problems of process planning have become simpler as the number of alternatives have decreased. The problems of production planning seem to become easier; however, as we delve deeper we find that not only do the same problems still exist, but there is also less flexibility in dealing with them.

From the recently published literature, it appears that the initial enthusiasm for the work-cell concept has to some extent been tempered by recognition of the above-mentioned problems.

S. Chandra and R. L. Shell (8) describe a SIMSCRIPT 115 simulation model for a multistation manufacturing system:

> The simulator could also be applied by the user to experiment after the system is operating, to see what improvements can be made in scheduling, machine utilization or product completion time as the product mix changes.

It is interesting to note that this work closely resembles the work in the production planning field on the problem of dispatching rules or scheduling policies. In the above-mentioned simulator, the following policies were applied:

1. Start the job with the shortest possible processing time.
2. Start the job with the longest possible processing time.

3.6 Group-Technology

3. Random selection.
4. First come, first served.

The following performance measures were evaluated:

- Average number of jobs in the system.
- Mean arrival queue length.
- Mean time taken for each job.

It is also interesting to note that:

> The performance of the system is relatively insensitive to the four scheduling policies tested . . . [it] also shows that the performance measure varies with change in the arrival rate.

It seems that the problems in the automated job-shop and their solutions are the same as the present-day problems and solutions.

Another approach to solving the problems created by the work-cell concept is reflected in the work of G. F. K. Purcheck (9). In this fascinating study, he introduces the Carafield system, which is an algebraic one, "by virtue of it being an application of Boolean Algebra." His concept is that a dynamic work allocation should be ensured:

> The control system should ensure that randomly-arriving job orders are placed into the least busy hospitality group under the principle of multi-group membership.

To achieve it:

> A similarity definition needs an object numbering system which has mathematical properties such that the class numbers form a closed set on which one or more ordering relations can be defined together with one or more combining operations.

Actually, this approach is a new method of production planning. There are many work cells, while the code number preserves the integrity of the original data and can be manipulated mathematically to create classes (or groups) according to the workload in each cell.

It must be emphasized that Purcheck's solution is a mathematical one, not a technological one. The process and its sequence are assumed to be given. The questions are: given by whom and how good is it?

Group-Technology is a human-oriented system. Classification and coding are manual work, even if carried out interactively with a computer. Retrieval is done by groups, and human intelligence is required to decide which of a group of drawings or processes is the desirable one. The human intelligence factor involved restricts the capabilities of the system, and thus a computer-oriented system would be much more flexible and powerful. Hal-Technology, described in Part III of this book, is such a computer-oriented system.

3.7 CAM/CAD

Computer-aided manufacturing (CAM) has many meanings and interpretations. On one extreme, it refers to the use of a computer to run an automatic programmed tool (APT) for programming numerical control (NC) machines, while on the other extreme, it refers to what technology forecasting predicts for the future—the "automatic factory." The automatic factory is a computer-integrated manufacturing system that controls all phases of the industrial enterprise: product design (computer-aided design—CAD), process planning, flow of material, production planning, positioning of materials, automatic production, assembly and testing, automatic warehousing, and shipping. No such system exists today, but the technology needed is feasible; in fact, computer-controlled work cells are already in use. Figure 3.9 shows the Ingersoll Milling Machine Company, Rockford, IL, system for machining tractor transmission housings.

Numerical control machines are built to cover most types of processing. The trend is toward computer numerical control (CNC) and direct numerical control (DNC). Adaptive NC systems have been introduced.

Industrial robots is another developed field. Computers and sensing devices increase the intelligence of the robots. Over 200 different models are available today.

The automatic factory will be an evolutionary process. The computers of the individual machine, cell, or device will be tied together in a hierarchy to form a complete network. Figure 3.10 shows such a network. Low-level computers will control the machinery and retrieve management information. The dynamic operations of the factory will be controlled by higher-level computers with on-line overall optimization programs. Computer-aided design will be part of the system; this will enable the designer to decide the best way to design the product such that all parameters are optimized. At present, the automatic factory is a very expensive project. It requires a national program and a massive pooling of resources.

The U.S. Air Force Integrated Computer-Aided Manufacturing (ICAM) Program is aimed at revolutionizing batch part-manufacturing by changing the concept of production from design engineering down. The program is backed by 75.8 million dollars.

Machine features and operating steps

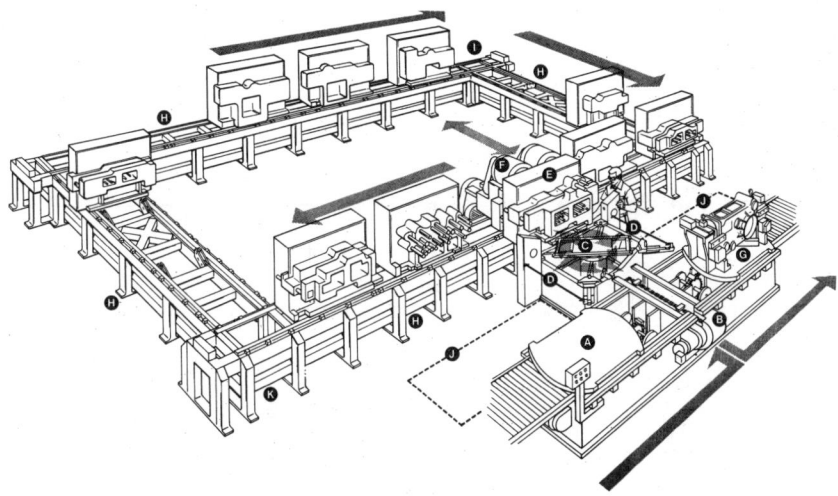

A. Pallet onto which machine operator loads and fixtures workpiece.
B. Shuttle conveyor moves palletized workpiece into position on indexing unit.
C. Four-position indexing unit supports pallet during machining operations and makes four 90° rotations to allow workpiece to be machined on four sides.
D. Hydraulic clamps (4) lock pallet into position, maintaining rigidity and level.
E. Change head is moved into operating position. The limit switch actuator cam on this head indicates that it will perform drilling and boring operations. Head is located by a retractable dowel and held in place by automatic hydraulic clamps. Change heads weigh approximately 3 to 6 tons.
F. Single-station feed slide with dual spindle drives. Four motors are associated with this component. The main drive unit at the rear of the slide has two motors, one for drilling and boring, the other for tapping. Each motor powers a separate male drive coupling which interfaces with an appropriate female coupling on the change heads. The slide itself is powered by two motors, an ac motor for rapid traverse, and a dc feed motor.
G. Workpiece is ready to be unloaded following machining operations on four sides. It will be moved to next head changer which will machine top and bottom. In this instance, the workpiece is an 1100-pound transmission housing.
H. Change head transfer system moves non-phasing heads clockwise around a 100-ft circuit. The heads move on skid rails and are driven by hydraulically actuated transfer bars.
I. Access station allows change heads to be removed from the transfer line for replacement or maintenance.
J. Optional machines (2) can be placed on either side of the workpiece. The optional machine would perform special single function operations.
K. Access station with safety interlock for tool change.

Figure 3.9 Computer-controlled automated production line at the Ingersoll Milling Machine Company, Rockford, IL. [From *Industrial Engineering*, Vol. 9, No. 11 (November 1977), p. 27. Copyright 1977 by the American Institute of Industrial Engineering, Norcross, GA 30092. Reproduced with permission.]

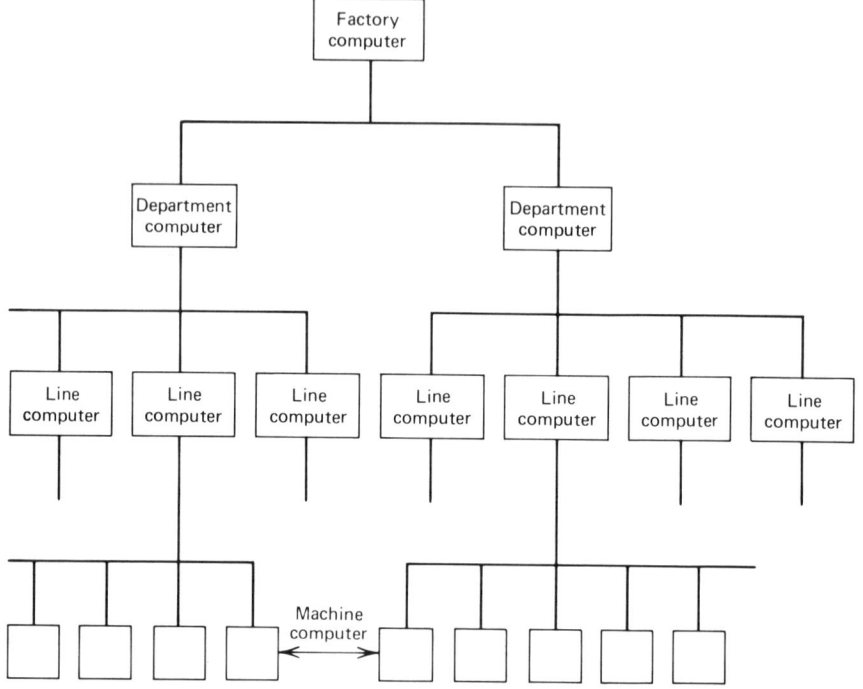

Figure 3.10 A computer network in an automatic factory.

At present, Japan has the most ambitious national program. Their plan, named "Methodology for Unmanned Manufacturing," is to develop a 200,000–300,000-ft^2 factory staffed by a control crew of only about 10 persons (compared to the normal 700–800 workers). The cost of the project will be approximately 100 million dollars; however, the predicted increase of productivity with this new technology is 7,000–8,000% compared with a conventional method.

The need for CAM and the corresponding economic incentive are greatest in the batch-type metalworking class of manufacturing. It is reported that 75% of such parts are manufactured in lots of 50 pieces or less, while about 40% of the total manufacturing personnel are engaged in this type of work. The savings in labor cost, capital tied down in production, and machine utilization could be enormous.

The common interpretation of CAM/CAD today is not as ambitious as the automatic factory. However, CAM and CAD are referred to independent systems in the following fields.

3.7 CAM/CAD

Computer-Aided Product Design and Drafting—CAD*

In this field, CAD is a human–computer interactive system, where the cathode ray tube (CRT) graphic display terminal replaces the drawing board, reference books, and the slide rule. The designer is able to "converse" with the computer and receive a direct response from it. This two-way conversation may be graphical or pictorial in nature. For example, the designer may generate a picture on the CRT graphic display terminal by using a light pen. As a result of previous programming, the computer "understands" the picture, makes calculations based on it, and presents answers or a revised picture to the designer within a few seconds.

The computer can carry out vast amounts of detailed work, tirelessly and without error. It can evaluate the consequences of an endless series of design alterations, performing both the engineering calculations and the graphical manipulations, and can file away each alternative for future reference. Remembering things is a task at which the computer excels. Optimum solutions for some problems cannot be obtained in close form, thus requiring the designer to resort to a tiresome trial-and-error process. For such problems, the computer can be instructed to increment a set of parameters and generate a family of solutions, from which the optimum one can be selected.

The synergistic effort of achieving this close coupling between the designer and computer has four important benefits:

1. Designers can immediately see and correct any gross errors in their drawings or input statements.
2. Designers can monitor the progress of a problem solution and terminate the run or modify the input data as required.
3. The designer can make subjective decisions at critical branch points which guide the computer in a continuation of the problem solution.
4. The graphic display may present data that cannot be readily understood or interpreted in a computer output listing or even in plotted output. Through clever programming, a computer-driven display can present multiple views, moving pictures, blinking lines, dashed lines, and lines of varying intensity.

Figure 3.11 shows a designer at work.

*This subsection is partially based on U.S. National Bureau of Standards publication No. PB 256 996—"Computer-Aided Design in Manufacturing" (February 1974).

Figure 3.11 A designer at work. (Courtesy of CAM-I, Inc., Proceedings P-76-BG-01.)

A light pen and function buttons are the input devices used by designers. Designers gain a distinct feeling of direct and personal control over the entire system because of actions they perform with the light pen.

The function buttons make available functions for creating lines, points, circles, erasing, reversing an image, and so on in an endless array. Many sets of functions can be called up, together with overlays for the function buttons that carry the corresponding designations. Users can call up "menus" of available actions, which will be displayed for their selection. Upon designating their choice with the light pen, an acknowledgment message is displayed. Users will also receive tutorial messages at the edge of the screen, reminding them of the next step they must take and the choice of actions available to them. Caution notices, such as illegal entry, impossible geometry, and capacity limit approaching, are also displayed.

The view and group functions are particularly useful. The view function enables the derivation of a third (or any auxiliary), or oblique, or perspective, or isometric view. In addition, one view (or any part of it) can be quickly transferred to any other drawing, with the scale or the orientation changed at will. The group function enables the instant translation of any group of lines or notes to any point on that particular drawing and also sets it up for transfer to another drawing in the same manner as a complete view.

The ideal computer application exists in a situation where, for a single input of repetitive elements, preferably oriented toward some discipline, the input data can be manipulated to produce a variety of useful outputs. For example, in the case of an electronic circuit, for the single entry of the elements of a schematic diagram, it is possible to generate a finished schematic drawing suitable for handbook usage, finished artwork for printed-circuit boards, NC tapes for printed-circuit boards, drilling or automatic wire wrapping, automatic testing machine programs, and information on various parts. This provides the potential for effective computer utilization. Most companies feel that costs do not, and will not in the reasonably near future, justify using the computer exclusively as a drafting instrument. Only where this multiple pick-off potential exists is the cost and time of getting the information into the computer justified.

Computer-Aided NC Part Programming—CAD/CAM

The NC process can be briefly described as follows:

1. The "part programmer" prepares a manuscript (the part program), describing the part geometry, machine geometry, and cutting instructions, using a language predefined to the computer. The most common language for this purpose is APT. The parts programmer regards this language as a dictionary of words, symbols, and rules for writing instructions that describe the part. The computer has programs that allow recognition of these written instructions, and these programs are called APT processors. The processors produce a general solution to the machine problem. This solution is often called the cutter location (CL) file.

2. The computer uses the CL file as input to a second computer program called the post processor; this program relates CL data to the format required by the particular machine control.

3. The processed control data is output to the machine on punched tape, magnetic tape, punched cards, or by direct wire.

The generation of part programs can be done as a component part of the CAD process. The geometric data base constructed in the computer by an interactive CAD system can be used to generate tool paths with a few extra commands. This minimizes the total design-to-production time, increases engineering efficiency, and improves quality control.

Checking of NC programming is aided by animation of the tool path on the graphic display terminal. This enables the part programmer to visualize tool motion.

Thus CAD integrates directly with CAM and can result in increasing productivity of both engineering and production personnel by factors of up to an order of magnitude or more, while improving quality control and reducing the design-to-production time.

The time reductions involved with CAD can be truly impressive. For example, McDonnell Douglas cites the following examples:

Task	Previous time	New time
Airfoil smoothing	6 weeks	10 minutes
Flight-path optimization	4 weeks	1 hour
Metering pin design	4 weeks	1 hour
Control system analysis	3 days	1 hour

Computer-Aided Manufacturing in Process Planning—CAM

The CAM process planning system consists mostly of retrieval programs based on GT methods. The machined parts are classified and coded for the purpose of segregating them into families. Each family is comprised of parts having attributes sufficiently common to prescribe a common manufacturing method. For each such family of parts, a standard manufacturing process plan is prescribed and stored in a data base.

The two prerequisites for such a system—classification and coding of parts and a standard manufacturing plan for each family of parts—are the responsibility of the user. It is up to the user to analyze the part spectrum and decide what level of refinement will be satisfactory. It requires a great deal of initial effort in these areas on the part of the user.

The CAM process planning system performs merely a retrieval function. It is a human–computer interactive mode of operation that uses a CRT graphic display terminal. The user keys in the classification code of the part to be process planned. The data base file is then searched in an attempt to locate the requested standard process plan. If the code number exists, the standard plan will be retrieved and displayed on the CRT graphic display terminal so that the user may modify or extend the plan to suit the specific part being planned. The modification and extension are the responsibility of the user. The CAM program provides the necessary software to handle such changes. Formatted handcopies of individual plans may be made for shop distribution, reference, and other purposes when appropriate.

The standard process plan is stored in memory, in several levels. Each subsequent lower level specifies the process in more detail, the lowest level specifying detailed work instructions and parameters.

The process planner works in a conversational mode with the computer. Initially, the general routine will be displayed on the CRT graphic display terminal:

```
010    Lathe
020    Grind
030    Burr
040    Inspect
050    Stock
```

Using a light pen or function buttons the process planner may request the extension of any of the above operations.

Computer-Aided GT—CAM

Group-Technology has been discussed in Section 3.6. Some of the GT concepts are utilized with the aid of a computer. We will not go into all of the applications, but only name a few:

- *Classifying and coding by computer.* Classification and coding by computer is a conversational mode. The user classifies a part by answering a series of logical questions posed by the computer. Based on the answers, the computer will then assign the code number to the part.
- *Retrieval of drawings.* Before designing a new part, it is advisable to scan the existing drawing file in a search for the required part. By assigning a code number to the desired part, this scan may be performed.
- *Standardization.* Classification and coding together with a computer retrieval program can be used to focus on standards. The application of these tools makes it feasible to examine the minute differences in parts and thus establish standards.

References

1. Petrov, V. A., "Flowline Group Production Planning," Business Publication Limited, London (1967).
2. Ham, Inyong, "Introduction to Group Technology," CAM-I Proceedings P-75-PPP-01 (1975).

3. Houtzeel, A., "An Introduction to the Miclass System," CAM-I Proceedings P-75-PPP-01 (1975).
4. Miclass System, Metaal institute TNO, Apeldoorn, The Netherlands.
5. Halevi, G., "Adaptive Production-Process Planning," Proceedings of CAM-I International Seminar, P-76-MM-02, Atlanta, GA (April 21–23, 1976).
6. Hancock, W. J., "Group Technology," CAM-I Proceedings P-75-PPP-01 (1975).
7. Halevi, G., "A Systematic Approach to Economics of Part Turning," Mecanique Industrie G.A.M.I. (France), No. 299 (November 1974), p. 33.
8. Chandra, S., and R. L. Shell, "Simulation for Automated job-shop Development" (1975).
9. Purcheck, Gunter F. K., "A Mathematical Classification as a Basis for the Design of GT Production Cell," *Production Engineer* (January 1975), pp. 35–48.
10. Ivanov, E. K., *Organization and Technology of Group Production*, Business Publication Ltd., London, 1967.

Chapter Four
Engineering Data Control

Engineering data, which include product structure and manufacturing routing, are generated in the engineering phases of the manufacturing cycle and then used, without question, as the basic production data. The engineering data are applied in various areas of the industrial enterprise. Each application requires these data in a different format and for different purposes. For example, manufacturing needs the engineering data to produce the desired product, material control needs them to plan the acquisition of raw material, and they are required by costing and marketing for the determination of estimated and actual cost.

Engineering data comprise a large volume of information that is constantly changing. If each application keeps a separate set of data, there will be a duplication of effort in maintaining them, and separate sets of data will most likely not agree with one another.

In many companies, bookkeeping, inventory, purchasing, and manufacturing develop their own names and codes for the same material. Each of these applications will often use a different unit of measurement and define the product structure to suit its particular needs. This interferes with daily communication between applications, since for a valid comparison to be possible, the reports of each separate application must undergo interpretation.

Thus the objective of engineering data control is to organize and maintain one set of data for the use of all applications, thereby eliminating the drawbacks of separate data sets. This set of engineering data control files becomes the official communication "language" throughout the company. The product, the materials, the operations, and the facilities are designated and given a "name" or code that all applications will use, thereby improving the communication in the company.

4.1 Classification and Coding

The primary objective of a classification and coding system for use in the IMS is to assign to each product, part, or material a unique name that will be used throughout the company—in engineering design, purchasing, inventory, production, costing, bookkeeping, and marketing.

To meet this primary objective, designation of any nature (shape or form) will satisfy the requirements. The code number does not have to represent any characteristic of the item; it must only be unique. An alphabetic name is as good as any code or set of digits. The IMS is not at all concerned with—and thus does not utilize—any characteristics of the code itself. The system refers to an item by the code name, while the relationships of part to product and of material to parts are given by links specified in the system files.

The secondary objectives of the classification and coding system are:

- Ease of use.
- Ease of detection and elimination of errors.
- Ability to select alternate materials.
- Ability to process groups of materials.
- Ability to compare the unit price of similar items.
- Ability to select sources of supply of similar items.
- Standardization of tools and materials.

As one can see, these secondary objectives do not correspond to those of GT. Hence, this classification and coding system is not the one needed for GT. A company will usually have different classification and coding systems, one for each purpose.

To satisfy the first two secondary objectives, the code must be numeric and short (see Section 2.5). Many companies use a classification and coding system of up to nine numeric digits, and there is general agreement that this produces a good compromise between ease of use and the ability to define the characteristic of the item.

To satisfy the remaining secondary objectives, the code number assigned to an item must have a meaning. The user has to specify the secondary objectives and the intended applications of the classification and coding system. Based on these specifications, a classification and coding system will be constructed. There are three types of classification and coding systems.

4.1 Classification and Coding

Meaningless (Running Numbers) Classification and Coding Systems

In this type of system, the code number assigned to an item is a chronological sequence number. This system is sufficiently satisfactory for the IMS. It is simple to install, does not require any investment of time and money, and does not require a specialist or trained personnel to maintain it. There are no wasted digits, and the number can be short. A company with one million items in stock can use six-digit code numbers. It reduces errors in copying the number, while also saving keypunching time and disk or file space.

The disadvantage of this system is that the last five secondary objectives cannot be achieved. For example, one cannot select an alternative item or material merely by the characteristics of the code number. (It should be noted that these features are seldom workable even with the sophisticated meaningful classification and coding systems.)

One difficulty that cannot be ignored is the editing of the catalog. How does one find the code number when the specifications of the item are known? This, however, is not a classification and coding problem, but rather one of publication. The catalog should be printed a few times, each with a different key. It is advisable to print it with the aid of a computer in cases where sorting is easily performed and key words used. Furthermore, the catalog should be first printed according to code sequence number, followed by the description, and then it should be printed and sorted according to key words, followed by the code number and description. In this way, the same material will appear many times in the catalog, each time under a different key word (e.g., steel, bar, cold-drawn, and tool steel). Once gotten used to, this type of catalog is often found to be more convenient than the conventional ones.

We are used to classification and coding systems in which we instantly have a general idea of what a code number is about when we hear or read it. This is of assistance and perhaps even essential in manual systems. However, in the IMS—a computerized system—the computer will interpret the code number for us.

Meaningful Classification and Coding Systems

In this type of system, the code number is based on a classification such that each digit of the code has a specific meaning built into the system in order to achieve the secondary objectives specified.

The construction of this type of classification and coding system is very difficult and time-consuming. The first step is to catalog (list and organize) all the types of items that the system should cover. Figure 4.1 shows such an organization; the cataloging system covers materials, machines, spare parts, products, and activities.

The second step is to assign a prefix that designates the item type. Since there are more than 10 types, the first two digits will designate item type; these two digits have to be assigned in such a manner as to formulate groups of items and leave enough room for future expansion.

The third step is in-depth analysis of each type of item. The purpose of this study is to learn the significance of the variables with respect to

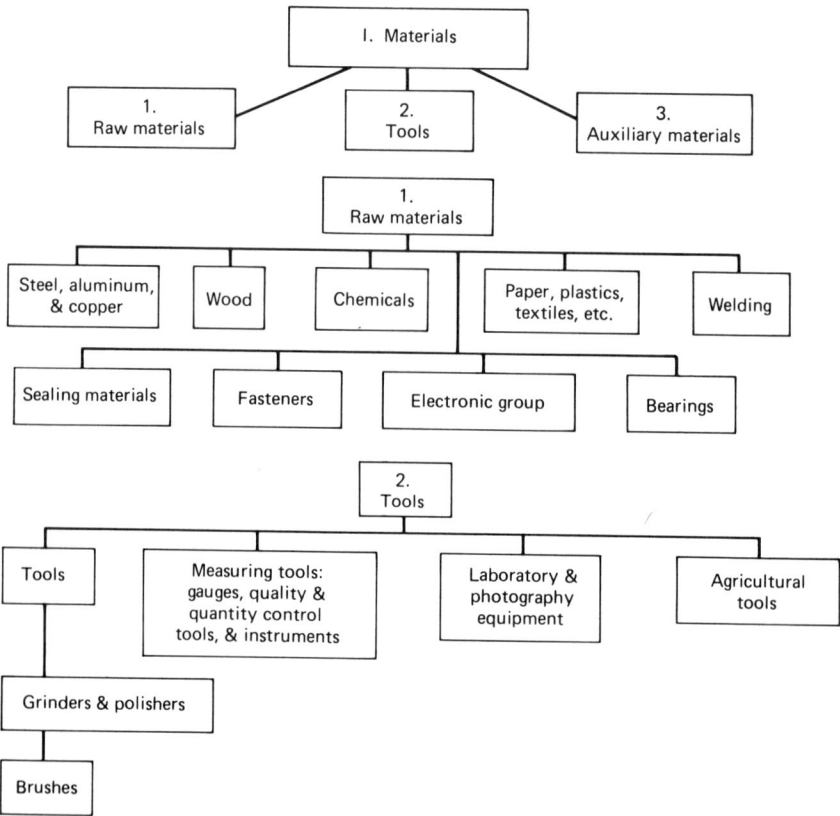

Figure 4.1 Preliminary organization of a meaningful classification and coding system. The three main groups of the cataloging system are: I, materials; II, machines and spare parts; III, products and activities.

4.1 Classification and Coding

uniquely defining the item type. In the case of fasteners, for example, one learns that one group is screws and bolts, which are specified by thread form, head form, material, class, tolerance, and dimensions; thus all types of thread forms, head forms, and materials, together with the classes, tolerances, and standard dimensions (e.g., diameter thread length and overall length), if any, should be listed. The relative importance of each parameter should be considered and a decision as to their sequence made.

The fourth and last step is classification. The frame, that is, the length of the code number, and whether the code number is to be numeric or alphanumeric must be decided on. Usually, one finds that 27–35 digits are necessary in order to both uniquely define an item and satisfy all the

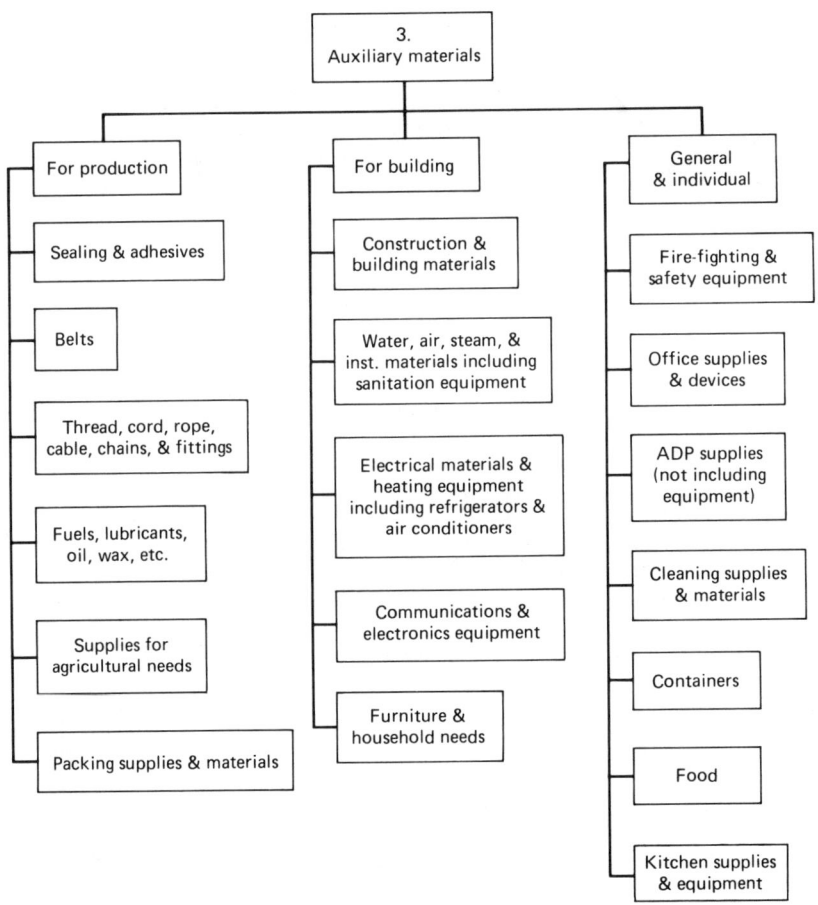

Figure 4.1 *(Continued)*

requirements. Since it is impractical to work with such a long number, a compromise must be made. For example, dimensions must be given by constructing tables and not by using the actual values. This will limit the use of the code number in the search for alternative materials (alternative dimensions). Material composition must be unified and coded by tables; this also limits use of the code number in the search for alternative materials. Some of the characteristics of the item that are not essential, such as the color of the screwdriver handle, will be dropped.

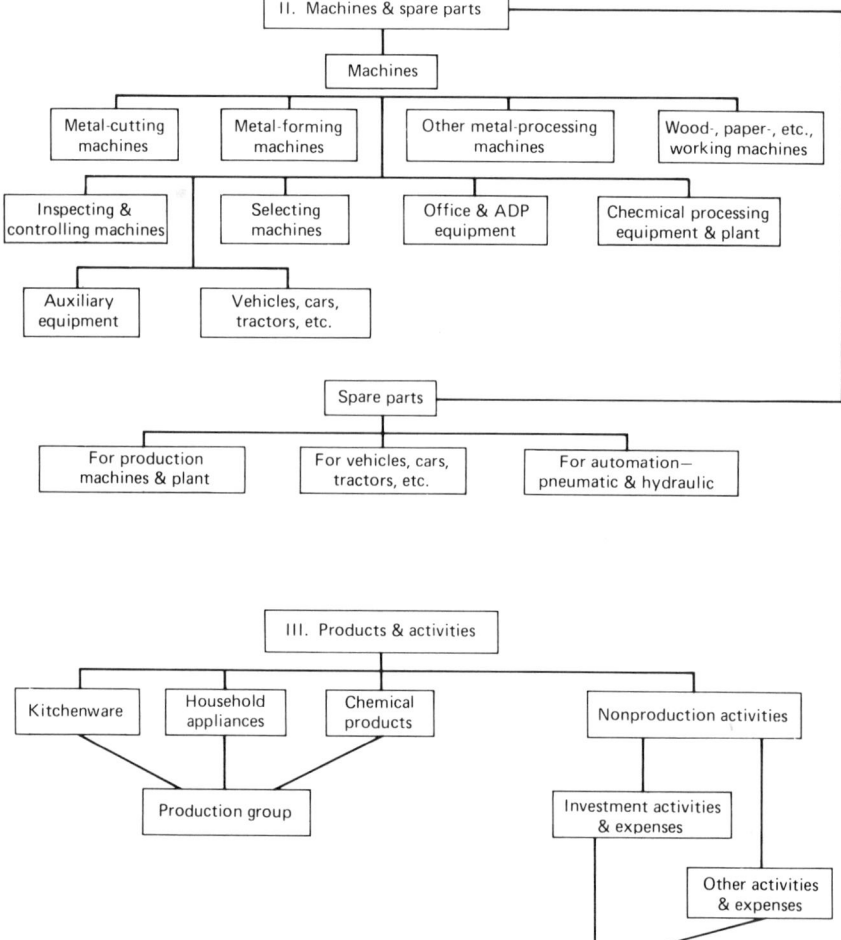

Figure 4.1 *(Continued)*

4.1 Classification and Coding

One common point of friction between the personnel involved in classification and coding and the production personnel is whether the code should be assigned by material specification or by material specification and supplier. The production people feel that buying material with the same specifications from different suppliers results in somewhat different materials. This argument has bearing on purchasing and inventory. Management must rule and state its policy on this matter. Figure 4.2 shows a meaningful classification and coding system that uses nine numeric digits.

This type of classification and coding system is very expensive to construct and requires continuous maintenance. New technologies are

A. Raw Materials

1. Steel, aluminum, copper, and other alloys.

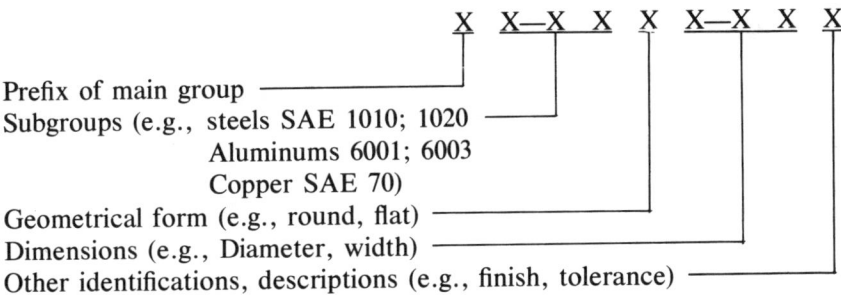

2. Fasteners.

Subgroup: screws and bolts

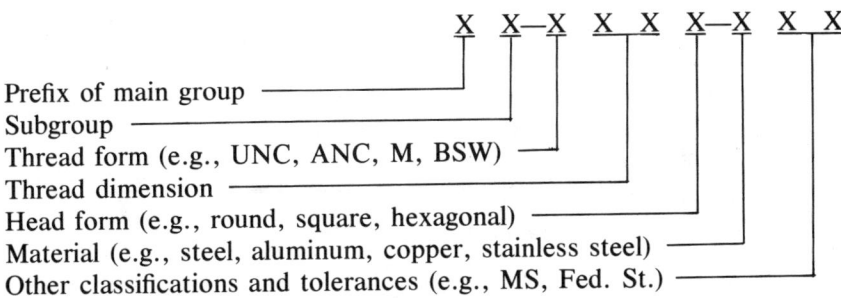

Figure 4.2 A meaningful cataloging system for materials—an example of classification and coding.

B. **Tools**

1. Cutting tools.

   a. Thread cutting tools

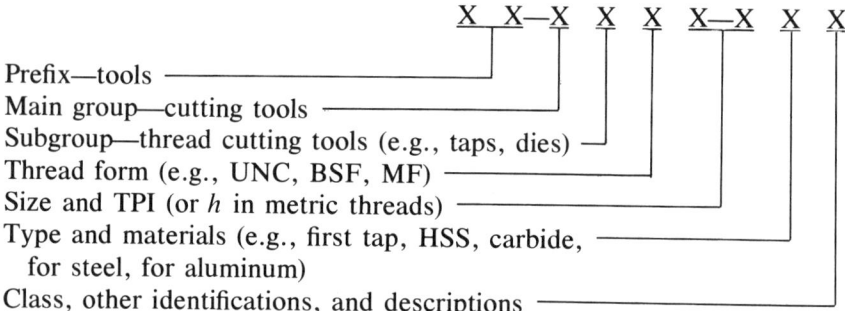

Prefix—tools
Main group—cutting tools
Subgroup—thread cutting tools (e.g., taps, dies)
Thread form (e.g., UNC, BSF, MF)
Size and TPI (or h in metric threads)
Type and materials (e.g., first tap, HSS, carbide, for steel, for aluminum)
Class, other identifications, and descriptions

Example

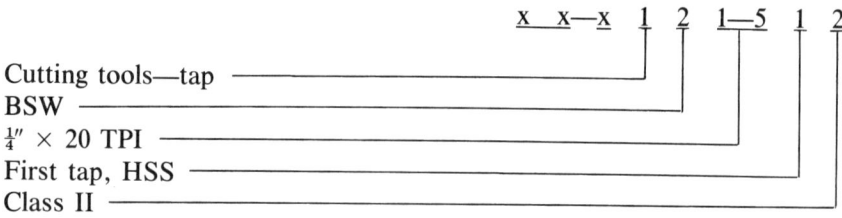

Cutting tools—tap
BSW
$\frac{1}{4}'' \times 20$ TPI
First tap, HSS
Class II

   b. Tools—cutting—Carbide brazed

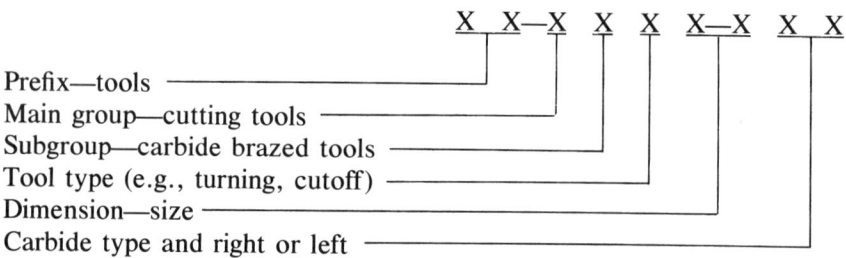

Prefix—tools
Main group—cutting tools
Subgroup—carbide brazed tools
Tool type (e.g., turning, cutoff)
Dimension—size
Carbide type and right or left

Figure 4.2 (*Continued*)

Example

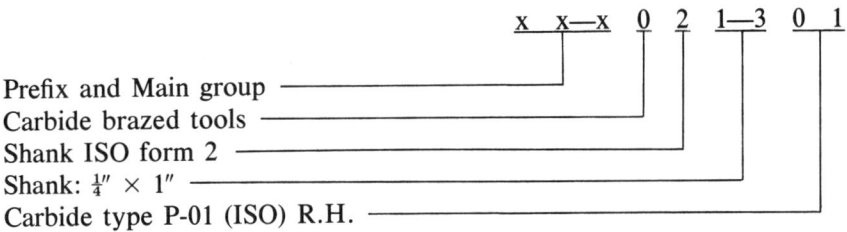

Prefix and Main group
Carbide brazed tools
Shank ISO form 2
Shank: ¼" × 1"
Carbide type P-01 (ISO) R.H.

Figure 4.2 *(Continued)*

introduced into the plant, hence creating a need for new types and groups of materials. Moreover, this system does not last forever. Reorganization and changes in the classification and codes have to be made every six to ten years. The limitation on code length also limits the extent to which the desired secondary objectives are achieved.

Mixed Classification and Coding Systems

This type of system is a combination of the meaningless and meaningful systems. There are many such combinations, but we shall briefly consider only three:

1. *External and internal classification and coding.* The idea behind these systems is to utilize the beneficial aspects of the meaningless and meaningful systems previously discussed. A meaningful classification and coding system will be constructed without code length limitation. The code number will be used internally in the computer software; for external use in inventory, production planning, costing, and so on, a short meaningless code will be used. The software will have a cross-reference table for these two codes. Hence, the limitations of the meaningful classification and coding system are removed, while the benefits of the short code are utilized. Changes in and reorganization of the meaningful classification and coding system will not result in changing codes in inventory and other files. The short code will remain unchanged, the only change being in the cross-reference table.

2. *Group-meaningful classification and coding systems.* This system uses a prefix number, which designates the item group, followed by a meaningless number. To construct it, only the first two steps of the construction of a meaningful classification and coding system must be

followed. It does not have the benefits of either system, but simply provides a general idea of what the coded item is.

3. *Semimeaningful classification and coding systems.* This system concentrates on one or two characteristics of the item and fully classifies them, the last digit of the code being a meaningless number. Only the absolutely essential characteristics of the item are controlled, but they are subjected to an in-depth study and then strictly classified. Other characteristics of the item are treated only as distinguishers and will result in a different short number.

Some of the secondary objectives and applications need not rely on classification and coding in order to be achieved. For example, such secondary objectives as the ability to process groups of materials or the ability to select sources of supply can be accomplished by other means.

4.2 Product Definition—Bill of Material

The purpose of the bill of material is to furnish administrative services to different applications. Figures 2.5 and 2.6 showed the interrelationships between the various applications in the IMS; Figure 4.3 isolates the bill of material and its uses from these two figures.

Each of the applications depicted in Figure 4.3 requires the data in a different structure: For the master production schedule, it is convenient to have a summarized bill of material, in which the total usage of each item is collected into a single list for the product; for requirement planning, it is convenient to have a single-level bill of material, in which only components used at one level are listed; for the costing of a product and its subassemblies, it is convenient to have an indented bill of material, in which the detailed makeup of the end product is shown; and for inventory control, where-used information (i.e., information of the location where an item is used in all assemblies and products) is required. For capacity planning the product structure is required, and for profit and loss report, a full product explosion is required.

The goal of the bill of material file organization is to serve all of the above-mentioned applications, while keeping only one set of files and including each item and assembly only once. This simplifies maintenance (i.e., addition, deletion, and other changes to the structure), File sizes are kept as small as possible, and all users are furnished with the same version of the product definition. These requirements call for random access file organization.

The bill of material regards the product as a "tree structure," and

4.2 Product Definition—Bill of Material

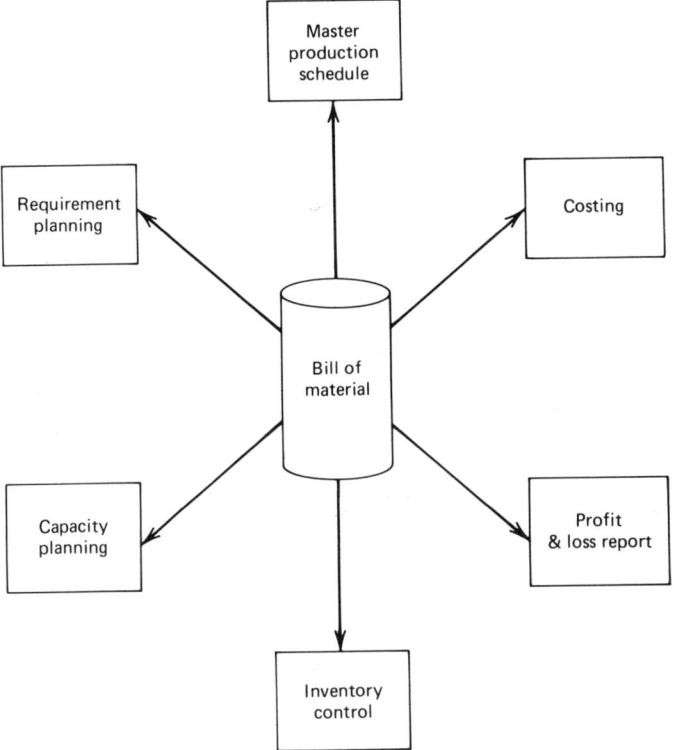

Figure 4.3 Applications based on the bill of material.

Figure 4.4 shows a product structure of this type. The end product is at the top—level 00. The items and assemblies that constitute the end product are one level lower—level 01. Each of the assemblies at level 01 is regarded as an end product for level 02, where the items and assemblies that comprise the assemblies of level 01 are found. The lines connecting the blocks indicate the structure—the items that are included in any assembly. The numbers in parentheses indicate the quantity of each item (assembly) that comprises the assembly one level up. For example, the assembly of one unit of product 2000 consists of two subassemblies 3024 and one subassembly 3991, while one subassembly 3024 consists of three items 3624 and two items 4873. The bill of material file organization is divided into two sections:

1. *Item master.* In this section (file) each item is entered only once, and any desired information regarding it is stored. This stored information might include name, unit of measure, drawing number, supplier code

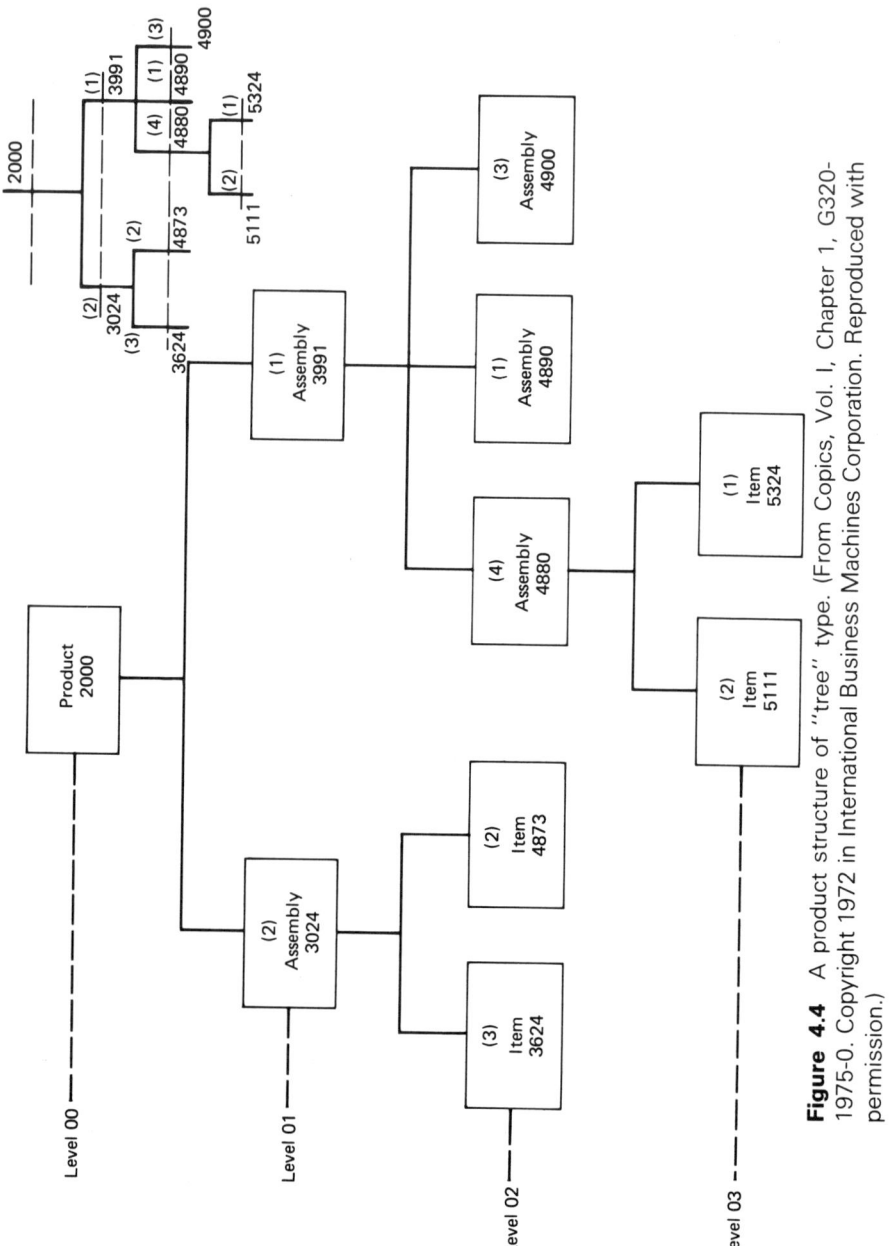

Figure 4.4 A product structure of "tree" type. (From Copics, Vol. I, Chapter 1, G320-1975-0. Copyright 1972 in International Business Machines Corporation. Reproduced with permission.)

4.2 Product Definition—Bill of Material

number, unit price of any type, lead time, make or buy code, scrap factor, and minimum order quantity. The bill of material system only requires a unique item number and some space in order to use this section.

2. *Product structure.* In this section (file) the relationship between items is specified; it is equivalent to the lines connecting the blocks in Figure 4.4. The minimum required data are parent item number, component (child) item number, quantity per assembly, and space for user's data. The user may add any information desired, such as assembly operation scrap factor, offset lead time, and alternatives.

These two sections (files) are connected by a chain address or pointer, depending on the program used.

By means of this organization, an item modification is introduced only once, and the bill of material program will update all products and assemblies concerned. The product structure connection record is given only once, and a connection to the top of the tree will retrieve the whole product.

Figures 4.5 and 4.6 show two types of IBM bill of material file organization—the DBOMP and DL/1. These and other available programs furnish:

- Protection against the definition of a product such that an endless loop will result. Figure 4.7 shows how this situation could occur.
- Low-level code assignments for each item. A low-level code is a number that indicates the lowest level at which a particular item is found for any product in the file. The low-level code is used to reduce processing time.
- Organization of the chain addresses (or pointers) in the files for explosion and implosion (where-used).
- Maintenance programs.
- Standard retrieval programs.

Figure 4.8 shows the logic and output of an indented explosion of a product; the request is for explosion product A (refer to Figure 4.6):

1. The item A record is retrieved and its information is available for output. In this record, there is a pointer to the first product structure (P/S) record.
2. The first P/S record indicates that the first item in product A is assembly item B and furnishes its address. However, it also indicates that there is another item in product A, whose name is stored in a given address in the file.

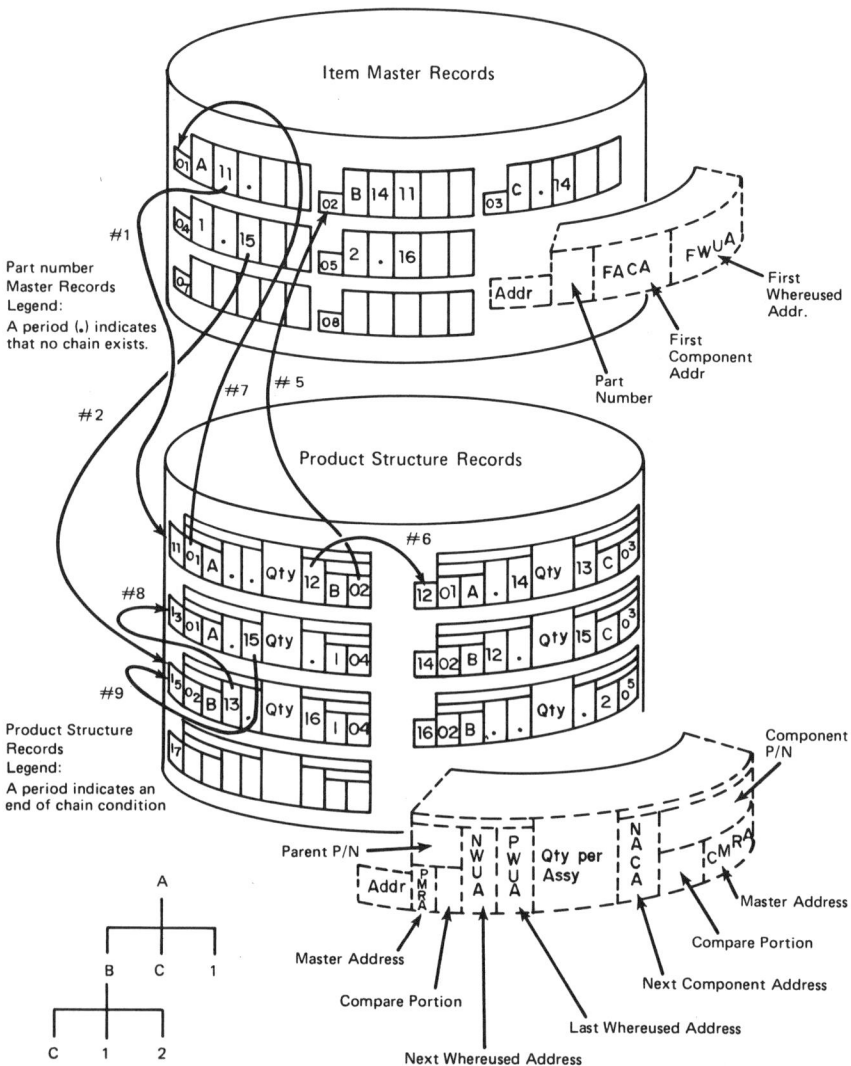

Figure 4.5 The IBM DBOMP bill of material file organization. (From *DL1 Data Base Techniques,* GE-20-0480-0. Copyright 1974 by International Business Machines Corporation. Reproduced with permission.)

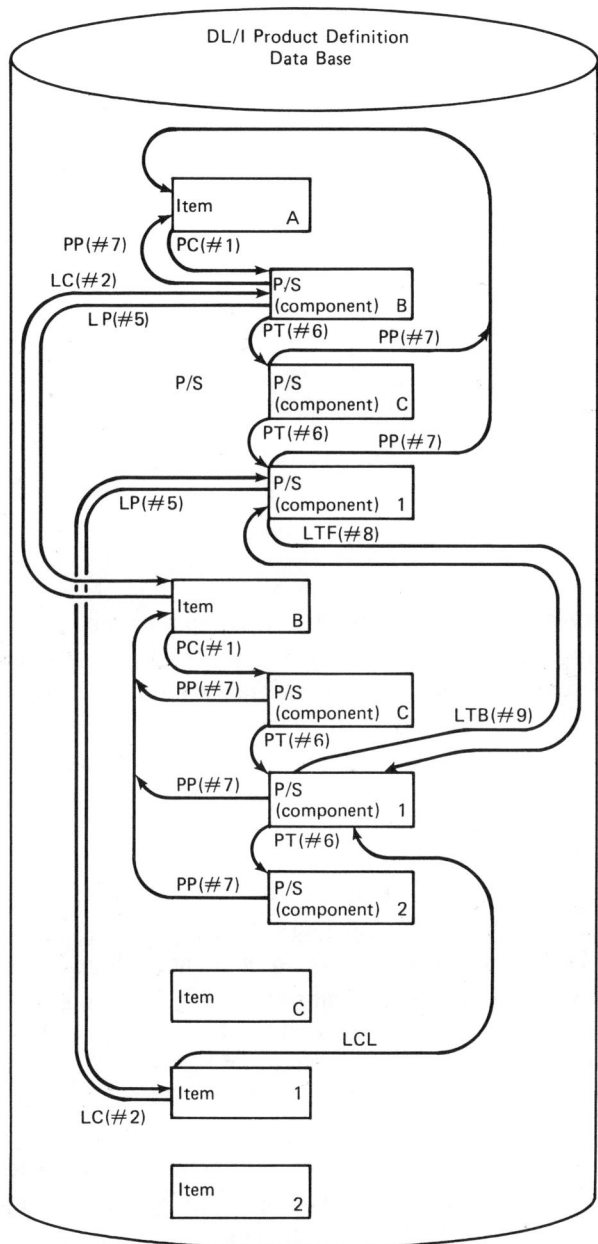

Figure 4.6 The IBM DL/1 bill of material file organization. (From *DL1 Data Base Techniques,* GE-20-0480-0. Copyright 1974 by International Business Machines Corporation. Reproduced with permission.)

Engineering Data Control

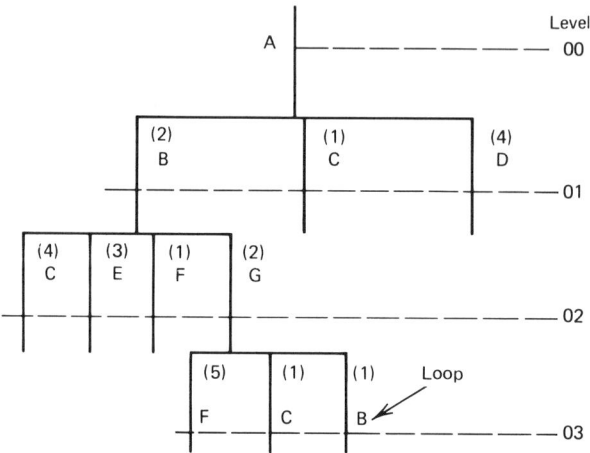

Figure 4.7 Loop definition.

3. This pointer address is stored in the memory table of level 01.
4. The program branches, retrieves, and records item B data. This record indicates that B is an assembly and points to the first P/S record of this assembly.
5. The program branches to the first P/S address. This record indicates that the first item in assembly B is item C and points to it. Moreover, it indicates that there is another item in this assembly, whose name is stored in a given address.
6. This pointer address is stored in the memory table of level 02.
7. The program branches, retrieves, and records item C data. This record indicates that item C is an elementary item on level 02.
8. The program looks at the memory table for the pointer of level 02 and branches to that address. This address is a P/S record that indicates that the next item in the current assembly is item 1 and points to it. Moreover, it shows that there is another item in this assembly, whose name is stored in a given address.
9. This pointer address is stored in the memory table of level 02.
10. The program branches, retrieves, and records item 1 data. This record indicates that item 1 is an elementary item on level 02.
11. The program looks at the memory table for the pointer of level 02 and branches to that address. This address is a P/S record that indicates that the next item in the current assembly is item 2 and points to it. Moreover, it shows that there are no more items in this assembly.

4.2 Product Definition—Bill of Material

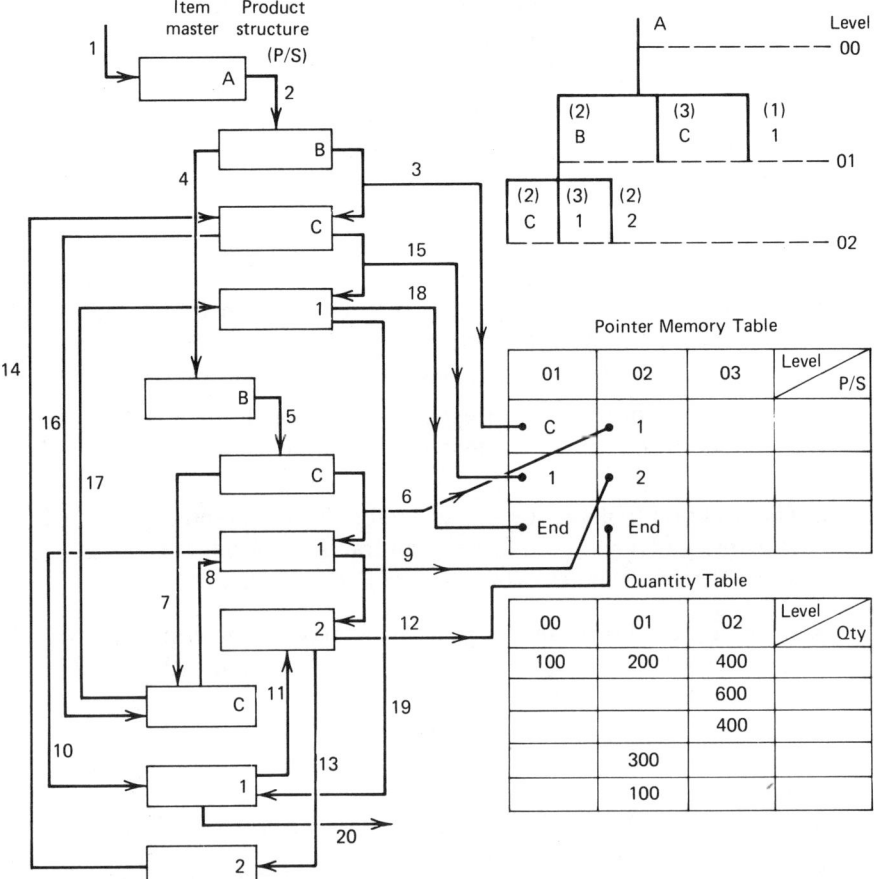

Figure 4.8 Logic and output of an indented explosion of a product. Given an order for 100 units of product A (level 00), the program gives the total required quantity of each component item: level 01—item B, 200 units; item C, 300 units; and item 1, 100 units; level 02—item C, 400 units; item 1, 600 units; and item 2, 400 units.

12. The end sign of level 02 is marked in the memory table of level 02.
13. The program branches, retrieves, and records item 2 data. This record indicates that item 2 is an elementary item on level 02.
14. The program looks at the memory table for the pointer of level 02. It finds the end sign and therefore subtracts one from current level (02), searches the memory table for the pointer of level 01, and branches to that address. This address is a P/S record that indicates that the next

item in the current assembly (A) is item C and points to it. Moreover, it shows that there is another item in this assembly, whose name is stored in a given address.

15. The pointer address is stored in the memory table of level 01.
16. The program branches, retrieves, and once again records item C data. This record indicates that item C is an elementary item, this time on level 01.
17. The program looks at the memory table for the pointer of level 01 and branches to that address. This address is a P/S record that indicates that the next item in the current assembly is item 1 and points to it. Moreover, it shows that there are no more items in this assembly.
18. The end sign of level 01 is marked in the memory table of level 01.
19. The program branches, retrieves, and once again records item 1 data. This record indicates that item 1 is an elementary item, this time on level 01.
20. The program looks at the memory table for the pointer of level 01. It finds the end sign and therefore subtracts one from current level 01. Since the result is level 00, the job is terminated.

Whenever a record is retrieved, processing may be performed on its data and on data stored in memory, such as product standard, actual cost, quantity needed, lead time, requirement planning, and inventory status.

Let us now consider a numerical example. Assume that 100 units of product A are ordered. An indented explosion with the quantity of each item is required. To achieve it, a quantity table with one location per level will be constructed. To follow the program logic, refer to Figure 4.8 and the previous example:

1. Since this is the starting point, it is level 00. The quantity 100 is entered in location 00 of the quantity table.
2. This P/S record indicates that the quantity per assembly of item B of level 01 is two. Multiply this by the quantity registered one level lower (level 00) in the quantity table. This results in 200 units for item B, which is entered in location 01 of the quantity table.
5. This P/S record indicates that the quantity per assembly of item C of level 02 is two. Multiply this by the quantity registered in level 01 of the quantity table: 200 × 2 = 400. This quantity is entered in location 02 of the quantity table.
8. This P/S record indicates that the quantity per assembly of item 1 of level 02 is three. Multiply this by the quantity registered in level 01 of the quantity table: 200 × 3 = 600. This quantity is entered in location 02 of the quantity table.

4.3 Organizing—Bill of Material

11. This P/S record indicates that the quantity per assembly of item 2 of level 02 is two. Multiply this by the quantity registered in level 01 of the quantity table: 200 × 2 = 400. This quantity is entered in location 02 of the quantity table.
14. This P/S record indicates that the quantity per assembly of item C of level 01 is three. Multiply this by the quantity registered in level 00 of the quantity table: 100 × 3 = 300. This quantity is entered in location 01 of the quantity table.
17. This P/S record indicates that the quantity per assembly of item 1 of level 01 is one. Multiply this by the quantity registered in level 00 of the quantity table: 100 × 1 = 100. This quantity is entered in location 01 of the quantity table.

Similar logic will be applied if the scrap factor of items in production and assembly must be incorporated, this also being true for computing the lead time and accumulated cost.

The chain (or pointers) is constructed by the maintenance program and works both from parent to child and from child to parent. In the previous retrieval examples, the pointers from parent to child were used. In the case of where-used inquiries, the child-to-parent pointer will be employed. For example, assume that information is needed concerning which assemblies and products use item C. The item C record will be retrieved from the item master file, whose data can be used for processing. In this record there is a pointer to the first where-used record, while in each consecutive where-used record there is a pointer to the next where-used record and to the parent record in the item master file. In our example, the item C record will have one where-used record pointing to item B and to another where-used record. The second where-used record points to item A and indicates that there are no more where-used records for item C. Item B will have a pointer to its where-used chain. That record will point to item A and indicate that there are no more where-used records for item B. These records and pointers will be used to write the where-use retrieval program.

4.3 Organizing—Bill of Material

The previous section dealt with the master product definition, and the emphasis was on file organization. This section is concerned with work product definition and the user's options in defining the product tree.*

---

*This section has been adapted from an IBM publication. Reprinted by permission from *Copics*, Vol. II, Chapter 1, G320-1975-0. Copyright 1972 by International Business Machines Corporation.

The engineering bill of material produced by the design department does not always meet the requirements of the production departments. An alternative method of specifying the data is often necessary because:

- The actual method of assembly may differ from the way an assembly is designed.
- Service (spare) assemblies are sold in a different form than the original production assembly (e.g., they may include parts for attaching the assembly).
- Engineering modifications are supplied for an installed product.
- Groups of parts issued to an assembly department may include portions of several different engineering bills of material.

The organization of these slightly different bills of material into separate data files can result in a large volume of duplicate data, and their maintenance is costly and time-consuming. Furthermore, errors in data maintenance would result in inconsistent bills of material, which could lead to the manufacture of obsolete items and products and an inability to determine the engineering modification level that was in effect at any given time.

When two bills of material contain assemblies, some of which are unique to one of them, it is possible to incorporate both bills into a single structure. For example, by coding assembly records to distinguish between engineering assemblies and production assemblies, one bill of material can be integrated with the other.

Figure 4.9 illustrates an engineering bill of material containing three assemblies. As a result of space restrictions, the gearbox must be fitted to the drive-shaft assembly before the motor is added, and so the production bill of material has to be different.

The combined engineering and production bill of material (Figure 4.10) is developed through the creation of two production assemblies. The first "removes" the gearbox from the motor drive unit, and the second "adds" it to the drive-shaft and base-plate assemblies. The assembly item records are coded to distinguish between engineering, production, and combined (i.e., effective in both engineering and production) bills of material. The original production or engineering bill of material can be retrieved as required. Service (spare) assemblies (and so on) can be handled in a similar manner.

The creation of different assembly types allows the requirements for multiple bills of material to be stored within a single product definition data base. Since much of the data are shared, this avoids the maintenance and control problems of separate bills of material.

The total quantity required for each item in a product should be the

4.3 Organizing—Bill of Material

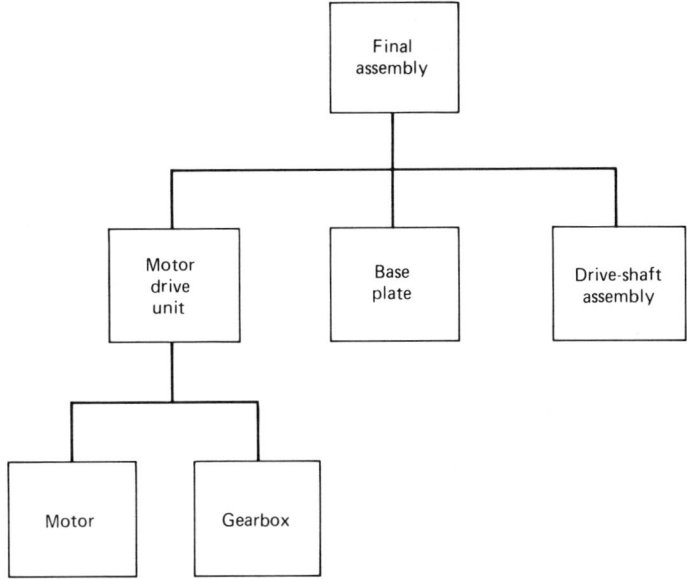

Figure 4.9 An engineering bill of material.

same, whichever bill of material is exploded. The system checks to make sure this balance is maintained.

The production bill of material is needed because the designer's view of how the product is made differs from the actual method of manufacture. The designer's bill of material may tie toegeher a group of assemblies which, although they make up a functional part of the product, are not collected or assembled to each other, but are added to the product separately at different stages of production. The complete fuel system of an engine is an example of a design assembly that is never manufactured as a unit.

Design engineering cannot always go back and alter their drawings to suit production requirements; some of these drawings may have been produced months beforehand. It is common, therefore, for production engineering to put production parts lists on the routing or operation sheets. The effect is that:

- The engineering specification is defined by a single engineering bill of material.
- The production specification is defined by both the engineering bill of material and the production routing. This increases the required file maintenance and, consequently, the probability of error.

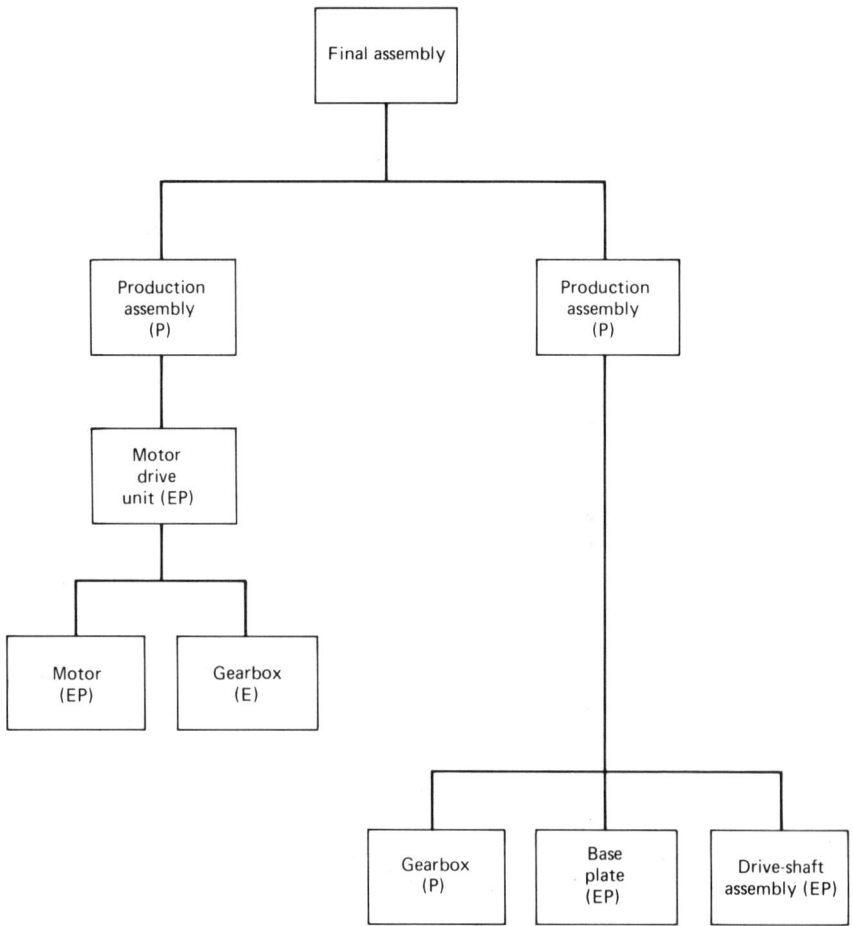

Figure 4.10 Combined engineering and production bill of material. Notation: E, engineering only; P, production only; EP, both engineering and production.

Specifying a production bill of material as a variation of the design bill, however, provides a complete picture of the way in which the product is manufactured, while at the same time reducing maintenance problems and the probability of error.

Two of the factors affecting the structure of the production bill of material are:

1. Which assemblies are stocked before incorporation into a higher-level assembly.
2. Which parts are issued together for assembly.

4.3 Organizing—Bill of Material

A key criterion is the stocking of partly assembled end products and assemblies. This policy is usually adopted to reduce the lead time from receipt of a customer order to shipping. The partially finished products or assemblies are usually stored at a level that provides the greatest commonality, that is, the level from which the widest range of end items can be produced. For example, if a pump can be stocked without feed and delivery connections, a bill of material must exist for this nonfunctional assembly. This allows common requirements from a wide range of end items to be summarized before production batches of the semifinished assembly are planned. Any item "created" by production and held semifinished should have its own unique item number and structure records.

The item numbers of assemblies that exist only for production reasons are identified as "production-only" assemblies in the combined bill of material.

A product *variant* is that part of the product which, while it must be present, may be specified to the customer's choice. For example, every automobile must have a transmission, but the choice of manual or automatic is open to the purchaser. On the other hand, the purchaser may or may not choose the *option* of a factory-fitted radio. Options and variants will be considered together, since the system's logic is identical for both of them.

Manufacturers who offer product variants and options often number the possible combinations in the millions. For each unique product made, a bill of material is required. However, there are certain drawbacks:

- It is often impractical to hold bills of material for every conceivable combination, since many of the possibilities will never be realized.
- If complete bills of material for only the common variants are retained in the system, the product definition data base becomes uneconomically large and contains many nearly identical end items.
- If a special, complete bill of material is specified and then discarded after use, the process of specifying it must be repeated whenever that unique product is required again.

Unfortunately, it is difficult to plan component part requirements if end item bills of material cannot be expressed exactly. A solution to this problem is to structure the bills of material so that the unique version that contains the variants or options can be specified easily and rapidly. This separate, unique bill of material is added to the product definition data only while the product is being manufactured; then, assuming it is no longer needed, it can be discarded.

The methods developed to solve this problem involve automatic conversion of an end item specification directly into a unique bill of material. While many methods are possible, two common ones are discussed here:

1. Coding of variants in the structure record into a compound bill of material.
2. Organizing bills of material by variants.

Where the number of choices is limited, alternative component items can be stored within a single product structure record. The end item specification is compared with each optional structure record as it is exploded, and the specified structure record is automatically included in the exploded bill of material.

Figure 4.11 illustrates the product specification and the compound bill of material for part of an end item in which there are three variants: panel color, left- or right-hand door, and voltage. When the panel section of the bill of material is exploded, the specified color—blue—is compared with the coded color in the structure record, and item number 146292 is selected. This process is repeated for each variant, and the generated bill

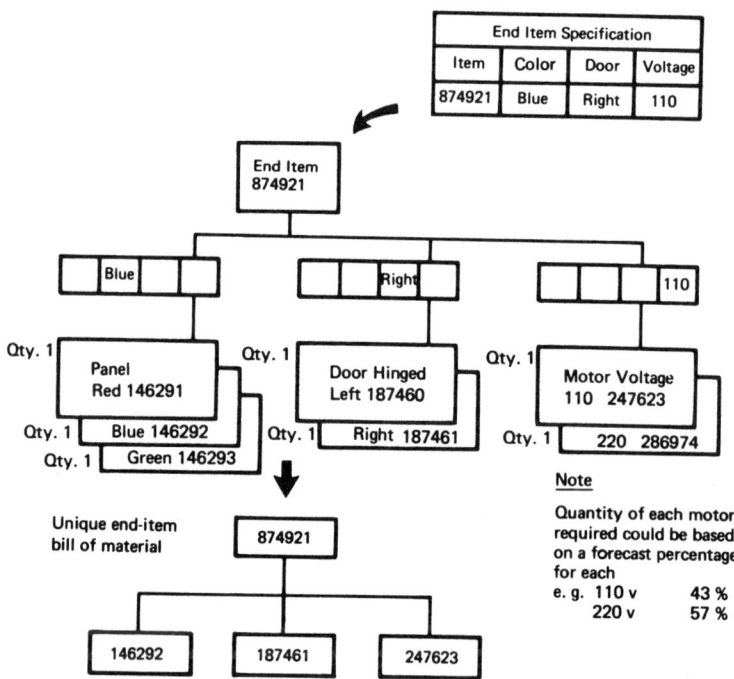

Figure 4.11 Development of an end item by a compound bill of material.

4.3 Organizing—Bill of Material

of material is illustrated in the figure. For the purpose of material requirement planning, the used quantities are expressed as a percentage of usage on the end product; for example, 110 V = 43% and 220 V = 57%.

This is a simplified example, but the techniques can be expanded to include larger sets of choices (e.g., requirements for control switches or instruction brochures in alternative languages).

The variants often occur at a low level in the bill of material and may depend on other variants; for example, a label may describe the specified motor voltage and be printed in a specified language. The huge number of possible end items can be generated from a single bill of material for each major end item.

When the number of available choices is large, a better solution is to organize the bills of material by variants. End item bills of material are specified without any variants (Figure 4.12), that is:

- Items common to all end items are collected into a main bill of material.
- Items that change with a specific variant are collected into a separate bill of material.

Using this type of modular bill of material significantly reduces the total number of separate bills needed to cover a wide range of variants. For example, Figure 4.13 shows an end item with only five variants (capacity, power, etc.) and a varying number of possible choices within each variant. With this combination, over 4,600 end item possibilities exist; yet only 33 bills of material are required in the data base.

In Figure 4.13, if the first two variants each contained 20 possibilities,

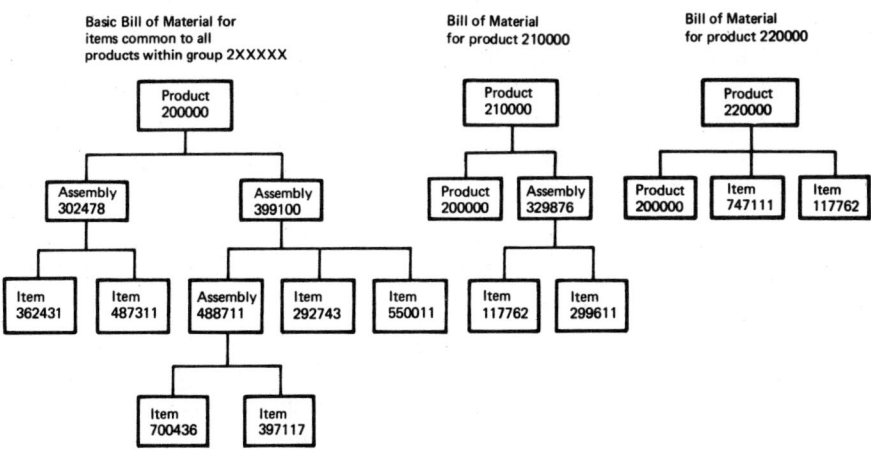

Figure 4.12 A compound bill of material.

Variant	Variant possibilities	Variant possibilities
Main bill of material	1	1
Capacity	12	20
Power	4	20
Material 1	2	2
Material 2	6	6
Switchgear	8	8
Total number of bills of material	33	57
End item possibilities	4,608	38,400

Figure 4.13 End item combinations.

the total number of end item possibilities would rise to more than 38,000, represented by only 57 bills of material.

The specific bill of material for an end item is generated for each product specification. If the total number of variants is low, this can be done manually. However, it is easy for the computer to develop the bill of material and so avoid the manual copying of item numbers with the attendant risk of error and waste of time.

Service assemblies usually fail to correspond with any of the existing bills of material. For example, a part may require an oilproof joint when fitted. This, in turn, may require a gasket or sealing material that is normally part of the next higher-level assembly in the production bill of material. When the item is sold, it must be identified as a service assembly containing the part and the gasket. Items sold together as a service assembly are combined under a spare assembly item number in a manner identical to that used for production assemblies (see Figure 4.10).

In many industries, there are particular requirements that are normally specified outside the bill of material, but ought to be included in it, since they are really part of the product definition. Examples where this applies include:

- *Field installation items.* Some items, for example, electrical connections and piping, can be fitted only after the product has been installed in the customer location. These should be shown as part of the end item bill of material.
- *Packaging.* "Super" levels of assembly, such as finished goods packaging, can extend the number of stocked assembly levels considerably. For example, pharmaceutical products can be stocked in tablet form, in various-size boxes, in boxes within cartons, and in special seasonal

packs. These can readily be held and maintained in the product definition data base.

- *Select components on the basis of a test.* In many electrical or electronic assemblies, the particular resistor used depends on test results. The possible resistors are held as variants in the product structure records. The quantity required is specified as a probability based on past experimental results, for example, 0.40 resistor A and 0.60 resistor B. Requirements resulting from product reconditioning will order the appropriate quantities of each resistor based on these percentages.

4.4 Manufacturing Routing

The goal of manufacturing routing is to prescribe the flow of work in the plant. It derives its information from process planning methods, time and motion study, and presents it to the production phases. Manufacturing routing describes how items are processed and assemblies produced. There are two types of data:

1. *Technical data.* These data instruct the operator how to perform the operation—the cutting speed, feed rate, depth of cut, inspection, and so on—but for the most part do not participate in the IMS. If this information is used at all, it is merely to print the instruction by the computer.
2. *Management data.* These data specify the sequence of operations and the information required to determine the work center load, manufacturing lead time, tools, jigs, and fixtures that will be used. This is the set of data used by the IMS in many applications, and which data will be stored in the files depends on the desired applications.

Figure 4.14 shows the applications that require routing data. Other applications use routing data indirectly, that is, after it has been processed by one of the above applications.

Not all applications treat the routing data in the same manner. Requirement planning, for example, needs only lead time, that is, summarized time, and disregards the single operation. Budget is concerned with a master routing and disregards the variations between batches. Capacity planning, on the other hand, needs the work routing specific to each manufacturing lot. Which data are stored in the routing file is a function of the desired applications. The minimum data required are:

- Operation number.
- Sequence number.

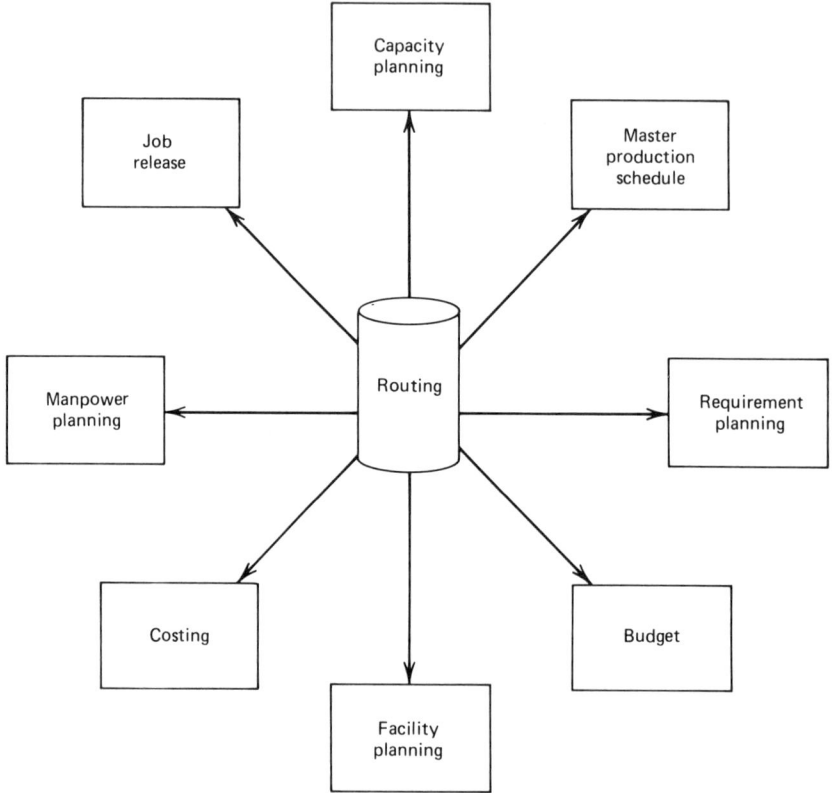

Figure 4.14 Applications that require routing data.

- Operation description.
- Work center.
- Machine type.
- Operation run time.
- Set-up time.

Different applications may regard the operation run time differently: Budget and costing probably need the standard time, while capacity planning needs the actual time.

Indirect operations, such as inspection, moving items, and raw material preparation, can be treated according to plant policy. It is useful to specify the first operation for each item as the issue of raw material from stock. A 100% inspection is regarded as an operation. An inspection

4.4 Manufacturing Routing

carried out by the operator is incorporated into operation time and is specified in the technical data. On the other hand, a random inspection carried on by quality control personnel is regarded as an overhead and is not specified as an operation. Moving items from one work center to another is also an overhead activity and thus is not included in the routing data; the IMS will handle it automatically through its inherent logic.

Specific data for an application, such as overlap code, split code, alternative machine, and alternative operation, are included in the routing data if capacity planning is used; the type of operator will be included if manpower planning is desired. The specific applications of these data will be considered in subsequent chapters.

The routing data are subject to a large volume of minor changes. Many of these changes are temporary and affect only some of the batches in production. They could be due to a change in the raw material, a broken tool, jig, or fixture, the use of an alternative operation, parts being rejected and requiring special correction operations, and so on. These changes have a short-term effect and do not create a permanent alteration in the master routing data.

Some changes are permanent. They may be due to the introduction of new facilities, change in product design, reexamination of the process plan because of a high rate of rejection during production, change of raw material supplier, and, of course, introduction of new items to be produced.

Some applications, such as master production scheduling, budgeting, facilities planning, and manpower planning, use the routing data for long-range planning. These applications are usually performed at random, and not very often. They are not concerned with the minor temporary changes of routing, but rather represent an overall general planning, and thus the standard master routing data are accurate enough for their purposes.

Requirement planning functions on an item level and is not concerned with the details of the operation level. The master routing data will suffice for this purpose.

The other applications are used frequently and require the exact routing data for each item. Capacity planning, job release, and costing are performed at the operation level (however, in some companies costing is kept at the item level) and usually keep their own files as a part of the application program.

Therefore, two sets of routing data are recommended: a master routing set and a work routing set. The work routing data will be created automatically from the master set.

When requirement planning indicates that a manufactured item is needed, or when a manual order for items is entered, a request for automatic routing data retrieval is generated. This request is directed to the master routing file. The routing data are retrieved (with the added information of lot number and size) and directed to the work routing data file. The engineer may examine these routing data and make any necessary changes. Temporary changes are directed to the work routing file, while permanent changes are directed to update work as well as to master routing data.

Requirement planning utilizes the bill of material file. Therefore, it is recommended that the lead time for each item be summarized from the master routing file as a separate task and stored as data in the item record of the bill of material file.

The master routing file can be organized as a separate sequential file. It will perform all its tasks and at the same time obtain all of the benefits inherent to this type of file organization. However, there are those who prefer to keep the master routing file as a part of the data base. In using this type of organization, a clear definition of master work routing files and capacity planning files should be made.

Chapter Five
Master Production Planning

The manufacturing process is directed toward meeting the objectives defined by management. The objectives specify what products are to be produced, in what quantity, and at what cost. The source of these data can be customer orders, forecasts of future demand, or a combination of both. These are nonengineering sources and are not concerned with plant capacity. In the master production schedule, the impact of alternate production plans on plant capacity and load balance is assessed. The result is a practical master production plan, which is the basis for the manufacturing activities, as well as the management and finance activities. The budget is the conversion of manufacturing standards to cost. The future development of the company—for example, its predicted profits, investments in research and development, and addition of new facilities—is reflected in the budget and the master production plan.

These activities and their interrelationship with the other IMS activities will be discussed in this chapter.

5.1 Customer Orders

Manufacturing activities start and end with the customer orders. The master production plan is based on these orders, and hence they initiate all of the subsequent activities of the manufacturing process, which are finally terminated by the delivery of goods. Customer orders also link manufacturing to sales, accounting, billing, management, and finance. Figure 5.1 shows these relationships.

The objectives of the customer orders application are to ensure on-time

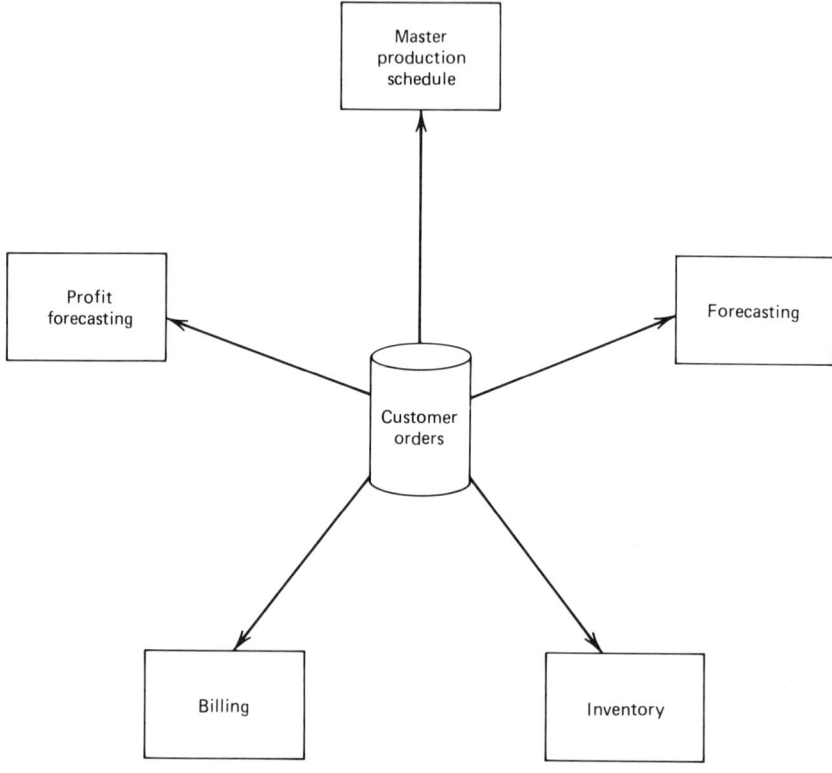

Figure 5.1 Applications that require customer orders.

delivery and improve customer service by rapid and accurate entry of orders into the system.

Customer orders data are organized in the IMS in such a manner as to enable quick response to inquiries, while at the same time eliminating data redundancy in such areas as customer name and address and item description. These inquiries may relate to the status of a customer order, items contained in the order, customer identity of a specific order or item, credit status, and so on. The inquiries can refer to any data available on file; what data are stored depends on the user and the intended application.

The data may contain the following information:

- Customer number.
- Customer name.
- Customer address.
- Bill-to address.

5.1 Customer Orders

- Ship-to address.
- Order date.
- Item number.
- Item description.
- Quantity.
- Unit of measure.
- Allowable quantity variations.
- Requested date.
- Priority.
- Delay penalties.
- Allowable delivery variations.
- Unit price.
- Cost.
- Delivery instructions.
- Inspection instructions.
- Delivery dates.
- Delivery quantities.

The customer orders file may also include data concerning the finance phases, such as:

- *Discount terms.* This can vary from the standard discount scheme based on the type of customer to a special arrangement in which each product class has a different discount for each type of customer and order quantity.
- *Payment terms.* This can include a down payment and a discount if payment is made within some designated period of time after the billing date.
- *Credit terms.* This can include credit by order or a total credit to the customer.

The company should have a system of assigning code numbers to customers and to orders, and each coding system should be controlled by one center. Customer order code numbers can be either chronologically meaningless or meaningful. In the latter case, the meaning may be determined as the type of customer, location of customer, and so on. It should be noted that this meaningful information can be stored as data in the customer orders file. Thus, processing customer orders by customer type or location can be done in either coding system. It is good practice to update customer files prior to entering orders.

The order number can be a serial number; it is used to ensure that orders are not lost between receipt and entry into the system. A control field that indicates the number of items in the order can be added. This will establish controls to ensure that all the lines on an order are entered.

Customer orders may be subject to such changes as cancellation of an order or of an item on order, change of delivery date, or change in quantity (e.g., add items). Prior to updating, it is possible to inquire about the order status in order to learn if the required change is feasible. When changes occur, the file is updated, and the system incorporates the new order into all of the other applications.

Many orders are preceded by a request for quotations, and thus it is good practice to keep quotations on file. This can be a separate file, or the customer orders file with a status code indicating a quotation record can be used. A quotations file can also provide an additional source of information for forecasting. It may serve as an automatic follow-up of response to requests and permit the study of unaccepted quotations, thus helping management to evaluate sales and product cost.

Upon customer acceptance of a quotation, the order is validated and given the revised status of a confirmed order.

The manufacturing process plans its activities through the basic engineering data, that is, bill of material and routing. A unique specification for each order must be available, and care should be taken (especially in the case of nonstandard products or of a standard product with variants and options) to define the product uniquely. This is difficult to do for nonstandard products that are not yet designed; however, the quotation, if detailed enough, can be used for initial planning, changes to the product definition being entered as the design progresses. If the quotation is not sufficiently detailed, the order should be entered in the customer orders file with a special status code, and the required activities will be incorporated into the manufacturing process as unplanned work. Since the customer orders initiate, predict, and plan the manufacturing activities, they can be used to increase the reliability of the manufacturing system by comparing activity taken with predicted activity.

The issue of items from inventory for delivery is validated against the customer orders file. The item catalog number, the quantity, the customer number, and the availability of the item in stock are checked before the issue transaction is accepted as a valid one. Additional validation checks may be made with respect to the required delivery date, the cost, and other items on order. If the transaction is found valid, the inventory records and customer orders file are updated. Shipping papers are prepared, and the handling of the order is transferred from the manufacturing system to the finance system for billing, accounting, and so on.

The receipt of a customer order is an unpredictable, unplanned event.

Therefore, it is difficult to validate the content of the customer orders file. Errors in this file are damaging and irreversible. For example, an error in the product item number will result in the production of products that no one may need, and hence they will become dead stock. Furthermore, these errors lead to delays in delivery and, consequently, customer dissatisfaction. It is a good practice to print the orders, as entered, and distribute copies both to the customers and to sales. The greater the distribution, the greater the probability of error detection.

Validation and error detection are very important, and all available means for checking should be employed: for example;

- Use the quotation file to validate confirmed orders. This is particularly important in the case of nonstandard products or a standard product with variants and options. The quotation was definitely read and inspected, and, consequently, any errors present were noted and corrected.
- Compare the customer code number on the order to the code in the customer orders file. Use a short name and the address for validation.
- Compare the ordered item number to those of standard items produced by the company. Make a cross check with past orders to see if this item fits into the line of business of the customer.
- Cross check the unit cost against the item number and its standard cost.
- Check the quantity by comparing the cost of the order as stated with the unit cost multiplied by the quantity. The unit of measure may also be of assistance.
- Check the delivery date against available stock or manufacturing lead time.

These are not absolute checks, but there is nothing better.

5.2 Forecasting

As stated earlier, the manufacturing process is directed toward meeting the objectives defined by management. The objectives specify what products are to be produced, in what quantity, and at what cost; they can be divided into two types: (1) short- to medium-range objectives, where the resources are assumed to be fixed, and (2) long-range objectives, which usually require the development of plans for substantially modifying existing facilities or obtaining new facilities. Manufacturing personnel act as consultants to management in planning the future development of the company.

It is management's responsibility to supply a workload that will completely utilize the manufacturing resources. Relying on confirmed customer orders may not supply a sufficient workload; management must generate orders and objectives in order to fully utilize plant resources. In setting these objectives, management might want to direct manufacturing to comply with its business and marketing policies in such areas as:

- *Delivery lead time.* For competitive reasons, the delivery lead time must be shorter than the manufacturing lead time, this being a standard procedure in the production of consumer items. A built-to-stock policy is used rather than a built-to-order policy, and thus no customer order exists at all. Producing some standard elementary items or even subassemblies to stock in order to reduce delivery lead time is used for certain specialized products; however, for products with options and variants, the standard items might be produced to stock.
- *Seasonal load balancing.* Products subject to seasonal demand can overload plant capacity, thereby reducing sales. Management may find it a good practice to start production at an early date in order to supply market demand.
- *Manufacturing lot size.* Inventory is a buffer between manufacturing and customers. Thus it can be profitable to manufacture greater quantities than required by customer orders.
- *Management control.* Trade shows, degree of competition, and product-style changes can have an impact on sales and pose business opportunities that management would like to utilize.
- *Financial control.* New tax regulations, bank interest, and new employee contracts can have an affect on management decisions concerning stock level and thus on manufacturing objectives. It is not uncommon for management to initiate a drastic inventory reduction just before the end of the fiscal year in order to make the financial statement "look good."

Manufacturing regards the orders generated by management as customer orders. However, these orders are not delivered until a confirmed customer order is received. Furthermore, the manufacturing system often considers them to be second-grade orders, that is, mainly fillers. This provides a certain degree of freedom in master production planning and thus throughout the manufacturing cycle. The variants, rush orders, and unforeseen rejects can be compensated for by using the resources originally planned for the management orders. It should be remembered that customer orders come first; management orders represent an inaccurate prediction into the future, and an overlap in the time between the confirmed customer orders delivery date and the final product assembly date will probably occur. This overlap will be used by manufacturing to adjust any deviation from the original production plan of these orders.

Forecasting is a tool that management uses to predict into the future. The forecast data should define the expected number of units of each product to be ordered in the future. Moreover, these data should also specify probable variations in future demand. There are three approaches to forecasting:

1. *Intuitive—expert opinion.* In this approach, people close to the market are asked for their estimate of future demand. An average or weighted average is then used to set the forecast. The primary advantages of this approach are simplicity, the use of human judgment, and the consideration of factors from different fields. The disadvantages are that if the number of items is large, the experts cannot afford the time required to do a good job; in addition, salespeople, although close to the market, tend to be either too optimistic or, if paid according to performance, too pessimistic.

2. *Statistical regression analysis.* In this approach, it is assumed that what has happened in the past will continue to happen in the future. The four basic steps in making the forecast are:

    a. Obtain historical data on demand per time period.
    b. Determine the curve equation that best fits past demand.
    c. Project the curve into the future.
    d. Revise the forecast as new demand data becomes available.

3. *Statistical with external factors.* This approach considers the factors that affect demand, such as:

- Distribution and promotion.
- Saturation of the market.
- Competition (share of the market).
- General state of business.
- Consumer taste and potential buying power.

The following four basic steps are involved:

    a. Obtain historical data on demand per period and associate them with one or more economic factors.
    b. Determine the curve equation that best fits past demand and, by regression analysis, assign weights to the factors considered.
    c. Establish the future relationship between factors (by forecast or other means). Using these factors and the weights, convert the historical data into a forecast demand.
    d. Revise the forecast as new demands, factors, and relationships occur.

This approach is more appropriate for groups of products than for individual products, since it is more expensive to develop and use.

Many mathematical models have been developed for forecasting. (See the references at the end of the chapter.) Simple forecasting techniques frequently provide results nearly as good as the more complex techniques, the use of which may involve greater costs than can be justified by the marginal improvement in accuracy.

The moving averages and trends technique is a good and widely accepted model. The historical demand data are represented as scattered points in a time (t) versus demand (D) plot, as shown in Figure 5.2. It is assumed that the curve that best fits the demand is a straight line. Any straight line can be described by an equation of the form

$$D = at + b \qquad (5\text{-}1)$$

where b is the D-intercept at $t = 0$, and a is the slope of the line.

The accepted technique for obtaining the best fit of a straight line to scattered points is the least squares method. By this method the straight line is defined in such a way that the sum of the squares of the differences

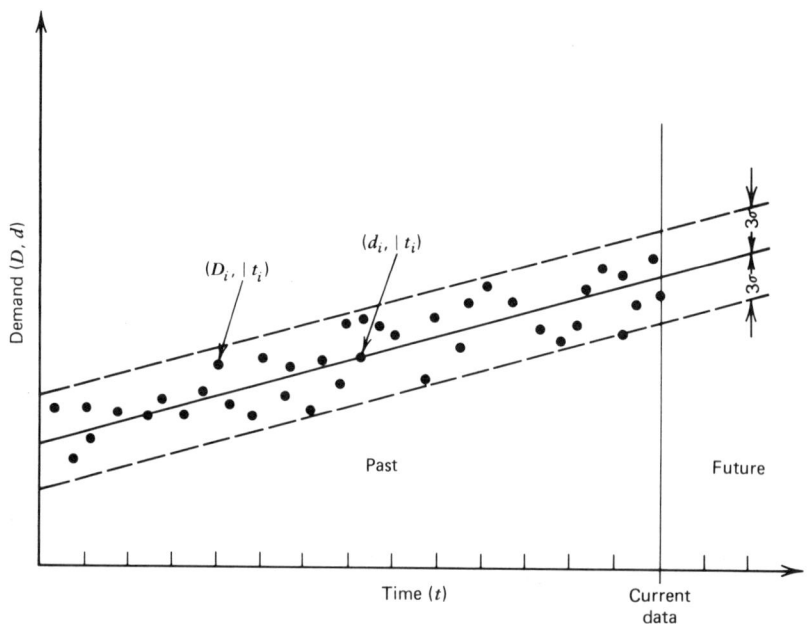

Figure 5.2 Forecasting by the moving averages and trends technique.

5.2 Forecasting

between the ordinates of the suggested line (D) and those of the given points (d) is a minimum. In mathematical terms, this is expressed as

$$(D_1 - d_1)^2 + (D_2 - d_2)^2 + \cdots + (D_n - d_n)^2 = \min \quad (5\text{-}2)$$

$$= \sum_{i=1}^{n} (D - d)^2$$

Substitution of eq. 5-1 into eq. 5-2 results in

$$\Sigma \,[(at + b) - d]^2 = \min \quad (5\text{-}3)$$

Multiplying out of eq. 5-3 gives

$$\Sigma \,(a^2 t^2 + b^2 + d^2 + 2abt - 2adt - 2bd) = \min \quad (5\text{-}4)$$

The variables are the unknown values of a and b. To find their values, the derivative of eq. 5-4 is found first with respect to a and then with respect to b; these two partial derivatives are then set equal to zero, thereby satisfying the curve fitting conditions.

The partial derivative of eq. 5-4 with respect to a is

$$\Sigma \,(2at^2 + 2bt - 2dt) = 0 \quad (5\text{-}5)$$

or

$$\Sigma \,(at^2 + bt - dt) = 0$$

or

$$a \,\Sigma\, t^2 + b \,\Sigma\, t = \Sigma \,dt \quad (5\text{-}6)$$

The partial derivative of eq. 5-4 with respect to b is

$$\Sigma \,(2b + 2at - 2d) = 0$$

or

$$\Sigma \,(at + b - d) = 0 \quad (5\text{-}7)$$

or

$$\Sigma \,(D - d) = 0 \quad (5\text{-}8)$$

Equation 5-8 means that the sum of the differences between the ordinates is equal to zero. In eq. 5-7 b is a constant. If the sum is for m points, $\Sigma b = mb$, and eq. 5-7 may be written as

$$mb + a \,\Sigma\, t = \Sigma \,d \quad (5\text{-}9)$$

The values of a and b can be determined from eqs. 5-6 and 5-9:

$$a = \frac{m \Sigma dt - \Sigma d \Sigma t}{m \Sigma t^2 - (\Sigma t)^2} \quad (5\text{-}10)$$

and

$$b = \frac{\Sigma d \Sigma t^2 - \Sigma t \Sigma dt}{m \Sigma t^2 - (\Sigma t)^2} \quad (5\text{-}11)$$

For simplicity, it is advisable to select $t = 0$ in such a manner that $\Sigma t = 0$; in other words, use an uneven number of periods and set $t = 0$ at the midrange. If this is done, eqs. 5-10 and 5-11 become

$$a = \frac{\Sigma dt}{\Sigma t^2} \quad (5\text{-}12)$$

and

$$b = \frac{\Sigma d}{m} \quad (5\text{-}13)$$

The general regression line equation becomes

$$D = at + b = \frac{\Sigma dt}{\Sigma t^2} t + \frac{\Sigma d}{m} \quad (5\text{-}14)$$

In eq. 5-14, b represents the average demand and a represents the trend. If the regression line is projected into the future, the forecast demand is predicted.

A measure of how well the model fits the data is obtained by means of the standard deviation, which is computed by the equation

$$\sigma = \sqrt{\frac{\Sigma(D - d)^2}{m}} \quad (5\text{-}15)$$

This standard error of estimate implies that 89% of the data are expected to fall within limits of $\pm 2\sigma$ of the regression line (see Figure 5.2).

The above model does not respond fast enough to changes in demand. Thus the use of exponentially weighted moving averages (also called exponential smoothing) is sometimes preferred. The basic idea is to start with the last forecasted demand and add a fraction of the difference between the actual demand of the latest period and the last forecast demand. This is called the smoothing factor, and it is denoted by the symbol α. This value applies to the average demand, the trend, and the standard deviation in the following manner:

New forecast = old forecast + α (actual demand − forecast demand)

A more familiar form is obtained by rearranging the terms:

New forecast = α (actual demand) + $(1 - \alpha)$ (old forecast demand)

The larger the value of α, the faster the forecast responds to a change in demand. A value between 0.05 and 0.4 is normally assigned to α and gives good results. Higher values of α will cause the smoothed average to react quickly not only to real changes, but also to random fluctuations.

This method is widely accepted because of its simplicity, consistency, and substantially lower requirements for data in storage. Furthermore, seasonal demand trends can be imposed on the regression line in the form of sine or cosine functions. However, an initial analysis of curve fitting is required.

In setting up the initial forecasting model, historical data are needed. As in statistical problems, the more data available, the better the model will be. The forecast period length selected has an effect on the forecast model and should match the manufacturing planning period—a period of one week to one month might be a good choice. The minimum amount of data needed depends on the importance of the item, the error effect, and the seasonal demand pattern; 7–12 points are a practical minimum for normal curve fitting. The seasonal demand pattern should be based on data from at least three to four seasons.

The data required for the exponential smoothing model are minimum. However, since no forecast model and parameters last forever, a constant tracking of the model performance is needed. This is done by measuring the deviation between the actual demand and forecast demand; as long as it falls within the normal standard deviation, let us say within $\pm 3\sigma$, the model holds. If the deviation increases beyond these limits, the forecast analyzer should reexamine the model. It is a good practice to track the deviations along a period of time, that is, to retain this information on file. Thus historical demand data for important items must be retained for a period long enough to allow the recomputation of a forecast model.

5.3 Master Production Schedule

The master production schedule transforms the manufacturing objectives of quantity and due date for the final product, which are assigned by the nonengineering functions of the organization, into an engineering production plan. The master production schedule is a coordinating function between manufacturing, marketing, finance, and management. It is the basis for further detailed production planning, such as requirement and capacity planning. Its main objective is to plan a realistic production program that ensures even utilization of plant resources—people and machines. This will be the driving input for detailed planning and will ensure, as much as possible, against overload and underload of resources at all periods of time. If formulated properly, the master production schedule can serve as a tool to marketing personnel in promising delivery dates.

The master production schedule is the phase where due dates are established for the production phases. Thus it controls the relative priorities of all open shop orders. If the master production schedule is unrealistic in terms of capacity, many shop orders will be rush orders with high priority, and the entire capacity planning system will not function correctly. To maintain valid shop priorities, the master production schedule must not exceed the gross productive capacity in any one period.

Planning the master production schedule is a difficult task, since it normally covers a wide range of products and represents a variety of conflicting considerations, such as demand, cost, selling price, available capital for investments, and company marketing strategy.

It is not purely engineering work. The engineers supply information and can simulate different strategies, but the final decision lies with management. In some companies, the sales department is responsible for preparing the master production schedule. In any case, production engineering must be involved in order to ensure a realistic program.

It is erroneous to talk about long-range capacity planning, since no capacity plan can last more than a few days—the conditions in the shop are very dynamic (the various reasons for this will be discussed in Chapter 8).

The new orders and changes in existing orders that occur continuously have impact on the capacity requirement. However, it is important to have a long-range master production plan, the main objective of which is to supply management with a "look ahead" tool—a tool that is needed in order to plan the future of the company. It provides simulation on capacity requirements for different marketing forecasts, on purchasing of new equipment, and on profit forecasts. In addition, it indicates the necessary requirement planning with respect to shop-floor space, warehouse space, transport facilities, and manpower.

No one actually believes that the master production schedule will be accomplished as predicted. However, it is a good starting point for planning; it does not really matter if product A or B will be manufactured some time in the future. The master production schedule represents a framework for the prediction of overall plant performance based on the planning for individual items. Preparing a master production schedule is not a one-time job, but rather a continuous process. The early periods are

5.3 Master Production Schedule

known with reasonable accuracy, while the distant future periods are rough estimates. As time passes, the estimates become confirmed customer orders, and the future is extended further. Every once in a while a new master production schedule is prepared, and it does not have to agree with the previous one.

The importance of master production scheduling is becoming more and more recognized. It is now recognized that it is the key to the success—or failure—of the detailed production planning. However, all of the mathematicians and economists that are developing economic models for production, such as sequencing, economic lot size, safety stock, and reorder point, just assume that there is a master production schedule. It is external to their area of interest, and how good the master production scheduling actually is does not matter to them—they are willing to build a whole theory on sand. It is a complicated problem, so let us leave it alone. The small amount of literature available on this subject merely states its importance and that it should be done; numerous articles have been published on inventory management, scheduling, and forecasting, but to the best of my knowledge not a single one has been devoted to the topic of master production scheduling.

The IMS is no different as regards the preceding discussion. It is a source of information that can display the capacity requirements for different combinations of product demand, lot size, and due date. It is recognized that it is impractical to try out all the combinations possible. Thus human judgment is necessary to predict the most likely combinations, and only they will be simulated by the system.

Basically, from the capacity point of view, the master production schedule represents long-range (infinite) capacity planning. Suppose that the company plans to produce certain products in certain quantities with different due dates. The impact of the plan in terms of production capacity is needed. The IMS files contain the product structure for each product and the routing for each item. The routing file tells in what work centers processing takes place and the sequence of operations; it also provides such lead time information as set-up time and standard machine time. By means of this information, we can explode each product in the order file, or the alternative plan under test, to its components and accumulate the workload at each work center by time periods. Figure 5.3 shows one rough way of doing this. (Refer to Section 4.2 and Figure 4.8.)

Product A can be produced in quantity Q_A at time T_{DA} as follows:

1. The product A record is retrieved and its information made available. This record points to the first assembly operation in the routing file.
2. The first operation is retrieved; it includes such general information as

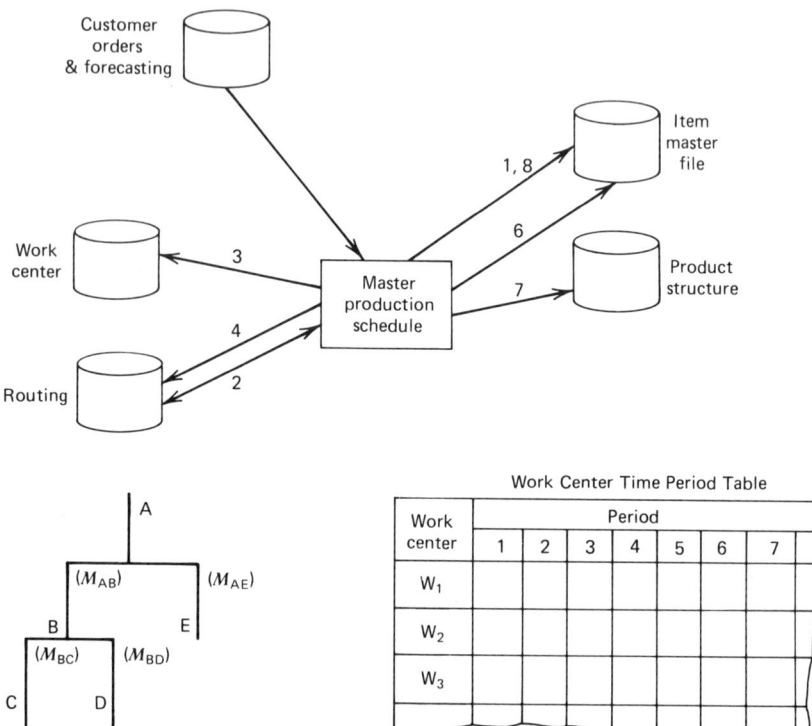

Figure 5.3 Master production schedule computation.

set-up time and machining time. The total lot size processing time is computed (t_{A1}). This record contains information on the work center (W_3) in which this operation is processed and points to its location in the work center file; it also contains a pointer to the second operation. This address is stored in memory.

3. The record of this work center (W_3) is retrieved. The record contains the time period table for the total planning period. Time t_{A1} (in which this work center is scheduled to perform this operation) is added to the table location of the starting period: $T_D - t_{A1}$.

4. The address of the second operation is kept in memory. Its record is retrieved, and computations similar to those in step 2 are performed. Operation time t_{A2} is computed, it is stored in order to update the work center time period table, and it is added to counter T_A, which gives the assembly lead time.

5. Steps 3 and 4 are repeated until a code in memory indicates that there are no more operations for product A.

5.3 Master Production Schedule

6. Controls return to the item master file and use its pointer to the product structure file.
7. The product structure record is retrieved in the same way as in the product explosion. This record indicates that the first item in product A is item B in a quantity of M_{AB} per assembly. This record points to item B in the item master file.
8. The item B record is retrieved. Its due date is the product A due date minus the assembly lead time: $T_{DB} = T_{DA} - T_A$; the quantity is $Q_B = Q_A \times M_{AB}$. This record points to the first operation of subassembly B.
9. Steps 2–8 are repeated until all items of product A have been exploded.
10. The next product in the order file is read and steps 1 to 9 repeated.
11. Steps 1–10 are repeated until all products in the order file have been processed.

We can now take cross sections along the different axes in order to obtain useful information. Figure 5.4 presents some of the important cross sections. Part a is an overall capacity profile of the shop, which shows the profile of normal available capacity (or planned capacity) and total capacity required per time period. This profile is for general knowledge only. If the required capacity is greater than that available, it indicates that the sales forecast exceeds plant capabilities. On the other hand, if the required capacity is equal to or less than the available capacity, it indicates that the plant, as a unit, is underloaded. However, in both cases there might be some work centers that are overloaded and some that are underloaded. To examine this, a work center load profile per time period, such as the one given in Part b, is developed for each work center in the shop. From these profiles, one can learn which work centers in the plant are overloaded and which are underloaded. One can also learn if the overload occurs at all time periods and to what extent, or if there is a mixture of overloaded and underloaded periods and what the average load is.

These profiles provide the information necessary for such decisions as whether to purchase new facilities for highly overloaded work centers; whether to work extra shifts or overtime in moderately overloaded work centers; and whether to balance the load by working overtime or extra shifts at certain time periods, changing due dates, changing lot size, subcontracting, or increasing the inventory buffer. Poorly loaded work centers can be eliminated by transferring their operations to other work centers.

Information on the capacity requirements at a given time period at

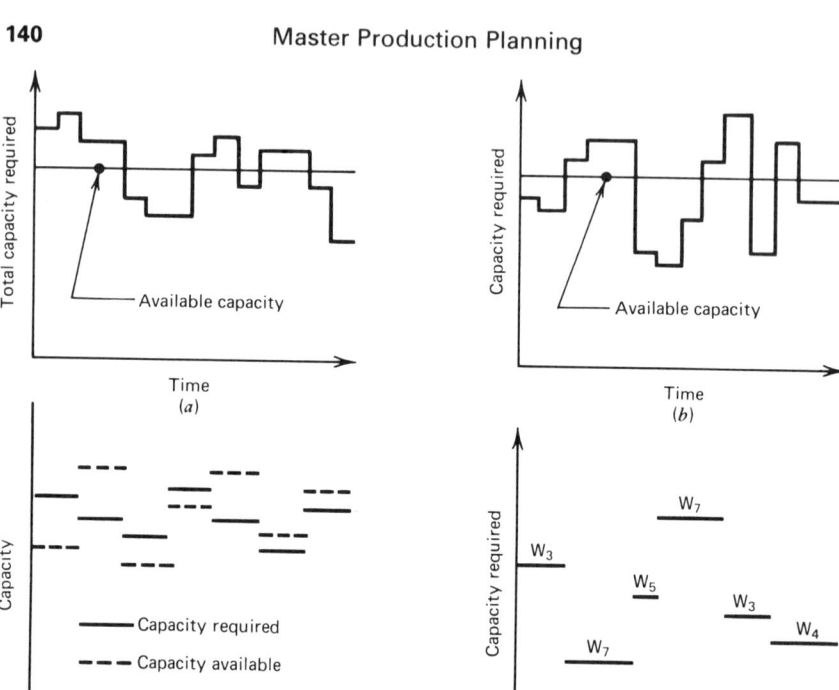

Figure 5.4 Cross-section information of the master production schedule. (a) Overall capacity profile. (b) Work center m load profile. (c) Work center load at time period n. (d) Work center load for product i.

different work centers can be obtained from a profile of the type given in Part c. This information is useful for balancing the load throughout the plant, not just in a single department.

If a decision is made to balance the load by changing product orders or lot size, information concerning the effect of each order on the load profile is needed. This information can be obtained from a profile of the type given in Part d.

It should be remembered that forecasting is not a precision tool and thus has its tolerances—the standard deviations. A trial fit of the master production schedule can be made by using the average, lower, or upper limit of the quantity of each product.

There is a general agreement that lot sizing should be made at the master production schedule level. The master production schedule considers the final product as a whole, not its components; the detailed

5.3 Master Production Schedule

planning is left to the requirement planning and capacity planning phases. The existing models for lot sizing are usually single stage, taking into consideration the set-up cost, but ignoring the capacity of the work center. Using these models, benefits gained due to economic lot sizing at one level of the product tree may be more than offset by the impact this has on other levels. Furthermore, these lot sizes are meddling with the master production schedule, since the previously balanced work center load is upset. These problems would not arise if the master production schedule took into account lot sizing.

Unfortunately, this is a complicated problem, and, to the best of my knowledge, at present there is no feasible mathematical model.

As one may conclude from the above discussion, preparing a good, realistic master production schedule is recognized as a must. However, no one actually knows how to do it scientifically, and thus it is usually done by intuition.

Actually, in the IMS the same scheduling is repeated three times (but for different purposes): master production scheduling, requirement planning, and capacity planning. In each one of these systems, the product is exploded to its components and scheduled by the lead time. The three systems function in series and, in practice, represent a loop problem. For example, master production scheduling ignores inventory and available open orders; consequently, there might be artificial loading in its planning. On the other hand, requirement planning considers inventory and open orders, but ignores capacity; this can create an artificial overload at some work centers and unnecessary rush orders.

One practical way to plan the master production schedule is to look at it as a continuous process in which only changes occur. These changes would include additional orders, cancellation of orders, and revisions of due dates and quantities. The master production schedule is composed of three time zones, as shown in Figure 5.5. The first zone is the "frozen zone." This zone covers job shop open orders or orders containing an item on which work has been started; it also usually covers the product lead time period. In this stage, the order is under the control of the capacity planning and dispatching (order release) phases. Changes in the master production schedule cannot be made during this period. The capacity load profile is obtained from capacity planning and is supposed to be accurate.

The second zone covers the confirmed customer orders that have been processed by requirement planning and entered in capacity planning. However, this zone is out of time range for job shop open orders.

The third zone covers the forecast orders and management filler orders, which represent estimates that enable plans to be made for the future.

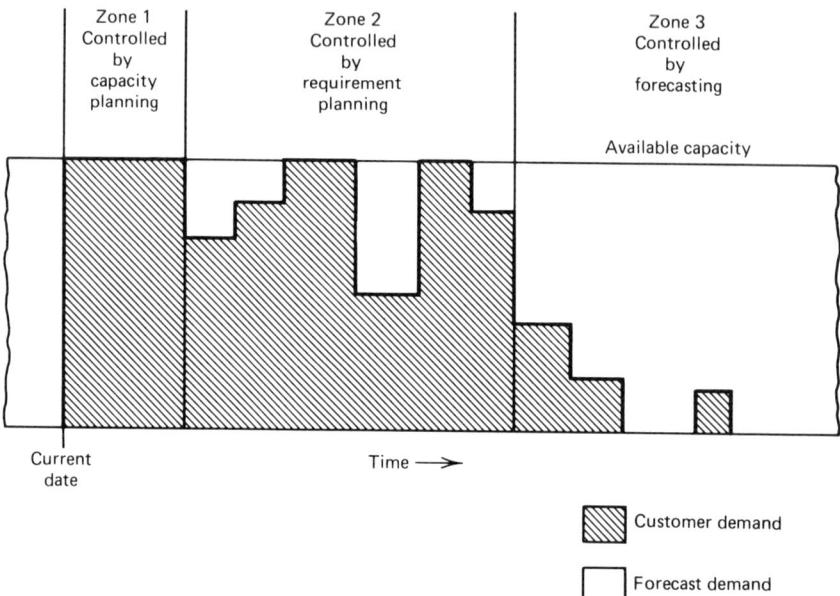

Figure 5.5 The zones of the master production schedule.

In a job-shop (or anyplace where many items, each with a different lead time, are produced) the zones overlap each other; in this case, instead of referring to them as zones, it would probably be better to refer to them as types of orders.

Items constantly advance through these three zones in time. Hence, the delivered orders become historical data, while items that were outside the production range enter as job-shop open orders. New customer orders are accepted, and each order received is checked against the master production schedule. If it was covered by a forecast or management filler order, the order type is changed and no further action is necessary. When the order does not exist in any form, a capacity trial fitting is made.

Initially, it will be attempted to supply the extra capacity required by removing some of the filler orders. If this is not sufficient, some of the forecast orders will be removed. When an order does not fit the schedule, a decision must be made either to change the schedule, to postpone the delivery date, or to split the ordered quantity into several delivery dates. The master production schedule system will, on request, simulate any course of action and supply data for a manual decision. If the order is accepted, the system updates the master production schedule with a reliable delivery date.

Every once in a while (monthly or quarterly) the master production schedule will be reviewed. Initially, only confirmed orders will be used to display the capacity load profile; eventually, however, the forecast and filler orders will be used to balance the load along the range of the planning horizons.

5.4 Management Control and Finance Planning

The master production schedule is the driving force behind further detailed production planning. However, it is also a management tool for controlling and planning the future development of the company. In this section, some of the uses of the master production schedule will be discussed.

Facilities Requirement Planning

In planning the master production schedule, different load profiles are generated, as was shown in Figure 5.4. The planner must schedule within the constraint of available facilities. However, management will use these profiles for decisions on facilities requirement planning.

Short-range or periodic overloads can be compensated for by subcontracting, working extra shifts, or working overtime. On the other hand, long-range or permanent overloads may make it necessary to buy additional machines (or even to build a new factory). The cross sections and profiles discussed in Section 5.3 indicate where a production bottleneck lies and what product creates it. Management must decide if orders are to be turned down and production restricted to available facilities, or if it wants to expand in response to demand.

The master production schedule can supply information for the simulation of different policies with respect to capacity, profit, investment, and manpower.

It is estimated that the life of a machine is about 20 years. Hence, about 5% of the facilities in a plant should be replaced every year. A machine load profile, whether indicating overloading or underloading, can assist management in deciding what type of new machine(s) should be purchased and what changes should be made in routing.

Manpower Requirement Planning

Manpower requirement planning is a conversion of facilities requirement planning. Direct labor needs can be specified according to work center.

(More accurate planning may be done by specifying the skill classification necessary for each operation in the routing file and computing it in a similar way to work center load planning. However, we do not believe that such accuracy is needed at this stage.) Thus the work center load profile can be converted into a manpower profile.

The historical job recording data can be used to arrive at some useful modifiers: A ratio of direct labor to indirect labor can be computed for each department; the efficiency (i.e., the ratio of standard to actual time per each operation) and its average for each department can be computed; and it is possible to compute the ratio of absent time to working time for each department or even for the whole plant.

The work center load profile predicts the total amount of direct labor required for each work center; this total amount can be expressed in terms of the needs of individual departments. The modifiers introduced above can be used to obtain an equation for the required manpower per department:

$$\text{Required manpower} = \frac{\text{(profile prediction)}}{\text{efficiency}} \times (1 + \text{absentee ratio}) \times (1 + \text{indirect labor to direct labor ratio})$$

When the manpower requirements of the individual departments are summed up and the general management staff is added, the result is the total manpower required in the plant per period. These figures can assist the personnel department in planning their recruiting and training activities; they may also be of assistance in preparing a budget.

The manpower planning at this stage is a rough approximation; it for the most part represents a prediction of future needs. A more accurate plan, but valid for a shorter period, is obtained by using the open order files as shown in Figure 2.6. Thus, detailed requirement and capacity plans have been made, the demands are confirmed customer orders, and inventory and on-order items have been taken into account.

Cash Flow Planning

Cash is an important resource, and the predicted cash flow per period of time can assist management in its decisions on when to invest and what commitments to make.

The master production schedule estimates what products will be produced, the quantities, and the delivery dates. These deliveries can be converted into costs and (by the company credit policy) into cash receivable.

5.4 Management Control and Finance Planning

The manpower profile, which was discussed previously, can be converted into salaries and wages. This conversion can be made in the form of a rough estimate (one that we believe is accurate enough at this stage) by using an average salary and wage multiplied by the number of anticipated employees. A more accurate estimate, if so desired, would use departmental averages or skill averages. The final result of either estimate will be classified as cash payable.

Subcontractors are considered as work centers in the master production schedule, but these work centers are not included in facilities and manpower planning. Their load profile indicates the amount of work per time period to be subcontracted. One may define one work center to cover all subcontracted jobs, or define many work centers according to the accuracy desired, or according to other planning purposes, and they can be organized with respect to suppliers, hourly rates, type of process, or any other leading variable. Accordingly, the load of these work centers can be converted into cost, and, upon consideration of the terms of payment policy, these costs can be offset and converted into cash payable.

Material (and purchased items) accounts for a substantial portion of the standard cost of a product—usually about 35%, although for some products it is likely to rise to 80% or more. The percentage of the total cost of any product or item that is contributed by the material cost can be computed using data from the bill of material file or the costing system. The average percentage contributed by the material cost over the complete product mix of the plant can be computed by using the balance sheet; it is a rough approximation, but one that is very easy to arrive at. We may assume that material purchasing is a continuous process that has no particular relationship to time, and that new orders are continuously released to the shop. Thus a rough estimate of cash payable, which is accurate enough for the purpose of predicting cash flow, can be obtained from the following equation:

Cash payable per period for material = total standard cost of master production schedule products × average percentage contributed by material to total cost of product mix ÷ number of periods that the master production schedule covers

If greater accuracy is desired, each product can be treated individually. The amount of cash payable is then equal to the standard cost of the product multiplied by the percentage contributed by material to the total cost of the product multiplied by the quantity. This pay will fall in the period given by the delivery date minus the production lead time minus the safety lead time plus the term of payment.

Other expenses, such as heat, energy, rent, and office, can be treated as if divided equally along periods. The main purpose of the predicted cash flow is to serve as a tool for management in deciding when to invest and what commitments to make. Commitments already made by management will serve to modify the cash flow predicted by using the master production schedule.

Profit Forecasting

Profit forecasting is an important source of information for management. It is essential in planning the future of the plant, since decisions on investment in expansion, new facilities, and research and development are based on potential profit.

The master production schedule can be used to forecast profits. The actual future (forecast) cost can be obtained by conversion of the master production schedule from time to cost, the sales prices are known, and the difference between these two is the predicted profit. A profit margin per product is computed. The master production schedule lists all of the products that will be produced, the quantities, and the delivery dates. Multiplying the product profit margin by quantity gives the predicted profit. Even though it may not be completely accurate, it is a good enough approximation to serve its stated purpose.

The profit estimate is one of the parameters in evaluating different combinations of master production scheduling. Here each product has its profit value, and the problem is to balance the work center loads under the constraints of the forecast orders and the goal of maximum profit. As the master production schedule is made, the profit forecast is known and car be used by management for decisions of various nature.

Budget and Management Control

The master production schedule represents a statement of management objectives in the form of production program that is the best mutually acceptable compromise between the conflicting requirements of the production and sales functions. The approved master production schedule is the yardstick by which management controls operations. If performance deviates from the plan, management must either take corrective action to overcome the deviation or reexamine other decisions and plans that are based on this schedule.

The master production schedule includes many variables with different dimensions. For management control, it should be expressed in a com-

5.4 Management Control and Finance Planning

mon denominator—money. In other words, the production plan is expressed in monetary terms, that is, in the form of a budget.

The conversion of the master production schedule to money values is best done by using standard costs. Standard costs are determined from carefully analyzing the two sets of cost elements—labor and material—for a given level of efficiency. Labor elements cover the whole sequence of operations and can be determined by means of a time study, if one is carried out. Material elements can be determined on the basis of the engineering specifications set forth in the bill of material and anticipated rejects.

A production budget based on standard costs measures efficiency in relatively absolute terms. This shows management how much improvement in performance is still possible and in which areas it could most profitably direct its activities.

Hence, the production budget should be a dynamic one. Whenever a new master production schedule is prepared, a new production budget must also be prepared. The budget is a zero-base budget—it is built on the basis of actual elements, not on the basis of the past budget.

The production budget should be realistic and a working tool. At this stage, we have only estimates; although these estimates are accurate enough for general information, they are not accurate enough for working and controlling purposes. It is recommended that a budget be prepared at a later stage, after requirement planning and capacity planning have been done in detail. At that stage the opening and closing stock of goods on hand, the net requirement, subcontracting, and shop orders are all known with good accuracy.

The budget should specify the amount of work that is planned to be done with a predicted amount of money. Controlling only the money is senseless. There is not much use in knowing, for example, that only 80% of the budget was used—it is very good if all the work has been completed, it is fair if 80% of the work has been completed, and it is disastrous if only 10% of the work has been completed. The figures must be reliable and accurate; otherwise, they cannot be used for controlling purposes.

There are elements in the budget that are not direct expenses and are thus independent of production. These elements can be controlled on a fixed yearly basis and organized in three additional separate budgets:

1. An indirect expenses budget, which includes such elements as general supervision, material handling, maintenance, security, light, heat, electricity, office supplies, engineering, depreciation, and tooling.
2. An investment budget, which includes the approved investment in

machinery, building, tooling, office equipment, laboratories, and so on.
3. A research and development budget, which includes the approved R&D projects of the company.

Budgets are prepared for the lowest controllable level of the company, such as the department. The budget for each successive higher level consolidates those of the level under it. This process continues right up to the general manager.

References

1. Adelson, R. M., "The Dynamic Behavior of Linear Forecasting and Scheduling Rules," *Operations Research Quarterly*, Vol. 17, No. 4 (December 1966), pp. 447–462.
2. Algorithms Supplement, *The Computer Journal*, Vol. 9 (February 1967), pp. 414–417.
3. Baer, Reinhold, *Linear Algebra and Projective Geometry*, Academic Press, New York, 1952.
4. Box, G. E. P., *Discrete Models for Forecasting and Control*, U.S. Department of Commerce Federal Clearinghouse Number: AD 641 935 (June 1966).
5. Box, G. E. P., *Models for Prediction and Control VII, Forecasting*, Clearinghouse Number: AD 650 276 (February 1967).
6. Box, G. E. P., *Models for Prediction and Control VIII, Forecasting Seasonal Time Series*, Clearinghouse Number: AD 650 277 (February 1967).
7. Box, G. E. P., *Models for Prediction and Control IX, Dynamic Models*, Clearinghouse Number: AD 656 684 (May 1967).
8. Box, G. E. P., *Models for Forecasting Seasonal and Non-Seasonal Time Series*, Clearinghouse Number: AD 656 685 (June 1967).
9. Box, M. J., "A Comparison of Several Current Optimization Methods, and the Use of Transformations in Constrained Problems," *The Computer Journal*, Vol. 9, No. 1 (May 1966), pp. 67–77.
10. Brown, R. G., *Smoothing Forecasting and Prediction of Discrete Time Series*, Prentice-Hall, Englewood Cliffs, NJ, 1963.
11. Fletcher, R. and Powell, M. J. D., "A Rapidly Convergent Descent Method for Minimization," *The Computer Journal*, Vol. 6, No. 2, (July 1963), pp. 163–168.
12. Grass, D. and Ray, J. L., "A General Purpose Forecast Simulator," *Management Science*, Vol. II, No. 6 (April 1965), pp. B-119–B-135.
13. Harris, Bernard, Editor, *Spectral Analysis of Time Series*, John Wiley and Sons, Inc., New York, 1967.
14. Kreider, D. L., et al., *An Introduction to Linear Analysis*, Addison-Wesley, Reading, MA, 1966.
15. Lasdon, L. S. and Waren, A. D., "Mathematical Programming for Optimal Design," *Electro-Technology*, Vol. 80, No. 5 (November 1967), pp. 55–70.
16. McMillan, C., and R. F. Gonzalez, *Systems Analysis*, R. D. Irwin, Homewood, IL, 1965.

References

17. Nelder, J. A. and Mead, R., "A Simplex Method for Function Minimization," *The Computer Journal*, Vol. 7, No. 4 (January 1965), pp. 308–313.
18. Papoulis, Athanasias, *Probability, Random Variables and Stocahstic Processes*, McGraw-Hill, New York, 1965.
19. Saaty, T. L., and Joseph Bram, *Nonlinear Mathematics*, McGraw-Hill, New York, 1964.
20. Spang, H. A., "A Review of Minimization Techniques for Nonlinear Functions," *SIAM Review*, Vol. 4, No. 4 (October 1962), pp. 343–365.
21. Spendley, W., G. R. Hext and F. R. Himsworth, "Sequential Application of Simplex Designs in Optimization and Evolutionary Operation," *Technometrics*, Vol. 4, No. 4 (November 1962), pp. 441–461.
22. Stewart, G. W., "A Modification of Davidon's Minimization Method to Accept Difference Approximations to Derivatives," *Journal of the ACM*, Vol. 14, No. 1.
23. Wilde, D. J. and C. S. Beightler, *Foundations of Optimization*, Prentice-Hall, Englewood Cliffs, NJ, 1967.
24. Zoutendijk, G., "Nonlinear Programming: A Numerical Survey," *SIAM Journal On Control*, Vol. 4, No. 1 (February 1966) pp. 194–210.

Chapter Six
Requirement Planning

The master production schedule sets the goals for the production phases of the manufacturing cycle. It specifies what products are to be produced, the quantities, and the delivery dates. Production activities are dependent on this master production schedule; hence, they can be planned and are predictable.

Production activities include plant shop manufacturing as well as subcontracting operations to other shops, purchasing items, subassemblies, assemblies, and raw materials from external sources. At any point in time numerous activities are underway in a working plant. There are open shop orders, open purchase and subcontract orders, and items in storage between operations and activities. All of these activities must be considered when converting the master production schedule into production activities.

A working plant is a dynamic environment, subject to many changes and unplanned interruptions, which may lead to the accumulation of unrequired stock; these changes and interruptions might include:

- Customer orders being added or deleted; quantities and delivery dates being altered.
- Purchasing being restricted by package size, economic consideration, lot size, and changes in delivery dates.
- Interruptions in shop causing early or late finish of jobs; reject rate being higher or lower than anticipated. These will cause unbalance in quantities of different items required for assembly, the controlling item being the one available in the smallest quantity; excess units of the other items are leftover after assembly.

All of these factors lead to the accumulation of stock. This stock can often be utilized later in manufacturing. The objective of requirement planning is to plan the activities to be performed in order to meet the goals of the master production schedule, while accumulated stock is taken into account.

6.1 Manufacturing Activity Planning

Requirement planning is not a new concept, having previously gone under such different names as items balance sheet, activity planning, and inventory management. The logic and mathematics upon which it is based are very simple. The gross requirement of the end product for a specified delivery is given by the master production schedule. This requirement is compared against on-hand and on-order quantities and then offset by the lead time to generate information as to when assembly should be started. All items or subassemblies (lower-level items) required for the end product assembly should be available on that date, in the required quantity. Thus, the above computation establishes the gross requirement for the lower-level items. The same computation is repeated level by level throughout the entire product structure, the net requirement of a level serve as the gross requirement for the lower level. Figure 6.1 shows an example of these computations.

The demand for product A is specified in the gross requirement row of the product A table. There are 40 units of product A in on-hand inventory, and there is an open order to assemble 40 units, which are scheduled for period 3. The demand for 20 units of product A in period 1 will be met from inventory. This will reduce the on-hand quantity to 20 units. The demand for 10 units of product A in period 2 will also be met from inventory, thus reducing the on-hand quantity to 10. An additional 40 units will be received in period 3, thus increasing the on-hand quantity to 50. The demand for 30 units in period 4 can again be met from inventory, reducing the on-hand quantity to 20. The demand for 30 units in period 5 will be partly covered by the 20 on-hand units, leaving a net requirement of 10 units. The demand for 30 units in period 6 is not covered; this results in an additional net requirement of 30 units. Since the lead time for assembling product A is two periods, the assembly of 10 units should start in period 3 and of 30 units in period 4.

Product A is composed of two units of subassembly B and three units of item C. Thus there is a gross requirement of 20 units of subassembly B in period 3 and of 60 units in period 4, while for item C it is 30 units in period 3 and 90 units in period 4. There are 90 units of item B on hand, which covers the demand. Thus there is no demand for items D and E. However, the 60 units of item C that are on hand, while totally covering the demand in period 3, will only partly cover that of period 4. This results in a net

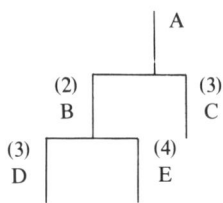

Product A
Lead time—two periods

Period	1	2	3	4	5	6
Gross requirement		20	10		30	30	30
Schedule receipt				40		
On hand	40	20	10	50	20	
Net requirement						10	30
Offset planned orders				10	30	

Subassembly B
Lead time—one period

Period	1	2	3	4	5	6
Gross requirement				20	60		
Schedule receipt						
On hand	90	90	90	70	10	10	10
Net requirement						
Offset planned orders						

Item C (purchased item)
Lead time—three periods

Period	1	2	3	4	5	6
Gross requirement				30	90		
Schedule receipt						
On hand	60	60	60	30		
Net requirement					60	
Offset planned orders		60				

Figure 6.1 Requirement planning computations.

6.1 Manufacturing Activity Planning

requirement of 60 items in period 4. Since the lead time for item C is three periods, the planned order must be offset to period 1.

Thus the activities required to meet demand are:

1. Issue of a purchase order for 60 units of item C in period 1.
2. Issue of an assembly order for 10 units of product A in period 3.
3. Issue of an assembly order for 30 units of product A in period 4.

Lot sizing is incorporated in the requirement planning system and will be discussed in a later section.

As one can see, the logic and mathematics amount to the simple equation

Net requirement = gross requirement − on-hand inventory − on-order units.

This approach tackles the problem of inventory management as part of the manufacturing system. Its reasoning is that the ideal situation would be not to carry any inventory, since from an investment standpoint it is simply a waste of money. However, from an operating point of view, inventory is a necessity; the manufacturing lead time alone accounts for an investment in work-in-process. Other types of inventory are required for various reasons:

- Finished product inventories are maintained in anticipation of demand and to absorb the difference between forecast and actual customer orders.
- Inventories of semifinished components, parts, and subassemblies are maintained in order to reduce the quoted delivery time for end products.
- Inventory is also carried as a protection against possible late delivery. To avoid shortages, delivery can be specified for some time before it is actually required. If delivery is not late, it will raise inventory because the stock must be carried until it is actually used.
- The ordered quantity of a manufactured item may exceed requirements so as to spread the cost of setup over many units. Excess quantities are also ordered to take advantage of supplier volume discounts and to reduce purchasing, receiving, and material handling expenses.
- In many industries the objective is to maintain a constant rate of production despite seasonal fluctuations in customer demand. During periods of low demand, inventory is created in anticipation of a greater demand later in the cycle.

For these and other reasons, the maintenance of inventory is necessary. However, these considerations can be taken into account in the master

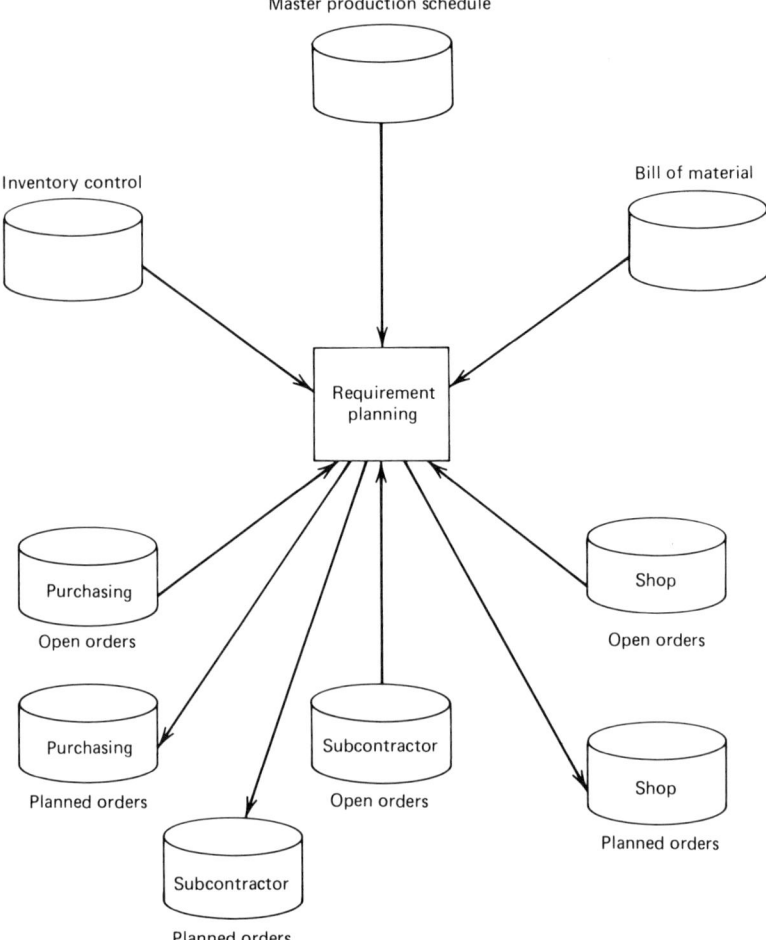

Figure 6.2 The relationship between requirement planning and other applications.

production schedule (as management policy) and not left to inventory management.

In spite of the fact that the logic and mathematics behind requirement planning are very simple, this phase of the manufacturing cycle is very difficult to implement. Figure 6.2 (which is based on Figure 2.5) shows the relationship between requirement planning and other applications. The implementation calls for data from many applications and requires discipline in reporting. The required handling of a large amount of data has been made possible by the computer. In fact, one of the first changes of concept that the computer introduced into manufacturing was in the area of requirement planning. The impact and benefits that accompanied the abandonment of manual systems will be discussed in Section 6.2.

6.2 Manual versus Computerized Systems

Inventory, for the reasons previously discussed, is a necessity in manufacturing. However, the level of inventory must be controlled. In many companies, management has little actual control over the level of this investment. Quite often, inventories tend to grow until some kind of crash program is initiated to reduce them. Investment in inventory is not usually planned—it simply happens. It is almost impossible to control inventory of dependent items (i.e., items that are not sold or ordered by customers as such, but are incorporated in products) without the use of a computer and requirement planning program. By this planning, inventory is controlled as an integral part of the manufacturing cycle. When the computer was not available, one had to apply practical manual systems. In these systems, inventory control was carried out as a stand-alone application, in which such terms and concepts as order point, safety stock, and economical order quantity were very popular. In the manual systems, the inventory recording procedure in most stockrooms, regardless of size, employed records of the perpetual inventory type. These usually show the movement of material in and out of stock as well as the current balance of each item carried. Descriptive and identifying information, such as name, item number, and location in the stockroom, is entered on the record card; such inventory control information as order point and order size expressed in terms of both the minimum balance quantity and the maximum balance quantity is also included. It is very easy for the storekeeper who has updated these records to decide if the balance has reached the minimum value, in which case a new order should be placed to replenish used stock. Each item is treated separately using statistical, economical, and service level considerations, regardless of the production schedule.

When components are forecast independently, their inventories will not usually match assembly requirements well, and the cumulative service level will be significantly lower than the service levels of the parts taken individually.*

---

*The rest of this section has been adapted from an IBM publication. Reprinted by permission from *Copics*, Vol. IV, Chapter 5, G320-1977-0. Copyright 1972 by International Business Machines Corporation.

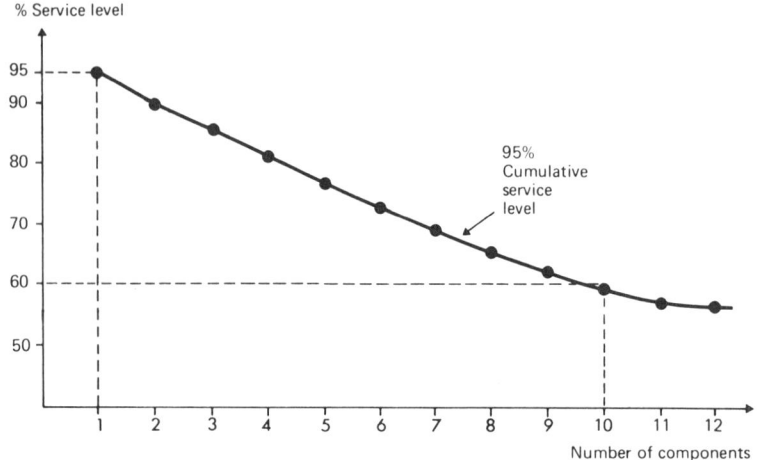

Figure 6.3 Service level as a function of the number of components in an assembly.

This is caused by combining the individual forecast errors of a group of components needed for a given assembly. If there is a 90% chance of having one item in stock when it is needed, two related items needed simultaneously will have a combined chance of being in stock of 81%. With 10 items, the odds of all of them being available are only 35%. Even with the service level set at 95%, the odds on 10 items would be no better than 60% (see Figure 6.3).

This kind of service would be unacceptable at the finished product level, but in many companies such a low service level does exist between component and assembly; expediting, rush work, and an increase in manufacturing costs compensate for this.

Another way of visualizing the timing problem is shown in Figure 6.4. Each time requisitions for components of assembly A are issued, a different component is out of stock. Requirement planning helps avoid this by using planning techniques to coordinate component delivery for all required parts, with the intent of achieving 100% availability of components.

Order point (statistical inventory control) techniques assume relatively uniform usage in small increments of the replenishment lot size. When this basic assumption of gradual inventory depletion is grossly unrealistic, the techniques of order point, safety stock, and economical order quantity will be invalid.

For components of assembled products, requirements typically are

6.2 Manual versus Computerized Systems

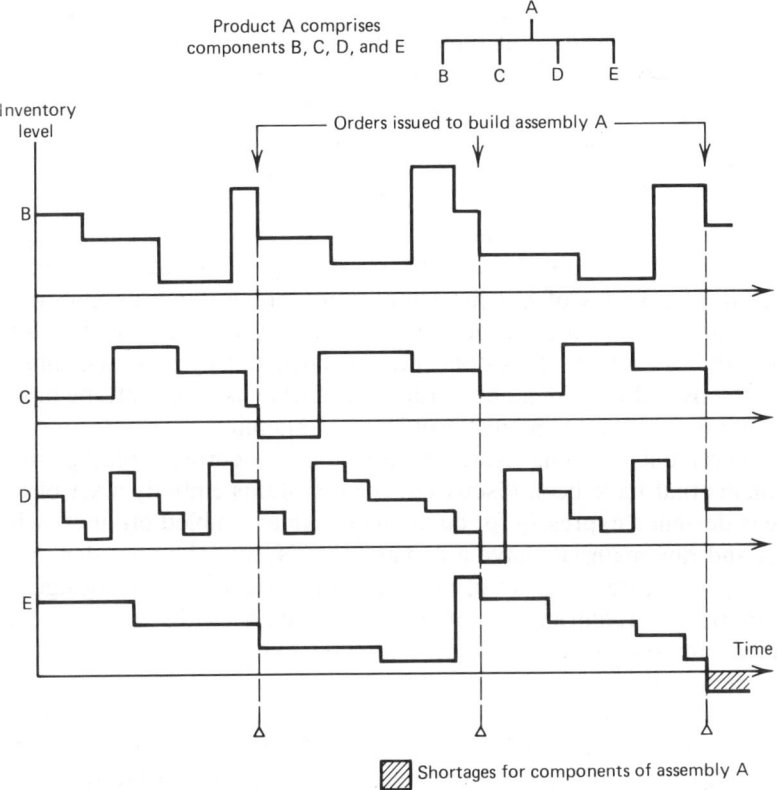

Figure 6.4 Coordination of the components used in an assembly.

anything but uniform and depletion anything but gradual. Inventory depletion tends to occur in discrete "lumps" because of lot sizing at higher levels.

Components are often not available when actually needed because they have been ordered independently of the timing of end item requirements. Even with high safety stock, if two or more different assemblies require an "order point" component simultaneously, it may not be available in sufficient quantity because order point techniques assume that annual demand will average out (typically on a weekly basis).

The fact that some manufacturing companies still get most orders shipped on time even though they use order point systems and maintain safety stocks on components may seem to contradict the above observations. However, such companies do so by carrying unnecessarily high inventories and doing a great deal of expediting. Expediters usually have some kind of "hot list" of assembly components for which there is a shortage and, regardless of the due dates that the inventory system has put on component orders, try to get the right items to the assembly floor. However, expediters face a dilemma with this hot list. If they expedite only those components for which shortages already exist on the assembly floor, it is obviously a case of too little too late. On the other hand, if expediters try to anticipate shortages and expedite components on this basis, they will have an extremely long hot list, and the foreman's logical question is, "Which do you want first?" To do an effective job, expediters really need a series of hot lists; they must break down assembly-floor requirements into time periods and indicate by period what the future requirements will be. This concept, extended through enough time periods to cover the entire manufacturing lead time, is, in effect, the basis of the technique usually called "requirement planning."

Requirement planning systems represent the correct solution to the problems that have been discussed. Such systems embody a set of techniques designed expressly for companies with assembled products whose parts and raw materials have a demand that is, by definition, dependent. This type of system is a set of procedures and decision rules designed to determine the requirements of inventory items—with respect to both quantity and timing—on all levels below the end product and to generate order action such that these requirements are met.

Safety stock is required to absorb a higher than average rate of demand during inventory replenishment. Figure 6.5 shows how safety stock is utilized. Starting from point 1 on the chart, inventory is reduced gradually until it reaches a level called "order point," at which time an order is released. Inventory continues to be depleted until, at point 2, the order quantity is received. However, if inventory is depleted at a higher than average rate, some of the safety stock is utilized (point 3). It is at this point, just before the receipt of the replenishment order, that there is the greatest chance of stockout.

Safety stock is normally not required when demand is solely dependent. The situation is illustrated in Figure 6.6. Only when an assembly order is placed for finished product A is demand for component C generated. The demand for component C is very discontinuous. Maintaining a safety stock of, for example, 20 units when faced with a periodic, "lumpy" demand of 100 units does very little good. Assuming a lead time of one period to replenish, the on-hand quantity of component C is kept unnecessarily high until the next shop order for finished product A is generated.

The ideal situation is represented in the bottom graph of Figure 6.6. The

6.2 Manual versus Computerized Systems

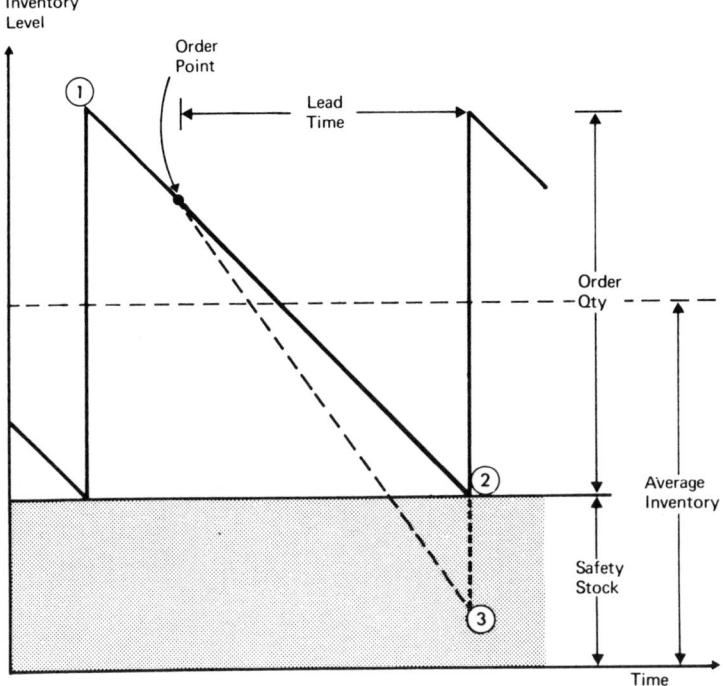

Figure 6.5 Utilization of safety stock.

object is to schedule the production of component C such that it arrives just before being needed in assembling finished product A. In this case, the inventory of component C is carried for only a short length of time. This result is exactly what requirement planning is designed to accomplish: It times the delivery of lower-level components so that inventory will be at a minimum. If the component is delivered on time, no safety stock is required and no stockout will occur.

The size of the order has a significant impact on the average inventory level (Figure 6.7). Through control of the order size policy, management can regulate the level of inventory. Control is exercised by changing either of the two cost elements that determine order size: inventory carrying cost and order cost.

The theory of economical lot sizing is illustrated in Figure 6.8. As the order quantity is increased, the average level of inventory rises, and the carrying cost therefore increases at a constant rate. On the other hand, as the order size increases, acquisition costs (e.g., set-up cost) can be spread

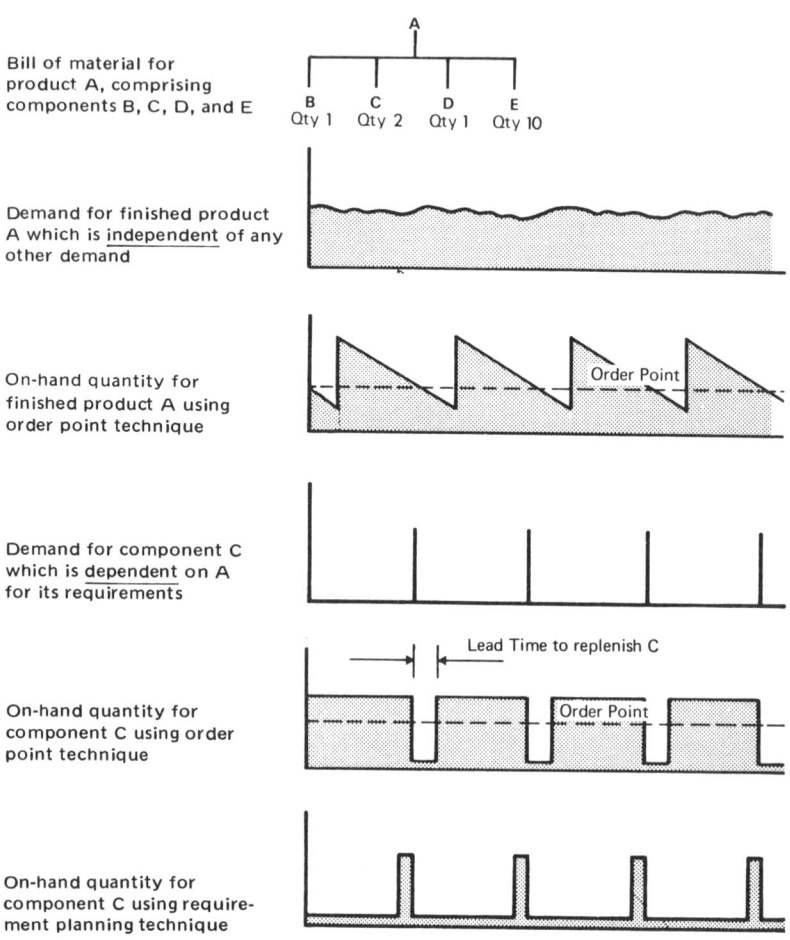

Figure 6.6 Comparison of on-hand inventory with order point technique and requirement planning.

over more units, and the unit cost therefore decreases. The total cost curve in Figure 6.8 represents the sum of the carrying cost and order (acquisition) cost curves. The point of minimum cost indicates the most economical order quantity.

Many techniques are used to calculate the economical order quantity (EOQ). The equation

$$EOQ = \sqrt{\frac{2 \times \text{order cost} \times \text{annual usage}}{\text{inventory carrying rate} \times \text{unit cost}}}$$

6.2 Manual versus Computerized Systems

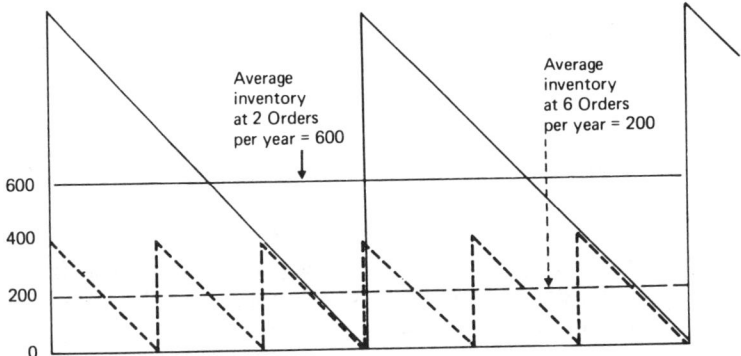

Figure 6.7 The impact of order size on the average inventory level.

which can also be written in the form

$$\text{EOQ} = K \times \sqrt{\frac{\text{annual usage}}{\text{unit cost}}}$$

is easy to use and works well for items subject to a fairly steady demand; therefore, it has found wide acceptance in manufacturing.

This equation assumes that the annual usage is known and that inventory depletion is gradual. In manufacturing, these assumptions are often

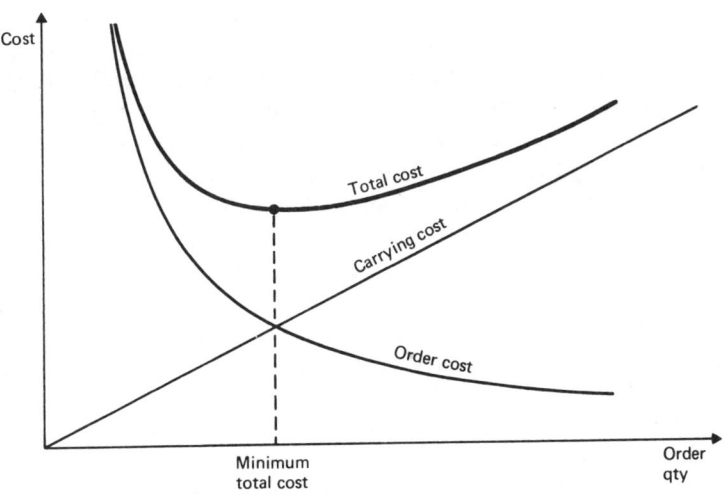

Figure 6.8 The theory of economical lot sizing.

not true and thus the equation ignores the timing of requirements. Therefore, the standard EOQ approach is not recommended for dependent items.

6.3 Requirement Planning Technique

The logic behind requirement planning was presented in Section 6.1, and it was demonstrated by means of an example of computing the net requirement for one end item (one master production schedule record).

The purpose of requirement planning is to plan manufacturing activities accurately by calculating the net requirement in conjunction with the production scheduling. However, if we treat each record of the master production schedule independently, the results will not be accurate: On-hand items will be allocated to orders that require them at a later date, while rush plan orders will be issued for the same items at an earlier date. This situation is demonstrated in Figure 6.9. Three end products are ordered: product A for period 9, product M for period 10, and product P for period 11.

If we calculate the net requirements for each product independently according to ascending order of due date, we will start with product A. This product requires 100 units of item B in period 8.

Suppose that there is a free stock of 100 units of item B on hand. This quantity will be allocated to product A, and no net requirement will exist. Next product M will be dealt with. This product requires 60 units of item B in period 7. Since the free stock of this item was utilized, a net requirement for this demand will result. Next product P will be dealt with. This product requires 40 units of item B in period 6 and 40 units in period 9. Since there is no free stock, a net requirement will result.

These calculated results are unreasonable, since they call for planned orders of item B of 40 units in period 5 and 60 units in period 6 while keeping in stock 100 units not required until period 8. This could result in rush orders or in not meeting the due dates for products P and M. One would expect the calculation to allocate the free stock of 100 units as follows: 40 units in period 6 for product P, 60 units in period 7 for product M, and a planned order of 100 units scheduled to start in period 7 for product A.

The above example dealt with only three orders that have common items. In practice, the number of such items might be much greater. In order to overcome this problem, the requirement planning calculations are carried out by using a low-level code and not by orders. The low-level code is an indication of the lowest level at which an item is used in any

6.3 Requirement Planning Technique

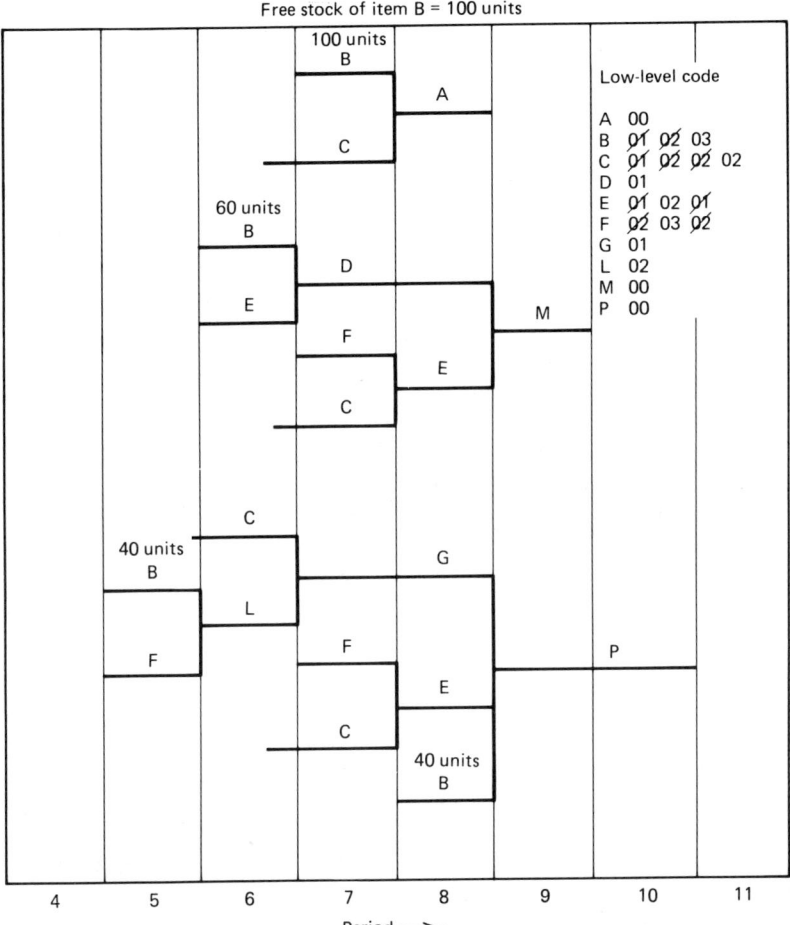

Figure 6.9 Requirement planning for a number of products.

product defined in the bill of material file. In Figure 6.9, item B is at level 01 in product A, at level 02 in product M, and at levels 01 and 03 in product P. Thus the low-level code of item B is 03. This code is kept in the item master record.

Requirement planning uses the bill of material file. A table is constructed for each item defined in the bill of material as shown in Figure 6.10. Initially, all orders of the master production schedule (the independent demand) are read, and the demand is entered as the gross requirement in the table of the appropriate item. These orders are for end

Item B

Free stock = 100 units

Period	4	5	6	7	8	9	10	11
Gross requirement			40	60	100	40		
For product			P	M	A	P		
Schedule receipt								
On hand		100		60	0			
Net requirement					100	40		

Figure 6.10 Requirement planning table.

products that have a low-level code of 00 or for spare parts, which might have any low-level code.

The free stock (on hand) and schedule of received orders are also recorded in the bill of material file in the table of the appropriate item.

The requirement planning starts by calculating the net requirements and planned orders (offset demand period by lead time) of items having a low-level code of 00. The required planned orders are recorded as the gross requirement in the table of the appropriate item; this is done by using the product structure file and the quantity per assembly. When all items on file with a low-level code of 00 have been processed, the calculation will handle items with a low-level code of 01. This process will continue, level by level, until all items in the bill of material file have been processed. By this technique, item B of the previous example will be treated at level 03, at which point the gross requirements of all orders are recorded in its table as shown in Figure 6.10. Thus the allocation of the on-hand inventory will be logical and according to expectation.

Requirement planning is designed to establish order due dates that correspond with the exact date of need, that is, when assembly is scheduled to begin. No safety stock is required, since it has been accounted for in the forecast and master production schedule. Some companies introduce the term "safety lead time"; this is a slight forward adjustment to the component order due dates. Safety lead time for manufactured items, if used at all, should be very short and, above all, uniform for all classes of parts; otherwise, relative shop order priorities become unrealistic.

The calculation of the lead time for manufactured items is based on routing data and modifiers, such as transport time, queue wait time, and inspection time. The formula is unique for each item, and no extra safety lead time is required. If the calculated lead time proves to be unrealistic, a change in the formula should be made. Thus, if so desired, the safety lead time can be incorporated into the calculation of the lead time and not considered as extra time to be added on separately.

6.3 Requirement Planning Technique

The lead time for purchased items is based on the suppliers' quotation, and the suppliers' rating should be taken into account in specifying lead time. Thus the "safety" may be accounted for at the stage when lead time is specified and not by adding extra safety lead time.

The timing of replenishment orders, as calculated by requirement planning, is based on the master production schedule, on the one hand, and on the lead time, on the other hand. The master production schedule balance load is determined with respect to work centers and not separately on specific machines; requirement-planning time period is longer than that used by capacity planning and might result in overloaded machines and inability to produce items on time, as computed by requirement planning. The requirement planning schedule, with its infinite capacity, serves mainly to set objectives for capacity planning and purchasing; It does not usually coordinate between these two types of items. Moreover, it schedules backward from the product due date and hence requirements may fall into the past for example, requirements are overdue. In requirement planning, there is no automatic feature to shift the product network by forward scheduling, starting from the current date.

These features that are lacking in requirement planning exist in capacity planning. The planned orders of requirement planning represent no commitment, since no action has yet been taken on them; they are an input to capacity planning and job release for the initiation of action. Hence, the lead time is for information purposes and there is no need for such safety factors as safety lead time.

Order quantity (or lot sizing) is an important economic factor. The standard EOQ approach assumes a gradual usage of items, which does not always hold true in manufacturing; therefore, its use is not recommended. A part-period balancing (also known as least total cost) gives better results for most items that are subject to fluctuating and possibly discontinuous demand.

A "part-period" is one part held in inventory one period. The total number of part-periods, then, is the number of parts held in inventory multiplied by the number of periods held.*

Each part-period incurs a certain carrying cost. Thus, whether three units are held for two periods or vice versa, the number of part-periods is six and the carrying cost is six times that for one part-period.

When the accumulated carrying cost exceeds the order costs, an order is planned. To facilitate calculation, an economic part-period is calculated. This is the point in time at which accumulated part-period costs

---

*The material from this paragraph to page 168 has been adapted from an IBM publication. Reprinted by permission from *Copics*, Vol. IV, Chapter 5, G320-1977-0. Copyright 1972 by International Business Machines Corporation.

exceed the cost of an order. The due date of the order is the earliest period's requirement included in the calculation. For example:

Item cost = $10.00
Order cost = $60.00
Carrying rate (24% per year) = 2% per period

$$\text{Economic part-periods} = \frac{\text{order cost}}{\text{inventory carrying cost}}$$
$$= \frac{60}{0.02 \times 10}$$
$$= 300 \text{ part-periods}$$

Figure 6.11 demonstrates how the calculations are carried. In this figure, the calculation starts with the first net requirement of 85 units in period 2. Any order delivered in period 2 will be immediately used for these 85 units and no costs are involved. However, if the next period's requirement of 220 units is included (for an order size of 305 units), the 220 units will have to be carried for one week or 220 part-periods. Similarly, the 175 units in period 4 would be carried two weeks and contribute a cost of 350 part-periods. At this point, the cumulative value exceeds the optimum value of 300. The calculation is therefore halted, and the order size is set as the sum of the requirements for all periods up to, but not including, the period causing the economic part-period quantity to be exceeded (period 4). Another calculation is started at this point and results in an order of 395 units. This is continued until all requirements are satisfied.

The planned order may be changed by the *look-ahead/look-back* fea-

Period	1	2	3	4	5	6	7	8	9
Net Requirements	0	85	220	175	145	75	145	110	210
Planned Orders		?							
Periods Carried		x0	x1	x2					
Part-Period Value		0	220	350					
Cumulative PPV		0	220	570					
Planned Orders	0	305	0	?					
Periods Carried				x0	x1	x2	x3		
Part-Period Value				0	145	150	435		
Cumulative PPV				0	145	295	730		
Planned Orders	0	305	0	395	0	0	?		

Figure 6.11 Demonstration of a part-period lot size calculation.

6.3 Requirement Planning Technique

Example: Economic part-period = 100

Period	1	2	3	4	5	6	7	8	9
Net requirements	50	35	30	5	35	20	5	30	35
Cumulative part-period value	0	35	95	110					
Normal part-period balancing									
Tentative orders planned	115				(x)				
Part-period value for next requirements					15	35			
Look-ahead									
Final orders planned	120				x				

Figure 6.12 Demonstration of a look-ahead/look-back calculation.

ture. This feature prevents planning receipt of an order when a period of abnormally low demand precedes a period of high demand. The basic concept is to avoid excessive carrying costs during periods of low demand. The solution results in an order being received in a period of high demand.

Figure 6.12 demonstrates how the calculations are carried. The look-ahead example illustrated attempts to move the next planned order to a period of high demand.

After an order has been tentatively planned, it "looks ahead." The cost of carrying the next period's demand with the first planned order is calculated (in the example, the demand of period 4 increases the cumulative part-period value of the first planned order by 15). This is compared with the cost of carrying the demand of the following period for one extra period (in the example, carrying the demand of period 5 in an order placed for period 4 instead of period 5 increases the cumulative part-period value by 35). The approach yielding the lowest cost is chosen. Since carrying a quantity of 5 for three periods is cheaper than carrying 35 for one period, the quantity of 5 is combined with the first order.

The calculated EOQ does not take into account certain practical limitations that affect actual order size. Limitations that are applied each time an order size is calculated are stored on the inventory segment of the product definition record. Some of the more common order size limitations are minimum and maximum quantities, package or container size, storage space, tool life, joint replenishment considerations, and rounding or multiple factors.

Lot size restrictions can be used individually or in conjunction with one

Order size limitation examples	Requirements Planned orders part-period balancing	0	85 305	220 395	155	145	95 255	145	110 370	210	130
1	Rounding factor (10)		310	390			260		370		
2	Maximum (350)		305	350			300		350		
3	Minimum (300)		305	395			300		325		
4	Pack size (50)		350	350			300		350		
5	Maximum (300) + pack size (100)		400	300			300		400		

Figure 6.13 Effect of different order size limitations.

another. The effect of some of these order size limitations is illustrated in Figure 6.13.

There are two basic approaches to requirement planning: requirement regeneration and net change requirement planning.

With requirement regeneration, the old plan is literally thrown away every time a new version of the master production schedule is authorized; the explosion starts again from scratch. This has the advantage of discarding old errors, plus data made invalid by changes, along with the old plan.

Net change requirement planning is a continuous system that must be continuously maintained. The old plan is retained and modified with current changes. This system presupposes that high data integrity can be sustained in both transaction entries and record maintenance. Before installing such a system, a company must be prepared to impose and maintain procedural discipline. Unless and until the management of a given company creates such a climate of discipline, that company can never attain a fully effective requirement planning system.*

Traditional requirement planning is an excellent tool for calculating the activities required in order to meet the goals of the master production schedule while maintaining a minimum level of inventory. However, it cannot bridge and automatically coordinate the conflicting timing scheduling of the manufactured and purchased items. Shop dynamics (which will be discussed in Chapter 8) and actual machine loading are out of the scope of requirement planning. Its logic is concerned with the overall orders as stated in the master production schedule and not with specific customer orders. Requirement planning calculates the net requirement and planned orders for each item separately; the product network is lost in the calculations, while the due dates for each planned order are based only on the lead time. Thus requirement planning is a rough scheduling. In manufacturing, one expects delays in deliveries, excess scrap, unscheduled stock issues, and overloaded machines. These factors offset scheduling and should cause a chain effect throughout the product network. Requirement planning is not capable of coping with this problem.

It is advisable to use requirement planning only for netting the requirements for each item, while controlling plant daily operations through the capacity planning and job release systems. These systems work with the product network and are able to shift networks backward or forward, to take into consideration the machine load, and to accomplish an accurate, realistic scheduling.

"Pegging" is a feature that can be added to requirement planning in order to link it with capacity planning. If the capacity planning system is not used, pegging can be of assistance in altering due dates manually. The "pegging" feature is discussed in Section 6.4.

6.4 Pegging*

The standard requirement planning procedure does not provide a connecting link upward to parent item records. Consequently, a display or printout of any item record shows requirement figures that are valid but anonymous, and this limits the usefulness of the standard system.

"Pegging" individual requirements to their specific sources is a significant feature that can be added to requirement planning; it provides an upward traceability from component to parent item record, all the way up to the end item requirement stated in the master production schedule.

Pegged requirements are used, for example, to:

- Check the source of requirements.
- Trace the effect of component delays on the delivery date of the finished product. For example, if a lot of castings is scrapped, pegging will specify which customer orders will be affected and what the extent of the delay will be.
- Examine the validity and significance of a system-generated request to change the due date of an order.
- Discover the effect of a pending engineering change on a customer's order or trace upward to the product serial number on which the change will become effective.

*The material in this section has been adapted from an IBM publication. Reprinted by permission from *Copics,* Vol. IV, Chapter 5, G320-1977-0. Copyright 1972 by International Business Machines Corporation.

- Maintain the customer's identity down through lower-level component orders for purposes of lot costing, inspection standards, and so on.
- Resolve contention between individual requirements for existing inventory and open orders.

When pegging is applied, two additional steps are appended to the logic of requirement planning. The first step is to retain the detail requirements in addition to summarizing them. This will establish effective upward traceability.

Figure 6.14 illustrates the concept of pegging and the basic data required. When planned orders for assemblies A and B are exploded, they generate individual (detail) requirements. If an item (E) is designated as pegged, its requirement is not merely summarized with other requirements, but, in addition, a detailed record is retained. The number of the planned order generating the requirement is associated with these data.

The second step is to allocate the existing and planned inventory to the detail requirements and to record this allocation. To do this, additional connection fields (segments) are created and maintained. These consist of the order number to which the order or on-hand quantity is being pegged and the allocated quantity. In the example of Figure 6.14, the released order 28040 is pegged to three different detail requirements records, while planned order 28777 is pegged to two requirements, leaving a quantity of 80 to cover future requirements.

In Figure 6.14, the additional data needed for the pegging of inventory and orders are indicated by the shaded areas (the assumption here is that requirements are already being held in date/quantity format). All of the data needed for pegging are generated at the time that component requirements are exploded.

Pegging can also be used to allocate available inventory and planned orders for finished products to customer orders. In a sense, the customer order becomes the top-level requirement.

The procedure just outlined is called "single-level pegging" because it identifies only the next higher level to which the item is allocated. Further tracing upward to the end item shop order uses the system-generated connection segments.

As an option, the end item shop order identity can be carried down to each component order. This can be achieved either by maintaining a record for every intermediate assembly order using the components or, to conserve file space, by maintaining only the end item identity and the next where-used shop order. This option is called "full pegging."

Whether to choose single-level pegging or full pegging depends on computer economics. Although the same information can be retrieved

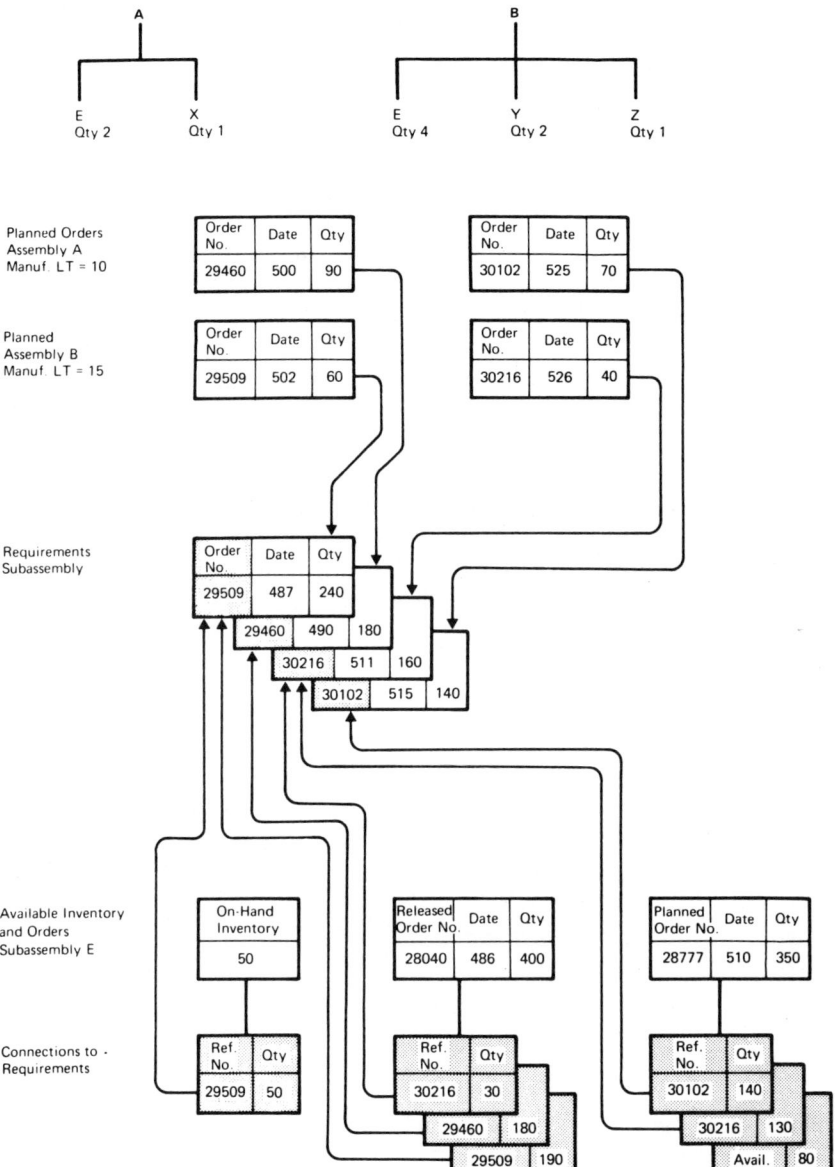

Figure 6.14 The concept of pegging and the basic data required.

either way, the where-used type of retrieval employed in single-level pegging reduces file space but increases the computer time required for retrieval; the converse is true for full pegging.

The requirement planning system can use pegged requirements for many different purposes, as previously listed. The following sections contain a discussion of some of these uses.

Tracing the Effect of Component Delay

Either single-level or full pegging allows the effect of a lower-level component delay to be traced to the customer orders or end item schedule that it affects. If delays are encountered, the system, using its pegging facilities, can reschedule the other related component orders accordingly, and capacity is not wasted trying to produce other components affected by the delay.

Figure 6.15 portrays this situation. Subassembly B is delayed, and this forces the delay of assembly A. The shop order for subassembly D is also covering a requirement for assembly X and therefore cannot be delayed. However, the shop order for subassembly C is assigned exclusively to assembly A, so its planned order due date can be revised to a later one. Its

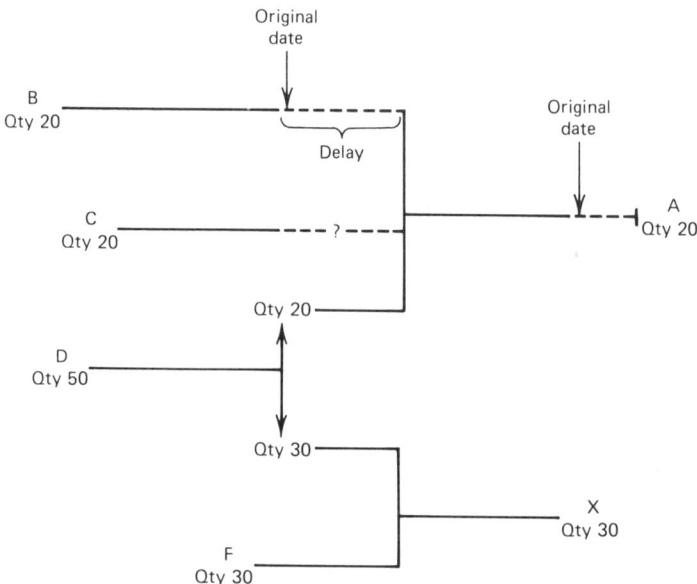

Figure 6.15 Effect of delay in one component on product assembly.

priority is lowered, and other, higher-priority shop orders competing for the same work center are given preference. If there are no other higher-priority orders, subassembly C is finished on schedule.

Checking the Validity and Significance of a Reschedule Message

Requirement planning has the ability to continuously reevaluate the validity of originally established due dates for released orders. In those cases where the date of actual need has changed and differs from the scheduled due date, the system generates a message to the inventory administrator, recommending that the order be rescheduled. Under another option, the system reschedules the order automatically.

In the former case, the inventory administrator can use the pegged requirements capability of the system to verify the validity—and significance—of the message. The inventory administrator may choose not to follow the system's recommendation. For example, in the case of a request to advance a component order due date, the inventory administrator might trace the requirement to an end product, learn that demand is running below forecast, and ignore the message.

Segregating Particular Product Orders

Some companies, especially those producing on government contract or making such very large finished products as turbines, must sometimes put restrictions on the requirements that can be grouped together into a planned order. The restrictions are usually imposed by the customer for certain products. The reasons for the restriction may include:

- A special lot-costing method, such as occurs with certain government contracts.
- The application of special inspection standards to particular orders—for example, military (or commercial) standards.
- The need to provide management with the capability of expediting or delaying items specific to a particular customer order.

6.5 Discipline Needed for Requirement Planning

A successful implementation of requirement planning depends heavily on the accuracy of the data used. Errors in data might cause an increase in inventory level and work-in-process, longer queues in shops, longer lead times, and dead stock. Furthermore, errors will be present in both the net requirement and the due dates. In addition, capacity planning, job release, and purchasing are based on the activities planned by requirement planning. These three phases function in series, each representing a more detailed planning than the previous one. Hence, an error in the requirement planning data will result in errors in all the detailed planning. Thus expediters and shop personnel very quickly learn not to trust the formal IMS.

Error-detection methods were discussed in Chapter 2, and the reliability of each phase of the manufacturing cycle is discussed in the chapter devoted to that phase. Requirement planning involves the data of many systems. In the following sections, aspects of the required interrelationship discipline between phases will be discussed:

1. The net requirement formula is very simple:

    Net requirement = gross requirement − on-hand inventory
    − on-order units

Requirement planning is performed on random dates—it might take place every week or every month, depending on plant dynamics. At the point in time when there is a need to replan requirements, the required data must be readily available.

On-order items are eventually received, and control is transferred from purchasing to the inventory system. Care must be taken that both systems be updated simultaneously; otherwise, the available quantity either will be taken into account twice—as on hand and on order—or will not be taken into account at all.

Since storekeepers are responsible for the physical inventory, they often may not wish to record the receipt of goods as stated in documents, preferring to count and check for themselves. This checking takes time, and incorrect requirement planning can result in the interim. One solution is to leave the items received under the control of the purchasing system; however, this might result in an incorrect late delivery report. Another approach is to keep a "shipment file" as part of the inventory system. In this file all receipts are recorded as soon as the goods are received (i.e., the file merely records the receipts as stated in the documents), and the storekeeper is not held responsible for these items.

A fixed amount of time is allowed for the storekeeper to count and check the receipts. This, in addition to the shipment file, can be used as a follow-up and reminder to the storekeeper if any shipment received has not been checked in on time.

6.5 Discipline Needed for Requirement Planning

On order	
Transactions	Balance
	50
− 50	0 (1)

Inventory	
Transactions	Balance
	100
+ 50	150 (3)
+ 40	140 (4)

Shipment	
Transactions	Balance
+ 50	50 (1)
− 50	0 (2)

Inspection	
Transactions	Balance
+ 50	50 (2)
− 50	0 (3)
− 50	(4)

Rejected	
Transactions	Balance
+ 10	10 (4)

Figure 6.16 Procedure required to account for the correct quantities of items.

In some industries, the received goods must pass inspection and laboratory tests. In such cases they are issued from stock, and the on-hand quantities must take these items into account—however, they are no longer part of the inventory control system. One solution to this problem is to keep an "inspection file," where all the items issued for the purpose of inspection will be recorded. Again, if a fixed amount of time is allowed for the inspection, this file can be used to follow up and remind the inspectors of any delay.

Thus, the on-hand quantity of an item is given by the following equation:

$$\text{On-hand quantity} = \text{inventory balance*} + \text{shipment quantity} + \text{quantity under inspection}$$

Figure 6.16 shows the procedure required to account for the correct quantities of items. Initially, there are 100 items on hand, and 50 items on order. The order is received (1), and the quantity is subtracted from on order and added to shipment. When these items are issued for inspection

---
*Inventory balance is later transformed into free stock.

(2), the quantity is subtracted from shipment and added to inspection. When inspection has been performed (3), the quantity is subtracted from inspection and added to inventory.

It is not essential to follow this procedure, and any shortcut from on order to inventory is valid. The transfer quantity between shipment, inspection, and inventory, as shown in Figure 6.16, need not necessarily balance, since, for example, count discrepancies and rejected parts can result in a quantity being subtracted from shipping different from the quantity added to inspection. If desired, another account, such as "rejected," can be used (4).

2. Manufactured items should be treated as on order; they are under the control of the job release or capacity planning system. Some operations required in the manufacture of an item, such as heat treatment, parkerizing, and grinding, can be performed by subcontractors. In many industries, shipment of items for any purpose, including the above mentioned, is carried out only through inventory.

Thus, it may occur that at a specific point in time three systems are handling the same item and quantity: an open order in the job release file, an open order in the subcontractor file, and the physical quantity of that item in inventory waiting to be shipped. Care and clear procedure must ensure that the quantity is accounted for only once in the requirement planning.

One solution is to code subcontracted orders for specific operations and to disregard them as on-order items for requirement planning. These orders should be recorded in the subcontractor file for purposes of follow-up, costing, billing, and so on.

A code should be used in inventory systems to designate transit. Thus, while the quantity is recorded as physical stock, it will be ignored as on hand by requirement planning.

3. The purchasing and subcontracting of items or assemblies are usually handled by the purchasing department. The purchasing system is initiated by a request for quotation or by issuance of orders. From that point on, the control of the "on order" comes under the purchasing system and files. This procedure leaves a "dead period" regarding information. The activity requirement planning is controlled by the production department; it issues the orders to purchasing. It might occur that at a specific point in time the production department has issued an order to acquire items, but this order has not yet been processed and recorded by purchasing—the quantity will be lost in the regenerated requirement planning. One possible solution is to use the old requirement planning output as a follow-up tool and coordinate purchasing from that point on.

6.5 Discipline Needed for Requirement Planning

4. There is a time lag between the moment a job is released to the shop for manufacturing or assembly and commencement of manufacture. The raw material or items for assembly will be withdrawn from inventory only when the actual manufacturing commences. Sometimes all of the items required for assembly (or the quantity as a whole) will not be issued simultaneously.

Thus it is possible that the same quantity will be accounted for twice in the requirement planning: once as on order from the job release system and then as on hand from the inventory system. There are at least three solutions to this problem:

a. On each job release order a code is used to indicate if stock has been issued from inventory. When data are retrieved for requirement planning, this code will be used. Orders that do not carry this code in the record will be disregarded.

This approach is especially suitable for those manufactured items where issue of only one item from stock is involved. For assemblies, where many items from stock are needed, this procedure is not appropriate, since these items are not issued as a kit. However, if a proportion of the quantity is issued as a kit, the job release order should be split into two orders—one for the quantity issue together with the code and the other for the remaining quantity without the code.

b. An auxiliary "assembly demand" file is used as an extension of the inventory file. This file records all of the allocated stock, and it contains two columns: allocated quantity and issue quantity. At the instant that a job release order is opened in file, a single-level explosion of the bill of material is made. Based on this explosion, the allocated quantity column is filled in. When the shop starts manufacturing, these data may serve as a condensed request for issue of items to inventory. As items are issued from inventory, the quantity is recorded in the issue column.

Thus, requirement planning considers the job release file as an on-order quantity, and the on-hand inventory term is transformed into a free stock term, where

Free stock = on hand quantity − allocated quantity + issue quantity

A comparison of allocated quantity to issue quantity reveals any scrap items in manufacturing. Furthermore, allocation of a greater quantity than the on-hand quantity is permissible; this will simply result in a negative free stock, which means that a net requirement exists.

c. An auxiliary "assembly demand" file (see solution b) that contains three columns—allocated, issue, and used quantities—is employed. The allocated and issue columns are used as in solution b, while the used column is used only for inventory control purposes. When the assembly is finished and enters inventory, the balance of the assembly record is updated. At the same time, the assembly is single-level exploded by the bill of material, and the used column is filled accordingly.

Requirement planning disregards the assembly job release orders as on order. The on-hand inventory is transformed into free stock by the equation

$$\text{Free stock} = \text{on hand inventory} + \text{issue quantity} - \text{used quantity}$$

The issue is regarded as on hand (located in the shop). Scrap and rejected items during assembly must be reported as used quantity to the assembly demand file. This approach is a compromise for cases where tight discipline cannot be maintained. In cases where the same item is used in many assemblies, it gives the foreman the freedom to decide for which assembly to use the items; this does not have to be declared when the items are issued from stock. Care must be taken to report scrap items; a physical count of items in the shop is usually required to compensate for errors in reporting.

The net requirements equation is misleading. Although it appears very simple, due to the situation in the shop it is one of the most difficult problems to solve in the IMS. Some, but not all, of these problems have been considered here. The main problem is to recognize and detect those situations that affect the requirement planning; once detected, solutions can be worked out. Discipline is thus essential for the successful implementation of requirement planning.

Chapter Seven
Inventory Management and Control

From an investment standpoint, inventory is commercially wasteful. However, from an operating point of view, it is essential. Finished product inventories are maintained in anticipation of demand and in order to absorb the difference between forecast and actual demand. Semifinished components and subassemblies are maintained in order to: (1) reduce the delivery time quoted for the end product; (2) balance seasonal demand fluctuations; and (3) take advantage of volume discounts in purchasing and manufacturing.

Inventory control is divided into two main parts: inventory management and inventory accounting. The objective of inventory management is to keep capital investment in inventory to a minimum while maintaining a desirable service level; this is the planning and controlling aspect of inventory. The objectives of inventory accounting are to keep track of inventory transactions and to supply information required by other systems.

The use of computers in industry made it possible to plan and control inventory as an integral part of the manufacturing system. The need for items and subassemblies is established to correspond with the exact date when assembly is scheduled to begin. These are dependent items—they depend on the master production schedule. The independent items are forecast and planned according to management policy in the master production schedule.

Conventional inventory management, with its theories of service level, economic order quantity, safety stock, and order point, was appropriate for manual systems; in spite of its unrealistic basic assumption of gradual depletion (in manufacturing, depletion tends to occur in discreet lumps because of lot sizing at higher levels), there was nothing superior. However, in the era of the IMS, conventional inventory management has become obsolete. Its objective of keeping capital investment in inventory to a minimum while maintaining the desired service level is met more satisfactorily by master production scheduling and requirement planning.

In conventional inventory management, dead stock is defined as items in stock with no issue movement for a predetermined period (e.g., two years), and a slow-moving item is defined as issue movement of no more than, for example, 10% of balance within a year. Whereas these terms were suitable for the conventional system, with the IMS the definitions should be changed. In the IMS we plan future activities and do not count on historical data; thus, for example, a better definition of dead stock would be: Dead stock is stock we don't plan to use for a predetermined period in the future. Requirement planning furnishes this information. In an extreme case, if an order was cancelled, dead stock might consist of items that have just arrived or that have not yet been received in inventory.

7.1 Inventory Objectives

Inventory control is central to the various manufacturing activities; in most industries, the activities start and end in inventory. Figure 7.1 shows this flow of activities. The received raw material is first entered into the storeroom, then issued to the manufacturing shops, and the finished items are entered into the stockroom; items are issued for assembly, and subassemblies and finished products are entered into storeroom; purchased components are entered into storeroom when received; finally, shipping to customers is carried out from inventory. This procedure places inventory at the juncture point of all activities, thereby making it a good source of information concerning the progress of manufacturing. Figure 7.2 shows the applications that require inventory data.

The objective of the inventory system is to supply information required by other systems; thus the inventory system is a dependent system—it depends on the applications desired and on the information required by the integrated system. The inventory system should be designed according to these specifications.

The following are examples of the above-mentioned applications and retrievable information that serve as the objectives of the inventory system:

- Control over plant properties.
- Supply data about on-hand stock to the requirement planning system.

7.1 Inventory Objectives

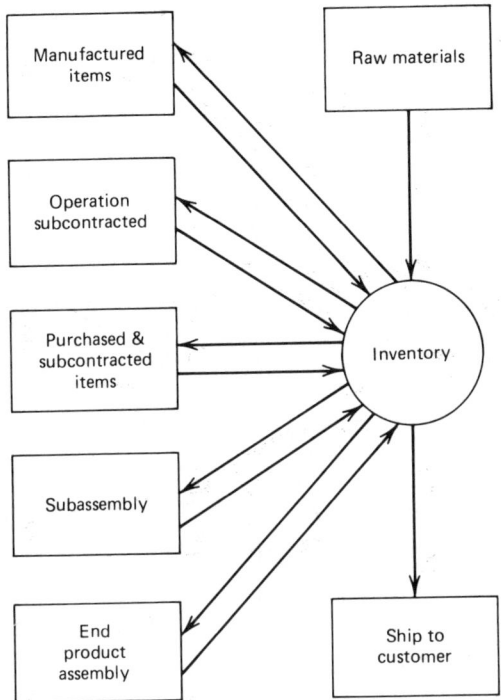

Figure 7.1 Manufacturing activities flow.

- Supply data to expediters on the availability of items required for assembly.
- Supply data for alternative materials.
- Approval of suppliers' bills.
- Supply data on the value of stock to the balance sheet.
- Supply data on material cost to the costing system.
- Control over indirect material usage.
- Supply data to estimate cost of products.
- Supply data on order delivery dates.
- Control over raw materials supplied to subcontractors (when the customer supplies the material).
- Control over quality control of suppliers.
- Supply data to suppliers' rating system.
- Supply data for calculating shop hourly rates.
- Supply data for budget preparation.

Inventory Management and Control

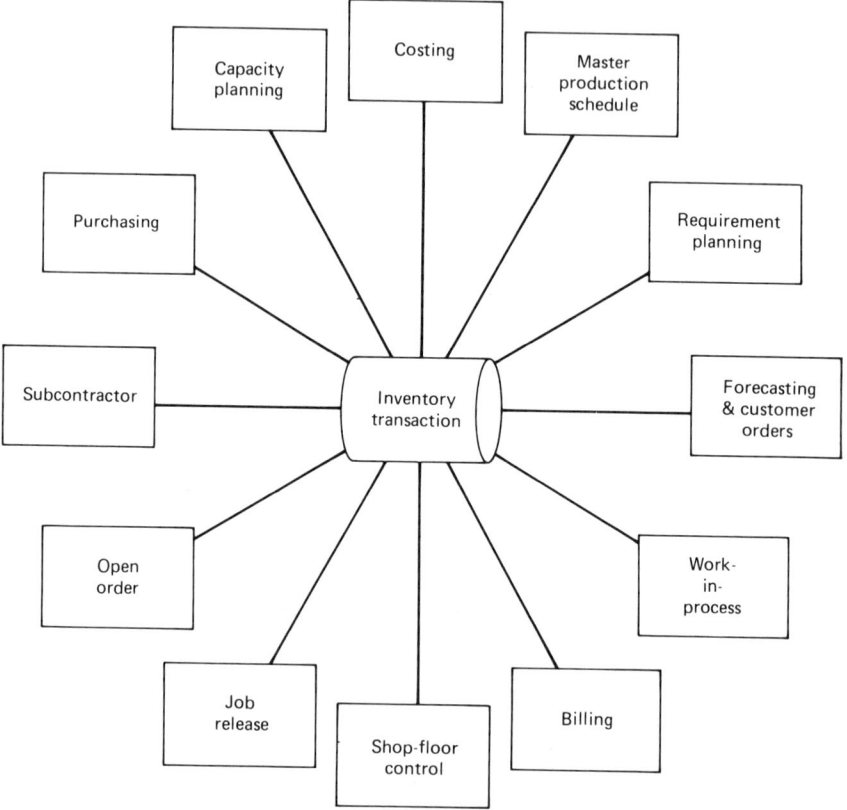

Figure 7.2 Applications that require inventory data.

- Supply data for forecasting future sales.
- Supply data for tax considerations.
- Supply data for evaluation of different price systems in inventory.
- Supply data needed for decisions on buying or expansion of plants producing required material.
- Control over dead stock and slow-moving items.
- Control over physical count of stock.

Each of the specified objectives should be analyzed *vis-á-vis* the required data and the way that they will be handled. In addition, the reliability and data processing technique requirements should also be considered. The inventory system is constructed with these objectives in mind.

7.2 Inventory System Technique—General

Inventory technique is basically very simple: It computes the inventory balance resulting from inventory transactions. To extend the scope of inventory to cover other objectives, one or more of the following codes and techniques are used.

Classification and Coding

From the data processing standpoint, each item in stock must possess a *unique name*. This name should correspond to the data processing technique employed and as far as possible be short and numeric. It should also serve as the communicating language between all applications using inventory data. A classification and coding system, such as the one discussed in Section 4.1, should be applied. The same code should be used in the bill of material, routing, job release, purchasing, inventory files, and so on.

Unit of Measure

All applications need information on the quantities of materials, such as the definition of quantity per part, quantity on order, and quantity in inventory. The quantity is not self-explanatory, and one must always indicate the meaning of the figure (unit, kilogram, meter, sheet, pair, etc.) describing quantity. It is not convenient to use the actual alphabetic name of the unit of measure, and it may also be a potential source of error. A code should be prepared for the use of the computerized system; Figure 7.3 illustrates such codes. The unit of measure assigned to an item usually corresponds to the shop-floor usage of that item, and the same unit of measure will probably also agree with the one designated in the bill of material. However, for purchasing, a different unit of measure must sometimes be used, but it is often of the same family (e.g., kilogram in shop and ton in purchasing or meter and centimeter, respectively) and can be easily converted to the other by calculation. On the other hand, sometimes the shop uses steel bars specified by length (according to the part length), but purchases it in kilograms or tons. The manufacturing system, and especially requirement planning, must speak the same language; thus, for example, since on-hand and on-order quantities are added and subtracted, they must use the same unit of measure. If this difference cannot be avoided, a conversion coefficient for each item must be applied.

Unit of measure codes

01	Unit	21	Centimeter
02	Pair	22	Meter
03	Dozen	23	Foot
04	Gross	24	Millimeter
05	Set	31	Square meter
06	Package	32	Square centimeter
07	Roll	41	Cubic meter
08	Box	42	Cubic foot
09	Bottle	51	Liter
11	Gram	52	Gallon
12	Kilogram	58	Cubic centimeter
13	Ton	61	Unit of thousands
14	Karat		

Figure 7.3 Table of unit of measure codes.

Transaction Code

The quantity in the inventory file represents the balance in stock, and the transaction represents the change, which can be a receipt or an issue. It is good practice to issue information on the nature and purpose of the change. This additional information is useful for other applications and for reporting on inventory usage. To designate these changes, a transaction code should be employed; Figure 7.4 shows a table of such codes. Code numbers 1 to 49 designate receipts of different origin, while code numbers 51 to 99 designate different kinds of issues.

The table given in Figure 7.4 was constructed for the use of a corporation with a centralized purchasing department. However, suppose that each plant is allowed to purchase goods for up to $1,000 and pay this from its petty cash account, while also being able to purchase goods of value up to $5,000 to be paid for by the central cashier. The code for each of these receipts is used to control and retrieve information on procedure violation; it is also used in evaluation of the purchasing systems and suppliers' rating.

The receipt from subcontractors code is used to control the raw material issued for the purpose of producing the ordered items. The issue transaction code indicates the purpose of the issue, such as transfer between storerooms or store location, for manufacturing, and for overhead. Note that for each issue transaction code there is an equivalent receipt transaction code.

Summarized Data

Historical information is usually retrieved from the transactions file and not the inventory file. However, in a medium-sized plant there might be from 30,000 to 50,000 transactions per month. Hence, to prepare a historical report for the year, about half a million records are handled. To avoid processing such a large number of records, some of the relevant information is kept summarized in the inventory file. The information, which is required for stock control, consists of the total amount of receipt (by purchasing), issue for manufacturing, and the last receipt and issue dates. However, if the information is to be meaningful, not all transactions should update these fields. For example, although the return of leftover stock from the shop floor is a receipt from the inventory standpoint, from the accumulation standpoint it should be subtracted from the issue and not added to receipt. Moreover, it should not cause any change in the last issue or last receipt date. Transfer of stock from the main store to a local store should subtract the quantity from the accumulated receipt of the main store and add it to that of the local store. In Figure 7.4, the codes in the body of the table designate the effect of the transaction on the summarized fields.

Status Code

Material issued for production and other purposes occasionally must be altered in form by sawing or cutting in order to meet requirements; this often results in some material being left over in production, and these leftovers can be scrapped (although it is the same material, it cannot necessarily be regarded as suitable for production) or used as indirect material and returned to store. For example, steel bars are purchased in 4-m-long units. In order to produce component A, 0.75-m pieces are required. Five pieces are prepared from a bar, a 0.25-m piece being thus left over. However, for production of component E, 0.25-m pieces are needed.

In inventory the purchased bars and the leftover pieces must not be controlled in a single inventory item. If they are controlled as one inventory item, an unrealistic situation might result. For example, four 0.75-m pieces, that is, a total of 3.0, are needed. In inventory, there are 20 pieces of 0.25 m length, which totals 5.0 m. Thus, the reply to the inquiry for available on-hand stock will be positive. To distinguish between this kind of leftover stock and the others, a *status code* is used. The key to the

Receipts

Transaction codes

			Date		Accumulation			
			Receipt (5)	Issue (6)	Receipts +(1)	−(2)	Issues +(3)	−(4)
	I.	*Purchases by "purchasing department"*						
01		Order after quotation	5		1			
11		"Small order"	5		1			
12		"Regular order"	5		1			
	II.	*Purchased by plant buyers*						
13		In cash	5		1			
14		On account	5		1			
15		On account from "annual order"	5		1			
	III.	*Receipts*						
20		From production	5		1			
21		From assembly	5		1			
23		From central storeroom	0		1			
24		From other plant storerooms	0	0	1			
25		From other storerooms within plant						
26		Tools and fixtures made in plant						
29		By lending from others	5		1			
	IV.	*Returns*						
30		Sundries from plant departments	0					
31		Manufacturing remains	0		1			
32		Manufacturing salvage	0		1			
36		Manufacturing surplus	0	0				
39		Receiving back of lent materials						
	V.	*Increase in balance*						
41		As a result of stock physical counting						
44		As a result of discrepancy in shipment						
45		As a result of catalog number changes						

Issues

			Date		Accumulation			
		Receipt (5)	Issue (6)	Receipts +(1)	-(2)	Issues +(3)	-(4)
	I. *Issues*							
51	From central to plant storerooms		0		2		
52	From plant to other storerooms		0		2		
54	From plant to another storeroom within plant						
56	Tools from storeroom to department						
	II. *Production*							
62	Materials for production in plant		6		3		
63	Materials for production by subcontractors		6		3		
64	Materials for overhead expenses		6		3		
66	Wasted tools		6		3		
68	Sundries to outside plants		6		3		
69	Fixtures to plant departments		6		3		
	III. *Returns*							
71	Sending back of lent materials	0	0		2		
72	Sending back of unsuitable materials	0	0		2		
	IV. *Sales & lendings*							
81	Sales inland						
82	Sales abroad						
89	Lending materials to others						
	V. *Decrease in balance*							
91	As a result of stock physical counting						
94	As a result of discrepancy in shipment						
95	As a result of catalog number changes						

Accumulation codes: 1 = + receipts, 2 = − receipts, 3 = + issues, and 4 = − issues.
Date codes: 5 = update receipt date, 6 = update issue date, 0 = no update, 7 = update receipt date in local store, and 8 = update issue date in local store.

Figure 7.4 (*Continued*)

inventory item record will be the original material code plus the status code. The status code might be as follows:

    1   new stock
    4   stock in cut pieces
    5   used stock
    6   not usable stock
    7   stock for sale
    0   scrap

Work Status Code (Transit Code)

Components and assemblies in the storeroom will be recorded in inventory under a record carrying the appropriate code number. Certain operations, such as heat treatment, parkerizing, or grinding, are sometimes performed by subcontractors. These items will be shipped and received through inventory. Furthermore, manufacturing program might have to be changed due to rush orders, excessive scrap, or missing tools and parts. In such cases, the production of certain items is stopped, and the foreman might wish to send these items to the storeroom. The above-mentioned items do not have unique names to be used in inventory; they are defined by code names and the last operation performed and are under the control of the job release or capacity planning system. Inventory is concerned with finished items and cannot usually handle operations of production. To solve the problem of storing such items in inventory, a work status code (transit code) is recommended. This code designates that although the item is identified by a code name, it has not reached its finished state, but is rather at a particular stage of manufacturing. This code is very important in the calculations of requirement planning, as previously discussed. Each plant should construct its own code table to correspond to its manufacturing procedures. The following is a sample of such a table:

    01   finished acceptable item
    05   finished part waiting for inspection
    10   before heat treatment
    11   after heat treatment
    14   before subcontractor operation
    15   after subcontractor operation
    31   before marking
    51   rejected part waiting for repair

7.2 Inventory System Technique—General

Thus milestones rather than exact routing are indicated. Production personnel usually press to increase the number of codes, but this should not be automatically granted. A clear distinction between items under production control and inventory control should be made. Solutions to problems and situations in manufacturing should be analyzed carefully before an extra work status code is granted. Although the use of a work status code appears to be a good, simple solution, in the long run it may turn out to be a source of complications. It should be remembered that the whole idea of the work status code is not a good one; it can jeopardize the manufacturing system if used loosely. However, problems are commonplace and solutions must be found. Therefore, the idea of the work status code must be used until a better solution is found.

Storage Location

Materials and items in stock occupy a definite physical space and location. It might occur that due to lack of space the same item is stored at several locations. To assist the storekeeper in his daily work, an indication of the storage location is incorporated as part of the inventory record. The storage location data might be either an informative or a working type. In the informative type, space is kept in the record, and these data (in addition to the variable data) are supplied by the storekeeper. These data may be either part of the transaction data or special update data. There is only one item record, one total quantity field, and as many storage location codes as necessary. The working type treats the storage location seriously and requires detailed information on each storage location. The storage location is part of the record key. (From the inventory standpoint, each location represents a different item.) Updating requires that each transaction—issue or receive—indicate a storage location. Furthermore, issuing of material from two (or more) locations requires two transaction documents. The information on storage location is not required by any application, and, on the contrary, it complicates the application programming and working procedures. However, it can be of assistance to storekeepers, if they so desire, and also if an automatic warehouse is used. Good storekeepers do not usually require this information, but rely instead on their memory, which generally proves accurate.

Batch Serial Number

Some companies must keep track of each manufactured or purchased batch and of the inspection certificates of each lot. These requirements complicate the system, but the problem can be solved by using the batch serial number and inspection certificate number as part of the inventory item key. Each batch will be regarded by the inventory system as a separate item. The transactions should specify this information on each issue or receipt.

From the data processing standpoint, introducing the batch serial number and inspection certificate number as part of the inventory item record key is a one-time job. It complicates programming, but, once done, allows as many batches as desired to be used. The user might find it cumbersome, demanding, and difficult to report and thus prefer to work with codes that indicate groups of inspection findings and codes that indicate the difference of batches, if any. The "blank" code means normal conditions, thereby simplifying the reporting to the inventory system and to the other systems that employ inventory data.

Material Pricing

There are several methods of pricing materials, the selection of a method depending on management policy. The following are some methods:

1. The *first in first out (FIFO) method* is based on the premise that the oldest material together with the price paid for it are issued first. As soon as the oldest lot is used up, the price of the material issued then reverts to that of the next oldest material. Under this method, since the material issued to the plant is charged to the current operating cost at the oldest price available, and also since the material in stock is valued at what most nearly approximates current market value, operating profits are exaggerated on rising prices. Company assets and capital tie-down are large.

2. The *last in first out (LIFO) method* is the reverse of the FIFO method: It is assumed that material most recently received is issued first. Under this method, the balance sheet will show lower profits and lower capital tie-down on rising prices.

3. The *cumulative-average method* is based on recalculating material price with each new receipt. The new price is calculated by the following equation:

$$\text{New price} = \frac{\text{old price} \times \text{quantity on hand} + \text{purchased price} \times \text{received quantity}}{\text{quantity on hand} + \text{received quantity}}$$

The price and stock value represent the actual amount paid. On rising prices, the average price will usually be lower than the current material

7.2 Inventory System Technique—General

price, resulting in increased profit, moderate assets, and moderate capital tie-down.

4. The *last price (current price) method* is based on adjusting the material price with each receipt. The value of on-hand stock is readjusted and debited or credited to an inventory-variation account. The issue of material is at the current price and therefore does not affect operating profit; the assets are at current value. At the end of the year, the inventory-variation account is cleared into a profit and loss account. On rising prices, the profit is clearly shown. The meaning of inventory value is not quite clear, since it does not represent the amount paid or actual value.

5. The *standard cost method* is based on charging materials issued with a fixed price. The fixed price is established by estimate or taken from past purchases. Operating cost variations may be clearly analyzed, since no change in material cost occurs. Operating profit and asset value are determined by management. Since the price of receipts of new material varies from lot to lot, an inventory-variation account is used to absorb the variations; this account is finally cleared at the end of the year into a profit and loss account.

From the computer program standpoint, the FIFO and LIFO methods are

Inventory record	Common to both	Transaction record
	Item identification	
	Plant number	
	Store number	
	Item code number	
	Status code	
	Work status code	
	Storage location	
	Batch number/code	
	Inspection certificate number/code	
	Unit of Measure	
Quantity balance		Quantiy
Unit price		(Unit price)
Last unit price		Transaction code
Date of last unit price change		Purpose:
Last physical count date		Product-part-batch
Number of updates		Order number
Accumulated receipt		Supplier number
Accumulated issue		Date
Date of last receipt		Sequence number
Date of last issue		

Figure 7.5 Fields in the inventory and transaction records.

more difficult to implement, since the price must be kept as part of the inventory record key or the receipt transactions used as the source of information.

In order to obtain information for pricing and costing, the transaction record must include, besides the record key and transaction code, the purpose of the transaction, that is, the product-part-batch for issues and the order number for receipts.

Figure 7.5 shows some of the information that should be available in the inventory and transaction records.

7.3 Design for Reliability of the Inventory System

Inventory is a passive stage in the manufacturing cycle; it does not plan or initiate any activity, but merely serves the active stages. This fact can be used to increase the reliability beyond the general reliability measures discussed in Section 2.5. Moreover, it may serve as a mini-PICS (production information and control system) for companies that do not wish to control manufacturing at the operation level and are satisfied with controlling it at the part level. This possibility will be discussed in Section 7.4.

Inventory transactions are not initiated by the storekeeper, but rather by one of the active stages of the manufacturing cycle. Therefore, each inventory transaction can be validated by comparing it to the planned activities. Figure 7.6 shows the inventory file as a nucleus with many reference files as satellites. These reference files contain all planned inventory activities. Each transaction is marked by a transaction code (see Figure 7.4) that indicates in which reference file the initiation of this transaction is recorded. Before updating the inventory file, a validation check will be made against the appropriate file. If the transaction is found to be valid, updating will take place; if not, the transaction will be marked as in error.

The numbers on the connecting lines in Figure 7.6 indicate the transaction codes as defined in Figure 7.4. For example, an inventory transaction with code 01 results from a purchasing order. The transaction indicates the item code number, the quantity, the supplier, the order number, and so on; furthermore, it must contain the key to the purchasing orders file. A validation check is made to ensure that the details on the transaction are correct. This is done by retrieving the appropriate record from the purchasing orders file. If all details match, the transaction updates the purchasing record with the quantity received, retrieves the unit price from the purchasing orders file and records it, and then the inventory file is updated. Receipt from the production floor will be validated similarly by

7.3 Design for Reliability of the Inventory System

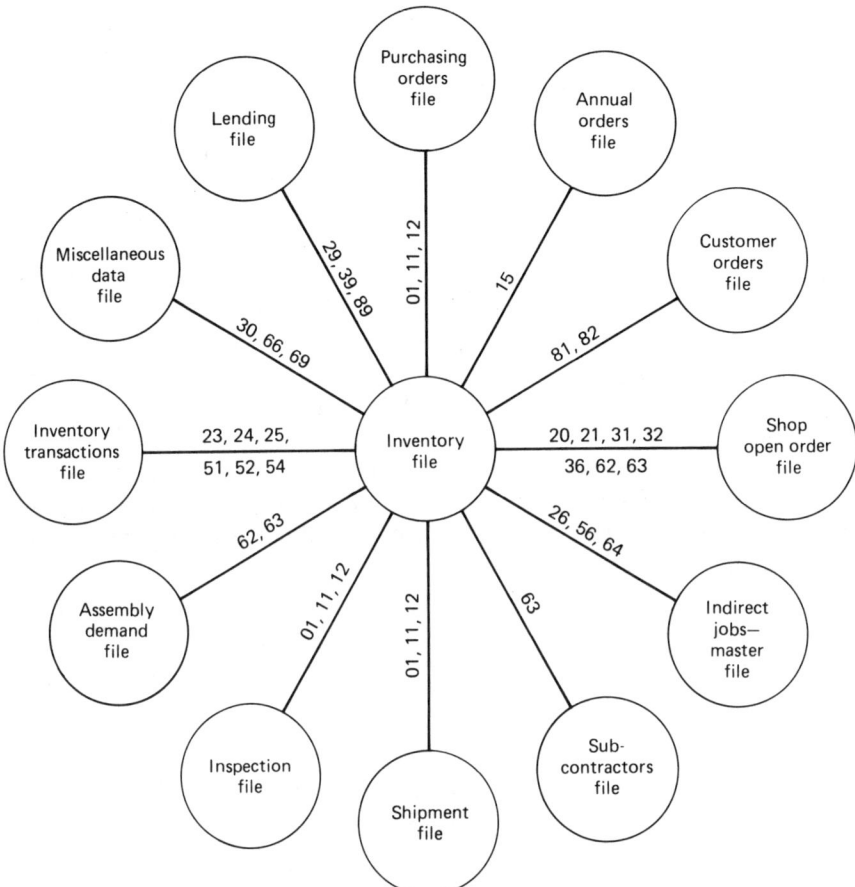

Figure 7.6 The inventory file as a nucleus with satellites (reference files). See Section 6.5 concerning the assembly demand, inspection, and shipment files.

comparison with the records in the shop open order file. Receipt from other company stores will be compared with the issues from the same stores, while issues to customers will be validated against the customer orders file.

Issues to assembly will be validated against the shop assembly order file and against the bill of material file to check if the issued item is required for the said assembly and if the quantity issued corresponds with the items per assembly. The principle of two-way data processing is applied; this saves reporting and thereby increases the reliability of the reference files. Although the reference files are used for validation checks, at the same time they can be updated if the transaction is found valid. The validation checks that the transaction was initiated by some phase, but at the same time it also checks the presence of the item in stock and of the reported quantity. Negative on-hand balance, for instance, is unrealistic.

This technique calls for retrieving all records involved from the appropriate files and bringing them into memory. It performs trial updating in the memory of all records while checking for validity. If found valid, the updated records are rewritten into their files. The use of the data base technique is very helpful from the programming standpoint.

Not all transactions can be so closely controlled. There are some unplanned activities for which no trace and backing can be found in any reference files (e.g., issues for overhead or miscellaneous use or receipts of items purchased and paid for in cash by plant personnel). The transaction code indicates this type of transaction and no validation against a reference file is carried out; however, some logical testing can be done. For example, the value of items received must be low and within company procedure. The issues can be checked according to item type and the department that made the request.

Other types of unplanned transactions include receipt of scrapped quantity, issue of quantity to replace scrapped items in assembly, and receipt of items due to production interruptions. These types of transactions are valid and should be controlled by the production phases, not by inventory. Inventory should serve production and not control its operation. Flexibility is therefore recommended in regard to quantities. Although the incidence of transaction errors is minimized through the validation tests, inventory discrepancies still occur. Some of the reasons for this are:

- Errors in count made in receiving or issuing.
- Errors in recording unplanned transactions.
- Entering a transaction twice or failure to enter it.

To establish confidence in the system, these errors must be corrected. They can be revealed and corrected by physical inventory count. Inventory counts are a legal requirement in some places and company regulation in others.

Traditionally, many companies have shut down manufacturing facilities in order to freeze all stock movements and take a count. Although it seems a perfect tool, errors still occur frequently. Counting is done manually, and even trained personnel make counting errors. Research on this topic indicates that when one counts more than a few hundred units of a single item, it is very probable that mistakes are being made.

7.4 The Inventory System as a Management Control Tool

Because of the inaccuracy and cost of shutting down production activities, it is recommended that "cycle counting" be used. Cycle counting is a rotating physical count at random intervals. The interval differs for each item. Some of the factors that will determine when to count are:

- *Stock level.* Count when the stock level is low. With fewer pieces to count, the job is easier, faster, and more accurate.
- *Number of transaction activities.* A dynamic item with many transactions is liable to be in error. Any item should be counted, for example, after 10 transactions.
- *Item value.* High-value items should be counted more frequently than low-value items. For example, high-value items should be counted every month, while low-value items can be counted only once a year.
- *Physical zero balance.* "Count" any item with physical zero balance. Storekeepers know when the physical balance is zero; they should record it on the appropriate transaction, and the balance will be compared to the recorded balance.
- *History of discrepancies.* Items that were found to be in discrepancy in the past should be counted more frequently.

These factors, and others, should be formalized into an algorithm and be part of the inventory updating program. Thus the system can decide when it is reasonable to physically count each item. The system can then prepare a list for counting.

Reporting physical count discrepancies is done as an inventory transaction with the appropriate "transaction code."

7.4 The Inventory System as a Management Control Tool

Inventory is central to the manufacturing activity. This fact can be used to increase the reliability of the inventory system, as discussed in Section 7.3. The proposed inventory system is demonstrated in Figure 7.6. The reference files contain most of the information required for management control. The information is on an item level and not on an operation level. The operation level is considered in shop open orders and open assembly shop orders. The inventory system does not know at what operation or stage of assembly the open orders are. However, it contains the information that an order has been issued, components taken for assembly, and raw material issued or not issued for manufacture. It also provides information as to whether the assembly or processing has been finished. Capacity planning or shop-floor control will contain the missing information as to the stage (what operation) the order is currently in and when it is scheduled to be completed. Chapter 8 goes into detail about capacity planning and shows how operations are controlled. Many companies find it adequate to work at the item level, leaving the detailed scheduling and capacity planning to the shop foreman. This section discusses the possibility of extending the inventory system to serve as a management control system as well.

The customer orders file contains the details of the orders. Shipment to the customer is an inventory issue transaction. This transaction will be checked against the customer orders file, and the latter will be updated if the transaction is valid. This file can then be used for both quantitative and financial open customer orders reporting.

The inventory file is also updated quantitatively and with respect to price (see below). The price value in the inventory file is the actual cost, while the customer orders file contains the selling price. These two files will be used to prepare a profit and loss report. The purchasing orders file contains the details of the orders. Receiving from a supplier is an inventory receipt transaction; this transaction will be checked against the purchasing orders file, and the latter will be updated if the transaction is valid. Thus, the purchasing orders file can be used to prepare financial commitment reports. The date when the goods were supplied is recorded on the receipt transaction, the promised date is in the purchasing record, and the quantity, number of rejects, and discrepancy between document and actual quantity is in the shipment and inspection files (see Section 6.5). This information can be used to prepare the suppliers' rating value. Upon updating inventory balance, the unit price is retrieved from the purchasing orders file, the actual price of items in inventory thereby being preserved. Receipt transactions of small orders were paid from the petty cash fund and will therefore carry the unit price on the transaction. This will be used as the actual price of the item in inventory.

A report of all transactions that did not follow company procedures, with respect to value and type of items, will be prepared and submitted to management. The unit price of receipts from the shop floor is somewhat of a problem. To solve it, it is advisable to extend the shop open order file to contain cost information. The record detail of the shop open order file is shown in Figure 7.7. The finance portion of the record contains the accumulation cost on the part level. Material and items accumulated cost is updated by inventory transaction. The issue transaction carries the shop order record key, that is, the purpose of the issue—product-part-batch. The transaction is checked against the shop open order file. If found valid, the issue quantity updates the inventory balance and the assembly demand file (see Section 6.5), and the cost value of the issue is added to the accumulated material and items issue value field of the shop open order file. This accumulation is in actual value, since the inventory

7.4 The Inventory System as a Management Control Tool

Product
Part
Batch
Ordered quantity
Standard time per part
Standard cost per part
Due date
Department
Code
Number of operations
Last operation number

Quantity reported in last operation
Quantity delivered to inventory

Accumulated material cost
Accumulated labor hours
Accumulated labor cost
Accumulated subcontractor cost
Number of operations performed
Number of last operation performed

Figure 7.7 Fields in the shop open order file record.

pricing system is actual cost. Subcontracted operations, including price information, are registered in the subcontractors file (see Section 6.5), and these items will be designated in inventory by a work status code. The price is transferred from the subcontractors file to the inventory file and, when issued to production, to the accumulated subcontractors cost field of the shop open order record.

The design for reliability and the concept of two-way data processing hold true for the job recording system (this will be discussed in Chapter 9). Each job-card is checked against the shop open order file. If found valid, the information on the job-card is added to the shop open order file record.

The hours reported are added to the accumulated working time. The department and cost center number is indicated on the job-card and checked against a table. If found valid, the hourly rate of that work center is known. This hourly rate multiplied by the hours worked gives the actual cost of the work reported on the job-card. This value is added to the accumulated working cost of the shop open order file record.

The sum of the three accumulated fields results in the actual cost—in labor and material—of the above shop order.

When the job is finished, the item, subassembly, or finished product is sent to store. Upon receipt, the transaction is checked against the shop open order file. If the transaction is found valid, the quantity updates the "delivered to inventory" field of the shop open order file. The sum of the actual cost is divided by the quantity received, and the actual unit price is computed. The inventory file record is updated by quantity and the inventory unit price.

By this method, the unit price of inventory items is always the actual cost. It starts with individual items, and, as manufacturing progresses, grows gradually to encompass subassemblies, assemblies, and the finished product. It covers all expenses—material, labor, and subcontracted jobs. The actual cost of a product is not computed by the bill of material, since it is theoretical and suitable only for computing the standard cost or estimated cost. The actual cost is not concerned with standard time per part, standard material, or items per assembly. The accumulated cost is retrieved from the shop floor by the job recording system and inventory transactions. When rejects occur, and extra material or items are issued, their cost is accumulated and divided by the actual quantity of good acceptable item received and inspected by inventory. If extra material and items have been issued, they are returned to stock and the accumulated cost value credited. This is one of the reasons why the assembly demand file regards the quantity data as information alone and does not restrict or constrain a transaction for which the quantity issue is greater than demand. The accumulated fields are not concerned with standards, but rather with actual occurrence.

A problem arises when partial quantities are delivered to inventory. The question is then what portion of the accumulated cost should be transferred to inventory value and what portion should remain in the accumulated fields. The solution depends on the method used for pricing, since one method will require more accuracy than another. The cumulative-average method, for example, allows, rough division of cost, since at the end of the batch the average will be balanced out anyway. In such cases, the standard cost multiplied by a coefficient may suffice. For more accurate results, the data included in the shop open order file, such as number of operations required, number of operations performed, quantity ordered, and quantity reported in last operation, can be worked into an algorithm to compute the estimated cost of the delivered quantity. It should be borne in mind that no special reporting system is required. The information will be updated by the job recording system, which is required anyway for salary, incentive, and other purposes. All of the computations and data transfer are done with no human effort if the system is correctly designed.

In preparing a monthly or annual balance sheet, knowledge of the value of work-in-process is essential, but determining it always constitutes a problem. Some companies take a physical count of the work near each machine. This provides data on quantity, which must be converted to cost. This is usually done by multiplying the quantity near each machine by its standard cost. Other companies assume that, on the average, all open shop orders are 50% complete. The value of work-in-process is thus 50% of the standard cost.

The extension of the inventory system can be used to furnish the required data at any given moment. The actual value of work-in-process is in the system files and is continuously being updated. The sum of the accumulated cost fields of all records in the shop open order file is the actual value of work-in-process.

Chapter Eight
Capacity Planning and Order Release

Requirement planning specifies the activities to be performed in order to meet the goals of the master production schedule. It plans both purchasing and production activities, taking account of requirements, but disregarding such manufacturing details as machine loading and shop dynamics. It sets objectives that must be transformed into a detailed loading plan for each machine or a group of machines in the plant. As distinct from this, capacity planning is the details planning phase; it is a scheduling and sequencing task. Finally, order release is the execution phase; on the basis of the scheduling, it initiates productive activities by the issuance of orders to the shop floor.

Requirement planning specifies the manufacturing requirements of individual items; it breaks down the order or the product into its components by the use of the bill of material. Although capacity planning might consider the individual items, this approach, sometimes called "build to stock," lacks the dynamics to overcome manufacturing divergencies. For example, if one of many items required for an assembly falls behind schedule, the due dates of the other items will be unaffected, they might still be a rush job in the shop, occupying overloaded facilities, only to have to wait in inventory for the missing item. A better solution is to use a network approach to capacity planning. The network in capacity planning need not be the same as the one specified in the product bill of material. The construction of a capacity planning network is done by using pegging and allocation of work-in-process, as assigned by requirement planning.

The time phase of requirement planning is a rough scheduling. Many features needed for accurate capacity planning are not available in today's requirement planning. It will suffice if one wants to plan and control merely at the item level, leaving the operation details to the foreman.

However, when planning and controlling at the operation level are desired, capacity planning is the tool to employ.

8.1 Capacity Planning Objectives

The major objectives of capacity planning are:

- Meeting delivery dates.
- Keeping to a minimum the capital tied down in production.
- Reducing manufacturing lead time.
- Minimizing idle times (*machine out of work*) on the available resources.
- Providing management with up-to-date information and solutions.

Some of these objectives conflict with each other. To minimize the capital tied down in production, the work should start as closely as possible to the delivery date; this will also reduce the manufacturing lead time. However, this approach will increase resources idle time in an environment in which resources are not continuously overloaded.

In the previous phases of the manufacturing cycle, the scheduling was done at the item level, disregarding available capacity. In such cases the foreman has a priority list of all the jobs to be performed. The priority is assigned either on the basis of the due dates resulting from the requirement planning or on the basis of some external parameter defined by management or sales. Foremen, with their skill and experience, can undoubtedly load the shop efficiently. However, they lack the basic information as to when the job is scheduled to arrive at their departments and what effect their jobs have on the item and the overall assembly of the order. If they could know in advance about high-priority jobs due to arrive in their departments, they would not tie up machines with long-term operations that later have to be interrupted for more urgent work. Programming at the item level assumes, unjustifiably, that the work on the shop floor runs smoothly with no interruptions, no part rejections, and no machine breakdowns. Therefore, the scheduling of items, each carrying its own due date, is sufficient to meet the objectives. In some types of industrial operation, such as small job-shops, line production, or process production, there is no need to schedule operations. However, where a large number of interdependent activities must share the same limited resources, the scheduling problem exists.

Scheduling is the assignment of target dates to operations in order to define when they must be completed if the manufacturing order is to be ready on time. Some people think of scheduling as a science, and many papers have been published on the theories of job-shop scheduling. On the

8.1 Capacity Planning Objectives

other hand, some people think of it as an art and believe that only the skill and experience of the foreman can effectively load the shop. In fact, scheduling is probably somewhere in between—a craft or trade, whereby the target dates are calculated according to certain rules, but the sequence of manufacture is determined by variable factors that differ as a function of local experience.

Scheduling is simply a forecast and as such will be often subject to errors. The capability of any scheduling system is measured by how well it can respond to changes, that is, how efficiently it can reschedule and reload the work in response to what is actually happening at any given time.

The tendency is to try to avoid operation scheduling by releasing work very early and then, with shortage lists, expediting the urgent orders. With this method, all orders become urgent at some time or else they are forgotten. The result is an increase in the capital invested in work-in-process because lead times are increased. With increased lead times, the priority of orders becomes vague, and, hence, much time is spent working on orders that are not currently required, which further aggravates the overload condition. This problem can be controlled only by considering all resources and analyzing decisions with respect to them.

The capacity planning system encompasses the following:

- Planning the capacity required at each work center and helping to allocate the machines and manpower required to meet the goals of the master production schedule.
- Controlling the level of work-in-process by regulating the rate at which orders are released to the shop floor.
- Helping to reduce manufacturing lead times by reducing the time a job must spend waiting for a machine.
- Planning and minimizing queue lengths to help ensure that machines and personnel will not run out of work.
- Determining how much work can be transferred to alternate work centers in an effort to reduce overloads or fill idle capacity.
- Analyzing remaining overloads and underloads to determine which orders can be subcontracted without causing idle time in other work centers.
- Assisting in making short-term capacity adjustments by planning overtime, adding temporary extra shifts, or releasing work to subcontractors.
- Leveling the planned load on each machine center (in certain instances), thus reducing idle time, overtime, subcontracting, and amount of manpower movement between work centers.

- Determining which orders should be released earlier to prevent idle time.
- Accurately estimating the completion time for every shop order and customer order.
- Planning the sequence of operations to be performed at each work center and providing a work sequence list for the foreman and for other phases.

Capacity planning is dynamic, since changing conditions call for new plans. The life of a production schedule depends on the environment in the shop. It is probable that after a week the schedule is no longer realistic, and thus a new plan is required. When a large number of operations is involved, the problem can rarely be solved satisfactorily by means of manual techniques. Although it can be analyzed by means of a manual system at any point in time relative to a given resource, under normal conditions this technique cannot be used to review the resulting decisions in terms of the full time range and with respect to all other resources. In addition, a satisfactory solution in one area may cause an unexpected problem in another area.

Thus a computer program should be employed for capacity planning, since it can reschedule all shop operations in a matter of hours. The logic of capacity planning computer programs today is similar to that of manual techniques—it merely utilizes the speed of the computer. Thus rescheduling that required a couple of days or weeks when manually planned is practical today as a result of the advent of the computer.

The following data are required for capacity planning:

- *Orders.* The manufacturing program as specified by the requirement planning phase or by management. This list includes individual items, product network, quantities, and due dates. In the case of a network, only the product or order due date is required.
- *Routing.* The operations and sequence required to produce each item or assembly. This includes the machine number and operation time.
- *Machines.* A list of all available machines, including available capacity time per period.
- *Parameters.* Depend on the option used. This topic will be discussed in Section 8.3.

Scheduling must consider many parameters, such as machines, toolings, jigs and fixtures, materials, and operator skill. It calls for sequencing in many dimensions, and this is very difficult to accomplish. Today's capacity planning programs mainly consider machines, whereas tooling and materials are treated as external data. Allocating machine operators to a

8.2 Capacity Planning Terminology

job is external to the program, unless they are defined as facilities to be loaded, for the purpose of the program.

8.2 Capacity Planning Terminology

Capacity planning normally applies backward scheduling. This term and others are illustrated in Figure 8.1.

Capacity planning uses a working day calendar instead of regular calendar dates to keep track of time, although the two are interchangeable. The working day calendar counts and assigns numbers to working days only. The day the scheduling is done is considered the current date (CD). In Figure 8.1, the CD is day number 60. Suppose that there is an order for an item requiring five operations; in this case, the due date (DD) is day 120. From that date backward the operations are laid out as shown in Figure 8.1. In order to finish this item on time, the latest date on which to start operation number 1 (OP. 1) is day 85. This is called the latest start date (LS) for OP. 1. The latest date on which the item can be finished is its DD, that is, day 120. This day is called the latest finish date (LF) for OP. 5. These LS and LF correspond to the item (or order); each operation is also assigned its own LS and LF. For example, the LS of OP. 3 will be day 97,

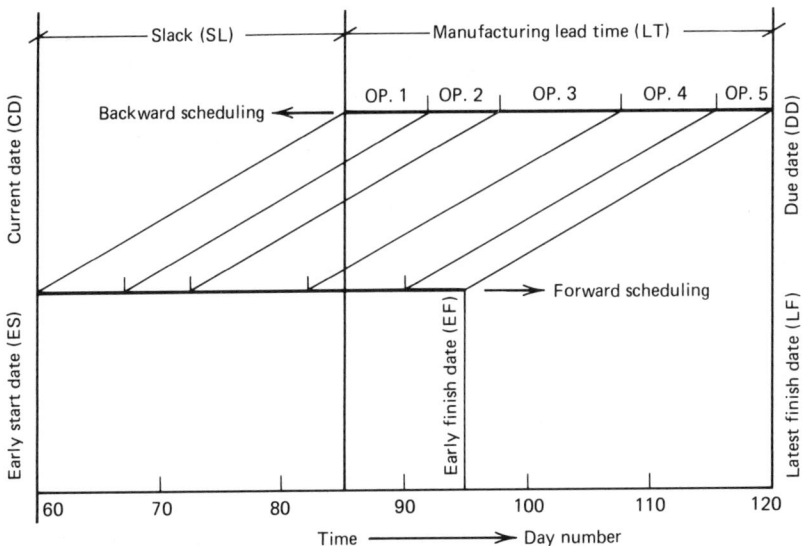

Figure 8.1 Capacity planning terminology.

while its LF will be day 107. The LF of an operation is the LS of the preceding one; the use of "latest" means that any delay in starting the operation on its LS will result in a delay in the DD; however in the example of Figure 8.1 there is a slack (SL). The starting date of the operation is in the future, but manufacturing of the item can begin on the CD; this day is also called the early start date (ES) of the item or of the first operation. If this is carried out, the item will be completed on day 95, which is called the early finish date (EF) of the item or of the last operation. Each operation has its own ES and EF, corresponding to item scheduling. For example, the ES of OP. 3 is day 72 and its EF is day 82. Thus the above item has an SL of

$$SL = LS - ES = LF - EF = 85 - 60 = 120 - 95 = 25 \text{ days}$$

This SL is for the item as a unit, but can be used on any individual operation.

Figure 8.2 shows the scheduling of three items (jobs). Each operation is marked by three digits: The first designates job number; the second, operation number; and the third, machine number. Initially all three items

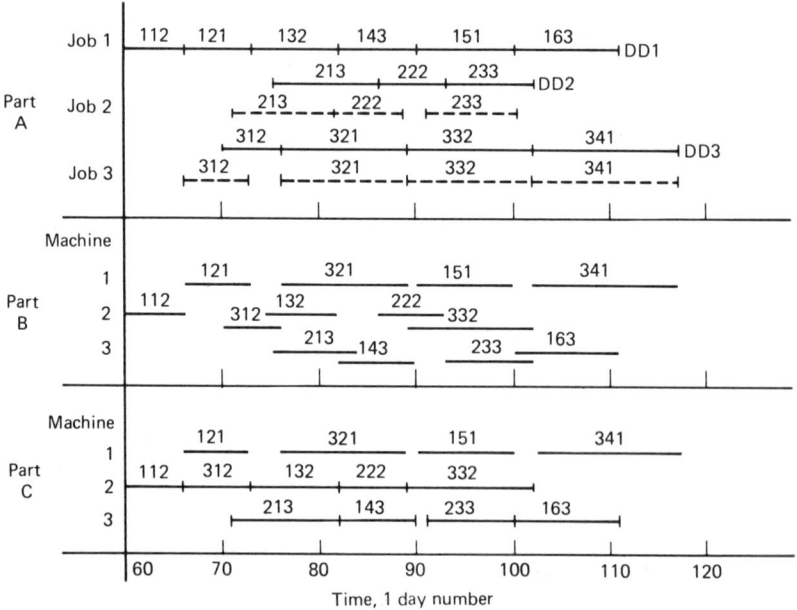

Figure 8.2 Scheduling of three items (jobs).

8.2 Capacity Planning Terminology

are backward schedule. The full lines in part A gives a cross section of order status.

Part B of the figure gives a cross section of the machine load for the backward scheduling of Part A. It shows that on days 73 to 76 machine 2 is overloaded: jobs 1 and 3 require the machine at the same time. A similar situation occurs on that machine on days 89 to 93 and on machine 3 on days 82 to 86 and days 100 to 102. To balance these loads, the SL time may be used. Since job 1 has no SL at all, any change in its LS will result in a late delivery. Job 2 has an SL of 15 days and job 3 of 10 days.

Part C of the figure shows the machine load cross section after the overload has been resolved by pulling jobs forward. It is based on LS loading and considers available capacity.

The dashed lines in Part A show the planned cross section of the jobs. Job 1 does not have an SL and is therefore unchanged. In such cases, ES = LS = CD and EF = LF = DD.

Job 2 used 4 days of its SL in order to balance the machine load. It has a new SL of 11 days, the LS being on day 71. The job is scheduled to be finished on day 100, two days before the DD.

In scheduling to finite capacity, all item operations and machines are linked together, and the meaning of the SL is changed: The item SL and operation SL do not coincide. In the example shown in Figure 8.2, the scheduled LS of task 213 is day 71. However, machine 3 is unoccupied, and this task may start on CD = ES = day 60. Hence, this operation has ES = 60 and LS = 71.

The second operation (task 222), due to the fact that machine 2 is occupied, can start only on day 82, which is its ES. It is scheduled to be finished on day 89, two days before the required date. Hence, this operation has ES = 82, LS = 84, EF = 89, and LF = 91.

The third operation (task 233) has a scheduled LS of day 91. However, by machine loading one can see that machine 3 is not occupied on day 90, and hence the operation can be pulled forward to start on day 90. The result is ES = 90, LS = 91, EF = 99, and LF = DD = 102.

The SL value may be positive, zero, or negative. A zero SL is sometimes referred to as critical, while a negative SL is called a delay. When working with networks, there is a third type of SL—network SL.

The overall manufacturing elapsed time is referred to as the manufacturing lead time (LT). Scheduling of the items in Figure 8.2 was done manually. One looks at the diagram and tries as many loading combinations as needed to obtain a satisfactory result. The terminology that has been introduced enables scheduling to be treated mathematically, thus allowing a computer to be employed.

8.3 Capacity Planning Technique

The capacity planning logic and programs are composed of several stages. The first stage is to examine the feasibility of meeting the DD (unlimited capacity scheduling).

For each item or network a DD and an ES are assigned. The ES may be explicitly defined by such constraints as the availability date of materials and tools. If the ES is not explicitly defined, the CD is substituted. Both the item LT and the available manufacturing time span (TS) are computed, and the two are compared.

Example

Due date (DD)	=	day 170
Early start date (ES) = current date (CD)	=	day 110
Daily working hours	=	8 hours
Manufacturing lead time (LT)	=	400 hours

Compute available time span (TS):
$TS = (DD - ES) \times 8$
$TS = (170 - 110) \times 8$ = 480 hours

Therefore, the slack (SL) is
$SL = TS - LT = 480 - 400$ = 80 hours

This indicates that the first operation should start, at the latest, 80 hours (10 days) later than day 110 in order for the last operation to be completed on day 170. In this case the part does not experience a delay, but an SL time of 10 days. That is to say, theoretically, the DD may be adhered to.

When the SL results in a negative value, that is, the TS is less than the LT, a delay might occur if normal manufacturing procedures are followed. An attempt is made to overcome the delay by the following methods.

Reduction of Indirect LT

Figure 8.3 schematically shows the elements of item LT. The LT includes interoperation time (IO) as well as operation time (TO). The IO covers the following time elements:

- Queue time (i.e., time spent waiting to be assigned to a machine).
- Preoperation time (i.e., time for marking out, cleaning, etc.).
- Postoperation time (i.e., time for inspection, cooling, etc.).
- Wait time (i.e., time spent waiting for transportation).

8.3 Capacity Planning Technique

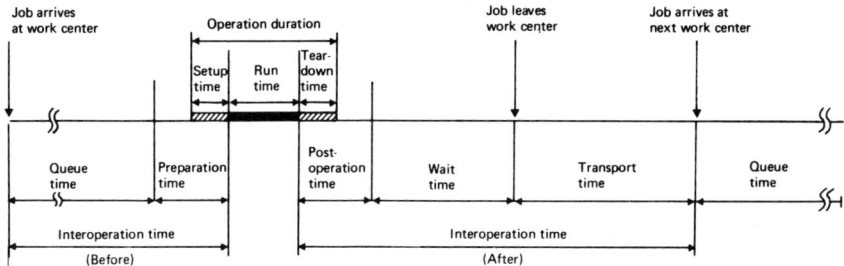

Figure 8.3 Elements of item LT. (From *Copics*, Vol. V, Chapter 6, G320-1978-0. Copyright 1972 by International Businness Machines Corporation. Reproduced with permission.)

- Transport time (i.e., time required for transportation to the next work center).

For normal manufacturing procedures, a generous allowance is made for these interoperations. It reduces expediting and preplans transport and inspection. However, in case of delay, expediting may be applied and initially allowed IO reduced; this reduction is shown schematically in Figure 8.4. The same effect can be expressed in mathematical terms and calculated on a computer.

The TO is composed of set-up time (SU), run time per part (T) times the batch quantity (Q), and tear-down time (TD):

$$TO_i = SU_i + Q \times T_i + TD_i$$

The IO can be expressed as a sum of individual operations (i). Each operation is allowed a certain amount of time that depends on its characteristics. A feature such as transport time can be based on a table designating the time as a function of department location and type of material handling equipment, whereas wait time can be assigned according to department and work center type.

Thus the LT per item is

$$LT = \sum_{i=1}^{n} (SU_i + TD_i + Q \times T_i) + \sum_{i=1}^{n} (IO_i)$$

or

$$LT = \sum_{i=1}^{n} (TO_i) + \sum_{i=1}^{n} (IO_i)$$

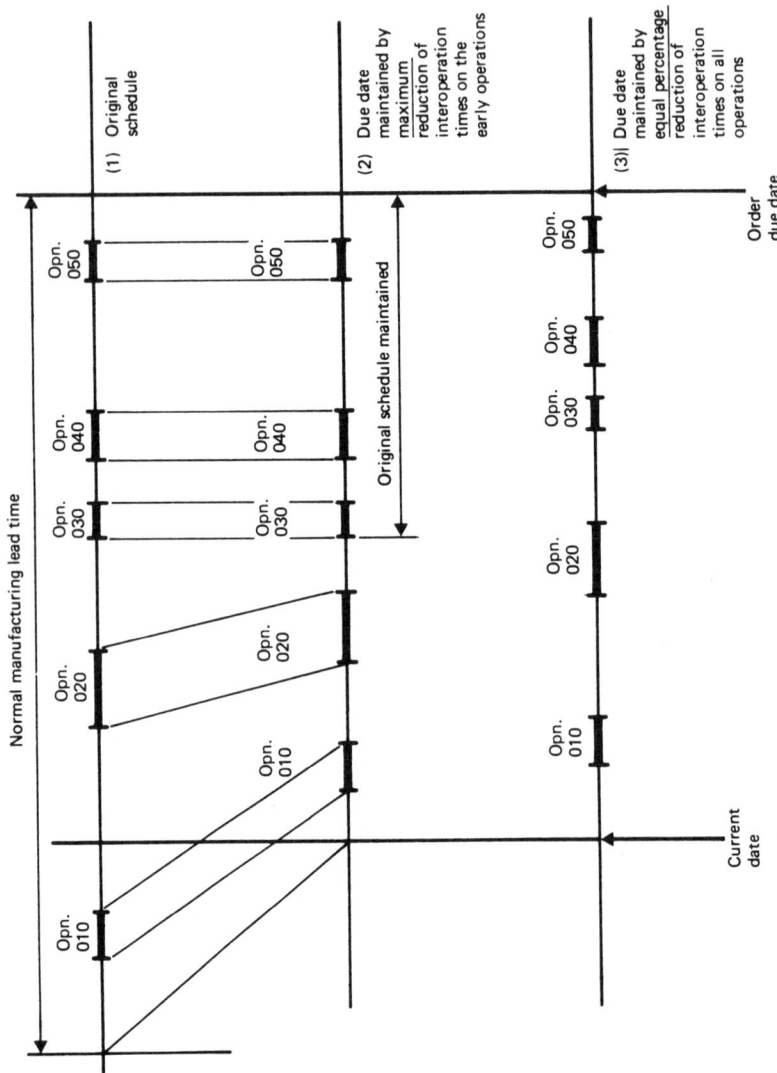

Figure 8.4 Effect of reduction in IO on LT. (From *Copics*, Vol. V, Chapter 6, G320-1978-0. Copyright 1972 by International Business Machines Corporation. Reproduced with permission.)

8.3 Capacity Planning Technique

The last equation can be succinctly expressed as

$$LT = TTO + TIO$$

where TTO is the *total* operation time, and TIO is the *total* interoperation time. The SL is computed by the equation

$$SL = TS - LT = TS - (TTO + TIO)$$

or

$$SL = (TS - TTO) - TIO$$

If the SL calculation results in a negative value, an attempt to equalize it to zero is made by reducing the TIO. (If the negative value is greater than the total TIO, this method is not applicable, and other measures, as will be discussed later, will be used.) The reduction is carried out by multiplying the TIO by a reduction factor (α), which under normal manufacturing procedures is equal to zero. Hence, the last equation can be expressed as

$$SL = (TS - TTO) - (1 - \alpha)(TIO) = 0$$

$$TS - TTO = (1 - \alpha)TIO$$

Solving for α, we obtain

$$\alpha = 1 - \frac{TS - TTO}{TIO}$$

Example

Due date (DD)	=	day 170
Early start date (ES)	=	day 160
Daily working hours	=	8.0 hours
Total operation time (TTO)	=	31 hours
Total interoperation time (TIO)	=	70 hours
Compute available time span (TS): TS = (170 − 160) × 8	=	80 hours
Compute manufacturing lead time (LT): LT = 70 + 31	=	101 hours
Compute slack (SL): SL = TS − LT = 80 − 101	=	−21 hours
Check for possibility of using reduction factor: TS − TTO = 80 − 31	=	49 hours

Possibility exists; compute reduction factor:
$$\alpha = 1 - (80 - 31)/70 = 1 - 49/70$$
$$= 1 - 0.7 = 0.3 \qquad = \qquad 30\%$$

One may assign a limit to the reduction factor in order to keep it practical. The value of this limit depends on the initial value assigned to the TIO and on the shop environment; a value of about 50% is normally used. Figure 8.5 shows this computation in a block diagram form.

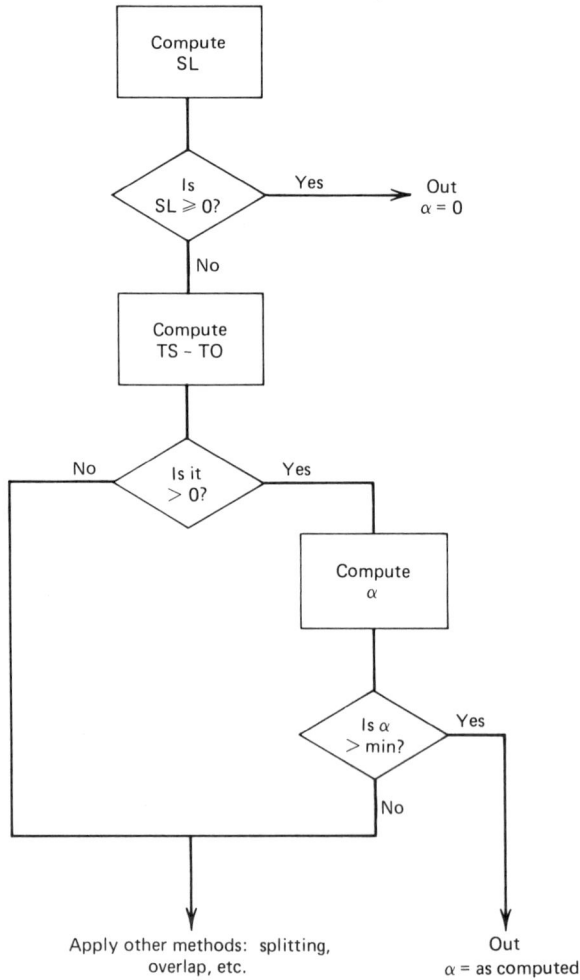

Figure 8.5 Logic of computing the IO reduction factor.

The reduction coefficient will be stored for further use in the scheduling phase with respect to priority consideration.

Splitting

Splitting is the simultaneous processing of an operation on several machines. By this means a reduction in operation duration is achieved. The operation time is expressed by the equation

$$TO_i = (SU_i + TD_i) + Q \times T_i$$

If N machines are employed simultaneously, the operation time will be

$$TO_i = (SU_i + TD_i) + \frac{Q \times T_i}{N}$$

The technical number of splits is determined by the number of similar machines or by the number of tooling sets available. This number is the upper limit for the number of splits possible. The economic number of splits is a function of set-up time and thus operation time. The longer the set-up time, the less economical the simultaneous use of more machines.

The plant must work out an economical algorithm. One practical algorithm is to specify combinations of set-up versus run time limits in the form of a table; this table will actually indicate the economical number of working hours per number of set-up hours. An example of such a table could be as follows:

$$S_1 \quad \text{and} \quad R_1$$
$$S_2 \quad \text{and} \quad R_2$$
$$S_3 \quad \text{and} \quad R_3$$
$$\text{with } S_1 \leq S_2 \leq S_3$$

where S_i is the set-up time limit and R_i is the run time limit. The limits are used as follows for the calculation of economic splits. The set-up time of an operation falls either between

$$0 \text{ to } S_1$$

or between

$$S_1 \text{ to } S_2$$

or between

$$S_2 \text{ to } S_3$$

or is greater than S_3.

If the last case is applicable, no splits are considered. In the other three cases, R_i will be selected accordingly and the economic number of splits is calculated as follows:

$$N = \text{number of splits} = \frac{\text{run time}}{\text{run time limit}} = \frac{Q \times T_i}{R_i}$$

The effective split is the computed value N with the constraint of the number of splits possible.

Example

Operation time (TO) = 200 hours
Set-up time (SU + TD) = 5 hours
Split table
$S_1 = 2$; R_1 = 25 hours
$S_2 = 6$; R_2 = 60 hours
$S_3 = 10$; R_3 = 120 hours
Given possible splits = 4

Since the set-up time falls between S_1 and S_2, the value of R_2 is used. Thus

$$N = \frac{200 - 5}{60} = 3.25$$

The effective split is therefore 3.

Overlap

Overlapping is starting the subsequent operation before the preceding one has completed the planned quantity. The result can be a considerable savings in LT. Figure 8.6 shows this effect and defines some of the terminology. Overlapping must be tightly controlled because it involves additional effort in coordination. One must consider at least three aspects that constrain the practicality of overlapping:

1. *Minimum overlap time.* This value ensures that a minimum overlapping of two operations can be achieved. If it does not, overlapping is not practical, since the savings in LT is not balanced by the additional coordination effort.
2. *Minimum time before overlap.* The overlapping operation may start only when the overlapped operation completes a given quantity and the required IO has elapsed. The SU of the overlapping operation may start in parallel, so that the TO can begin immediately after items are received.

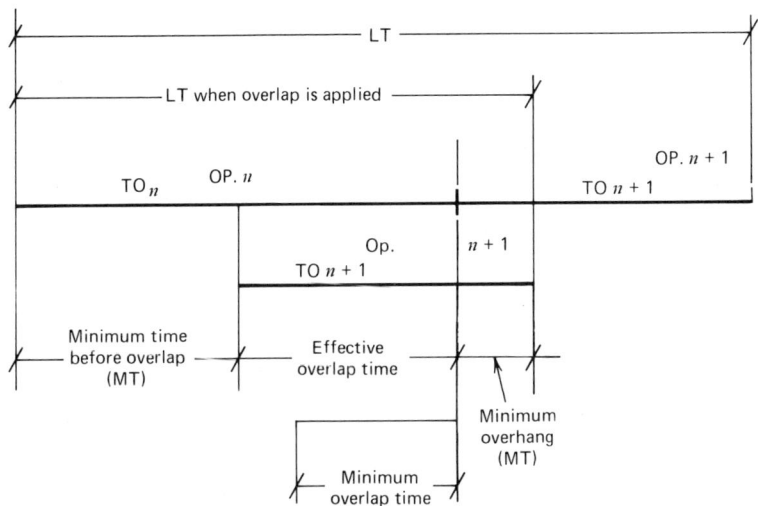

Figure 8.6 Effect and terminology of overlap.

3. *Minimum overhang.* This is similar to the minimum time before overlap except that the data refer to the end of the overlapping operation instead of the beginning of the overlapped operation.

Computations can be performed in order to decide whether and how to plan overlapping. If overlapping is worthwhile, the preferable form can be computed by the two methods shown in Figure 8.7. In case A the overlap is computed forward by using as a constraint the minimum time before overlap. In this case the key day (KD) to start OP. $n + 1$ is computed by the equation

$$KD_{n+1} = KD_n + SU_n + MT_n + (1 - \alpha)IO - SU_{n+1}$$

Case B uses as a constraint the minimum overhang and is computed by the equation

$$KD_{n+1} = SU_n + TO_n + (1 - \alpha)IO_n + MT_{n+1} - TO_{n+1} - SU_{n+1}$$

The effective overlap (EOL) must be equal to or greater than the minimum overlap (MOL). This can be computed by

$$EOL = KD_n + SU_n + TO_n - (KD_{n+1})_B^A \geq MOL$$

The method that will give the best results will be used. A practical rule is that if OP. n is longer than OP. $n + 1$, backward overlap should be used, while if OP. $n + 1$ is longer than OP. n, forward overlap should be applied.

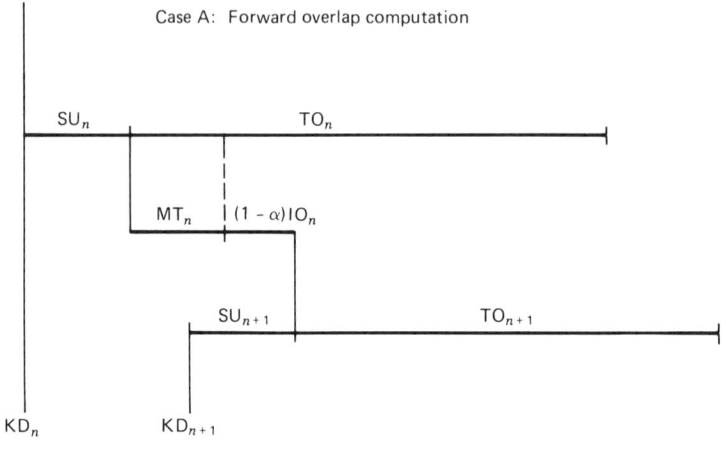
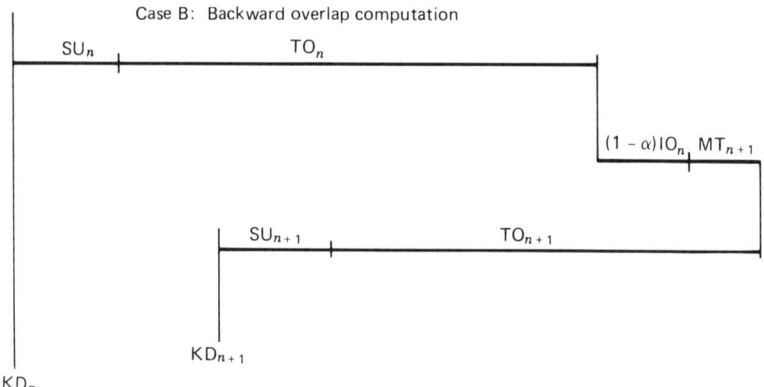

Figure 8.7 Two methods of computing preferable overlap. Notation: KD, key day.

These cases are illustrated in Figure 8.8. When one wishes to overlap three operations in series, with the middle one being short, special treatment is required, since the results achieved by normal overlapping are not satisfactory. In such a case it might be advantageous to split the middle operation into a few batches (see Figure 8.8).

If all three methods so far discussed (reduction of indirect LT, splitting, and overlap) cannot reduce the LT to fit the TS, a delay will occur. The lot size might be reexamined, split, or, possibly, reduced. Otherwise, the job is scheduled forward, starting from the CD. The EF − DD value gives the delay time, and the job is marked as critical. This step provides the required information concerning the jobs, that is, it indicates if a job is

8.3 Capacity Planning Technique

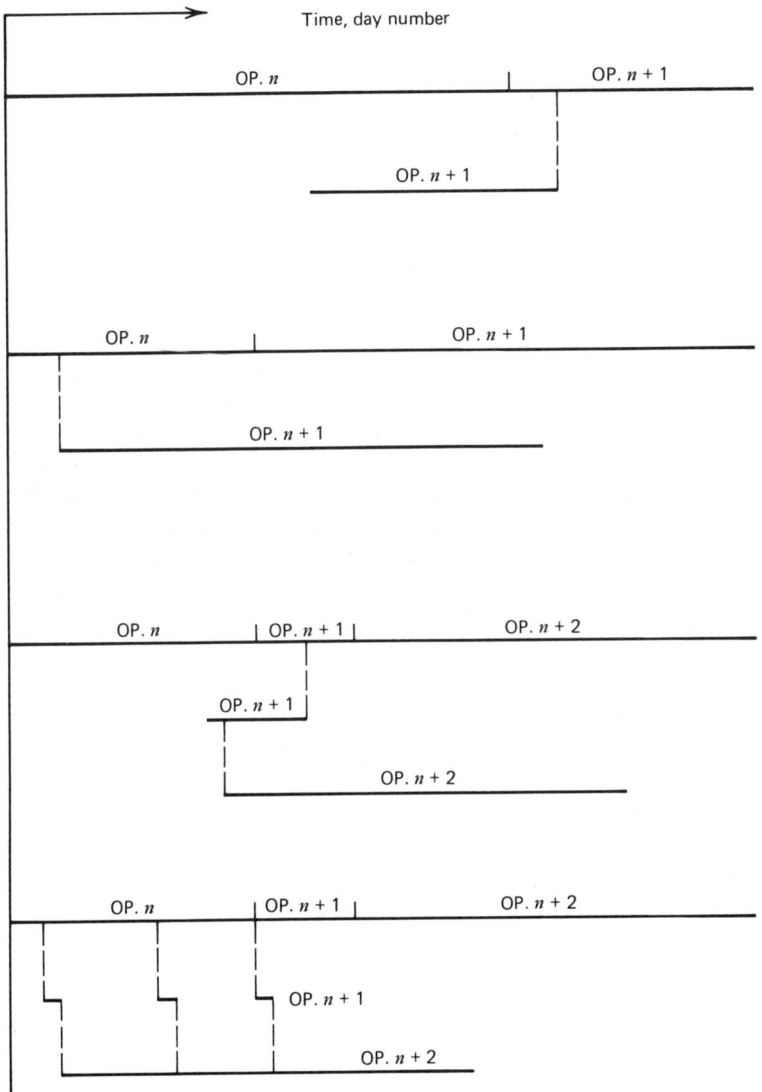

Figure 8.8 Overlap combinations.

normal or critical; its SL, ES, EF, LS, and LF; and if nonnormal manufacturing procedures are required. Furthermore, it gives a loading curve for each work center.

A load profile based on the LS will look different than one based on the ES. These profiles are used to decide if additional machines are required or if the daily working hours should be increased by approving overtime or an additional shift. The routing might specify alternate machines that can be used to overcome overloaded machines, work might be subcontracted, due dates might be changed, or lot sizes might be altered.

Thus, the second stage of the capacity planning and logic is an interruption to allow manual judgment and basic data modification. This is done in order to ensure a practical capacity planning. One should remember that up to now the constraint of available capacity was not applied. Jobs that seem to pose no problem with respect to meeting the due date might suddenly do so when exact scheduling takes place. This results from an overload on certain machines. The modifications in this stage serve to overcome bottlenecks on some machines and idle time on others while scheduling to available capacity at the next stage.

The third stage is long-term, finite capacity planning. If the load profiles of all work centers are smoothly balanced in the first stage, there is no need for further scheduling. Thus there will be no delay resulting from the constraint of finite capacity. Unfortunately, the load will not usually be smooth: There will be overloaded and underloaded machines together with periods of excess load and periods of machine idle time. The purpose of this stage is to balance the load and plan a practical scheduling with the goal of meeting due dates while keeping to a minimum the capital tied down in production and load balancing. To resolve the problem of periodic overloaded machines, work must be pulled forward or backward. The question is which job should be performed as planned and which should be pulled. To solve this problem, a priority rating is introduced. The basic purpose of rating orders by priority is to resolve competition for production capacity in attempting to meet order due dates. Some plants leave this problem to be solved on the basis of foreman or expediter judgment in recognizing the important jobs. Unfortunately, their decisions are usually based on insufficient facts.

The priority rating is a combination of two elements:

1. *External (order) priority.* This is a management decision and is probably based on the importance of the order, the customer, or the penalty clause in the contract. It is a network priority, covering all its items and operations.
2. *Internal priorities.* These priorities are assigned to each operation based on the planning parameters established in the first stage. They are dynamically recalculated as scheduling advances, some of the factors taken into account being:
    - The priority of a job that subsequently goes to an overloaded work center may be reduced.
    - The priority of behind schedule operations is increased.

- An operation already on a machine is automatically given a high priority.
- Jobs with successive operations on the same machine are given priority to ensure that another job will not intervene.
- A short queue in the next work center will cause the priority to increase.
- The bigger the expected delay, the higher the priority.
- When IO has been reduced, the priority increases.

In determining capacity loading, the system performs the following basic steps:

1. Order priority is determined.
2. Starting with the highest-priority order, each work center is loaded on the basis of the operation start dates assigned in the first stage. Loading is continued until an overload occurs.
3. When an overload is detected, the operation is moved to the closest period that is not overloaded. The number of periods it can be moved from its planned date depends on the SL between the earliest and latest dates on which the operation can be started. If an operation date is moved, the other operations of the order may also be moved in order to reduce the overall LT.
4. If the operation cannot be scheduled by the LS, check for the existence of an alternate work center or routing (e.g., a subcontractor or another plant). If an alternate is designated and available, it is loaded.
5. If alternates are not available, reschedule the release date and, if necessary, the order due date until the order can be loaded. If end item orders will be delayed, the master production schedule should be changed accordingly.
6. Check which orders can be released early in order to fill up any underloads.
7. Estimate completion dates for all orders based on when operations are scheduled.

8.4 Order Release

Order release is the link between planning and implementation. It initiates the production activities by the issue of orders to the shop floor according to the program prepared in the finite capacity planning stage.

The previous phases of the manufacturing cycle—requirement planning and capacity planning—do not impose any actual activity—they are planning phases. Order release, on the other hand, releases jobs for both the shop floor and also for purchasing and subcontracting. (It is good practice to replan these activities through capacity planning as well. Thereby, if a network is shifted, more accurate due dates than from the requirement planning system will result.) These planning phases do not impose any commitments. Each period (one week), a new requirement planning and capacity planning are prepared, while the old ones become obsolete. In the implementation phases, an order is a commitment of resources and material and cannot be overlooked in the next period. The shop floor and (probably) purchasing phases are dynamic, since reality seldom resembles planning: Suppliers fail to deliver on time, machines break down, tools break, and items do not pass inspection. Therefore, before authorizing release, a check should be made that material and tools are available.

Capacity planning schedules periodically to available capacity. Each operation, item, and assembly is assigned a starting date. Jobs should be started in shop on these dates.

The actual starting date is up to the foremen. They know best what is going on in their departments: the open job orders, the workers and their skills, the machines and their foibles, material and tools available, and the setup of each machine. Their responsibility is not only to complete jobs on time, but also to keep their operators occupied; they try to optimize operations in their departments, but at the same time keep operators satisfied. From their standpoint, sequencing operations by similar setup, for example, is highly economical. They are also exposed to pressure from the operators to assign "good" or easy jobs if incentive systems are employed. However, the foremen can function and make good decisions only within the range of the information they possess. They do not know which operations will arrive in their departments; moreover, they do not know, and thus cannot possibly take into account, the effect of their decisions concerning job selection on the delivery date of an order. Therefore, the amount of work released to a department should be kept to a minimum. Theoretically, jobs should be released on their starting date. Thus, optimal decisions (from their standpoint) made by the foremen and bad decisions from the network standpoint cannot cause much damage. Practically, there are some preparatory actions to be taken before the job starts on the machine: Material and tools have to be issued and transferred from stock; the machine has to be set up; and alternate jobs should be made available for emergencies. Thus, the foremen should be notified in advance of jobs due to arrive in their departments as well as of the approximate (or planned) arrival date. Plant management should specify rules on when to release orders to shop and what offset to use for the

8.4 Order Release

capacity planning program. These rules have to correspond to the plant environment, discipline, and the nature of production.

One should bear in mind that the act of job release serves two purposes. It transfers the control of this job from the planning stage to the implementation stage (open orders or open purchasing) and is an early warning to the foremen stating that this operation will, in the near future, be scheduled for production. The short-term scheduling (one day) will sequence the above operations (see Chapter 9).

From a systems standpoint, it is advisable to release items and not operations. For example, job release rules are as follows:

- Jobs for which the first operation starting date falls within production range (e.g., three weeks) are released.
- Critical items should be allowed extra time.
- Items with a large number of operations should be released at an earlier date.
- The production range for items scheduled by early start is reduced.
- The production range for items with postponed operations is reduced.

The basic functions of job release are:

- The physical availability of material and tools required for initial operation is checked. Only those orders that have all material and tools available are released.
- Shop order documentation is prepared. A printed order that includes item routing is sent to all departments involved in manufacturing the released item. Drawing number, special inspection demands, and additional descriptive text are printed on the job release form.
- The job-card for job recording is attached (for operation level of control, one card per operation).
- The material and tool requisition forms are attached.

The order is recorded in the shop open order file (see Figure 2.5) and assembly demand file (see Figure 7.6). Both files are required for requirement planning and inventory control. The data in the open order file are used for operation scheduling and shop control (see Chapter 9).

Chapter Nine
Shop-Floor Control

Capacity planning is a simulation of what is likely to happen on the shop floor. It attempts to schedule the jobs with respect to the current production plan and existing manpower and machines. Its design is good for the purpose of releasing orders to the shop, and its planning horizons are for several weeks or months; however, it cannot take into account any unplanned interruptions that occur. To overcome this problem, the available capacity can be reduced and two hours a day, for example, allocated for unplanned interruptions and urgent jobs. Although this holds true as a rule, it is not sufficiently accurate for actual job assignment in the shop. Usually, the life of a schedule is no longer than a day. After a week, a capacity plan will probably not resemble reality at all. In the shop, unplanned occurrences take place: machines break, tools break, operators do not show up, actual operation time doesn't work out as planned, the previous operation is not finished on time, a lot is rejected, or the foremen, for their own reasons, change the planned sequence. All of these cause changes in the implementation of the capacity planning program. The capacity planning simulation disregards the unplanned interruptions, that is, an operation is available for scheduling when the latest finish and the interoperation time of the previous operation was due. In practice, an operation can be loaded only when all previous operations have been completed and the components are available in the queue of the machine. The actual allocation to the individual machine is made by the foremen. They know best what is going on in their departments, the particular skill of each operator, the tolerances on the machines, and so on. Many companies, therefore, leave the daily scheduling to the foremen. On the other hand, some companies do daily scheduling by computer as a guide to the foremen, while others might establish dispatching rules to guide the foreman. In any case, the system must know what is going on in the shop. This information is vital for daily sequencing and for capacity planning. It is the basis for many other applications, such as costing, salaries, incentive pay, and absentee control. The frequency of receipt of this feedback is a function of the application. For daily scheduling it must be processed daily or hourly, while for salary and costing a week or a month will probably suffice. However, for reliability purposes it is recommended that it be processed daily or even in real time by the data collection equipment.

9.1 Short-Term (Daily) Capacity Planning

Capacity planning considers all company orders over a long period of time, its main purpose being to trigger order release to the shop. These orders to the shop are considered as information or as an early warning to the foremen with respect to what jobs might arrive in their departments in the near future. The foremen must make all the necessary arrangements to handle these jobs on short notice. Thus these orders do not represent a signal to begin processing.

The short-term capacity planning provides the actual order to start manufacturing. To be practical, it may, for example, be scheduled daily, but it normally covers a period of two or three days. This is done to compensate for computer down time and any unplanned interruptions that might occur. Input to the short-term capacity planning comes from the open order shop file, which establishes a realistic start date for each operation based on order priority and capacity limitations. Medium- or long-range capacity planning is processed at greater intervals—for example, every 10 days—and updates the open order shop file.

The method of sequencing is based on operation priority together with some special considerations, such as grouping jobs with similar setup and tool availability.*

The sequencing of operations is performed in three steps:

1. Establishing the hours available at a work center (on a machine).
2. Sequencing work from the queue.
3. End-of-shift routines.

The basis for establishing the hours available at each work center (on each machine) is to specify the particular hours of the day that it will be available. Staggered working hours can be shown by varying the start/stop times for a work center (machine).

---

*The rest of this section has been adapted from an IBM publication. Reprinted by permission from *Copics*, Vol. V, Chapter 6, G320-1978-0. Copyright 1972 by International Business Machines Corporation.

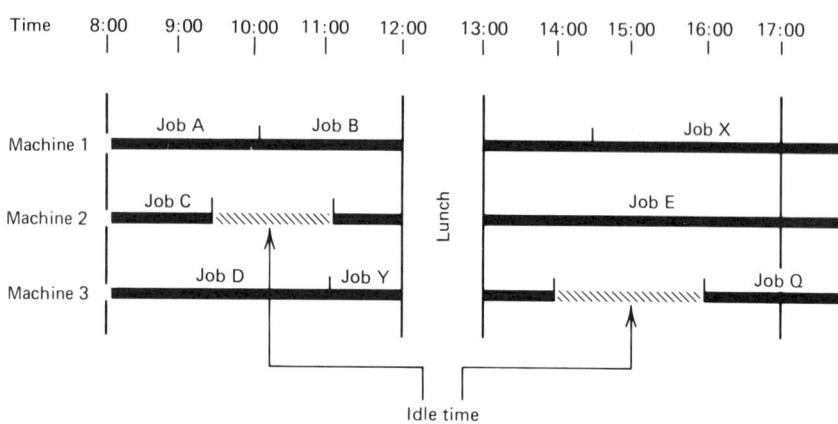

Figure 9.1 Idle time expected at each work center.

When sequencing operations, the system uses an internal 24-hour clock to simulate the actual scheduling of operations in the plant. Specifying actual start/stop times for each machine enables better sequencing decisions to be made. For instance, the sequencer would not plan to do a heat-treat operation at the end of the shift if the subsequent operation had to be performed within two hours and the next work center was working on a one-shift basis.

During processing, the clocks simulate the time of day and indicate when each machine will be available for the next job. The system keeps track of idle time expected between successive jobs on a machine because of unavailability of another job (Figure 9.1). The total idle time expected for the day or shift can be shown on the work sequence list illustrated in Figure 9.2.

The principle of sequencing is that all work centers are cyclically processed in turn. During one cycle, all jobs queuing at a work center are considered for assignment. Work is assigned for a specified period of time ahead (20 minutes, 1 hour, 2 hours, 4 hours, etc.). When all work centers have been processed, the clock is moved forward and processing repeated for the next cycle.

During each cycle, work expected to be completed at a work center becomes available for processing at the next work center after interoperation time, shift length, and so on are considered (Figure 9.3).

Length of Time Increment

The length of the cycle period can vary from minutes to days. The shorter the period, the more accurate the results. A longer increment, however, reduces computer processing time.

WORK CENTER	75205	RADIAL DRILL		NUMBER OF MACHINES	1	AVERAGE DAILY CAPACITY	16.0 HRS

DAY NO.	AVAIL HRS	LOADED HRS	PERCENT	IDLE TIME HRS	PERCENT	0 20 40 60 80 PERCENT 100
881	14.0	14.0	100	0.0	0	:XXXXXXXXXXXXXXXXXXXXXXXXXXXXXXXXXXXXXXXXXXXXXXXX
882	16.0	12.3	77	3.7	23	:XXXXXXXXXXXXXXXXXXXXXXXXXXXXXXXXXXXXXX
883	16.0	16.0	100	0.0	0	:XXXXXXXXXXXXXXXXXXXXXXXXXXXXXXXXXXXXXXXXXXXXXXXX
TOTAL	46.0	42.3	92	3.7	8	

SHOP ORDER NO.	ITEM NO	DESCRIPTION	OP. NO.	TEXT OF OPERATION	AVAIL -ABLE	QTY	OPN. HRS	SET-UP HRS	LAST WC-NO END	NEXT WC-NO START	PLAN WC
05510	003204	SCRAPER	020	CHAMFER	::	1000	3.5	0.0	61001 880	75310 883	
05530	220121	GRIP UNMOUNTED	080	DRILL AND CHAMFER	::	500	10.1	0.4	75204 880	75207 886	
05509	003204	SCRAPER	020	CHAMFER		1000	3.5	0.0	61001 881	75310 885	75637
05526	003210	ADDITIONAL SUPPORT	060	DRILL		400	7.8	0.2	75303 882	75301 889	
05501	220121	GRIP UNMOUNTED	080	DRILL AND CHAMFER	::	100	2.0	0.4	75204 881	75207 886	
05515	003216	ADDITIONAL SUPPORT	060	DRILL		400	7.8	0.2	75303 881	75301 884	
05516	003217	ADDITIONAL SUPPORT	090	DRILL		800	16.0	0.2	75303 882	75201 888	
05517	003210	ADDITIONAL SUPPORT	090	DRILL		800	16.0	0.2	75303 884	75201 890	

JOB QUEUE											
05502	220121	GRIP UNMOUNTED	080	BORE AND CHAMFER		500	10.1	0.1	75204 887	75207	

Figure 9.2 Work sequence list.

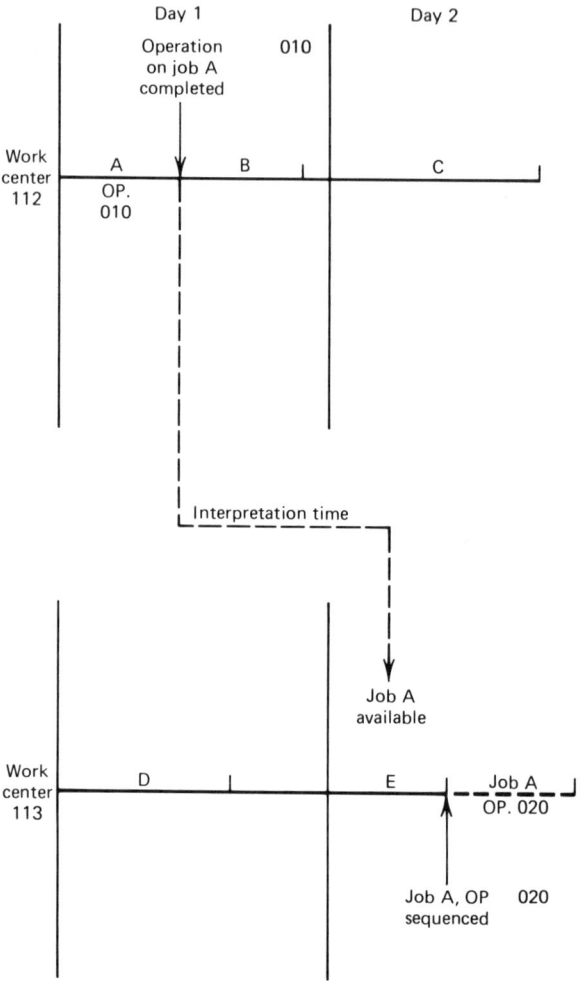

Figure 9.3 Jobs available for sequencing.

Work Center Sequence

The work centers themselves can be loaded in any sequence. In most industries there are certain "gateway" or first-operation departments (material cutoff, foundries, turret lathes, etc.) that ought to be sequenced first because they release work to later work centers.

At the start of processing, the first work center queue is checked for all operations available for sequencing. An operation is available for

9.1 Short-Term (Daily) Capacity Planning

sequencing when all previous operations have been completed and components are available.

The operation priority is now calculated for all jobs in the queue, and the job with the highest priority is loaded onto the first machine. The next highest priority job is loaded onto the second machine and so on, until every available machine in the work center has been assigned one job.

The clock for each machine is incremented by the operation time (set-up time + run time per part + tear-down time).

The next job is now loaded onto the first available machine in the work center (machine 3 in Figure 9.4) and so on, until the capacity within the cycle period is used up or until all available jobs meeting the operation priority criteria are sequenced.

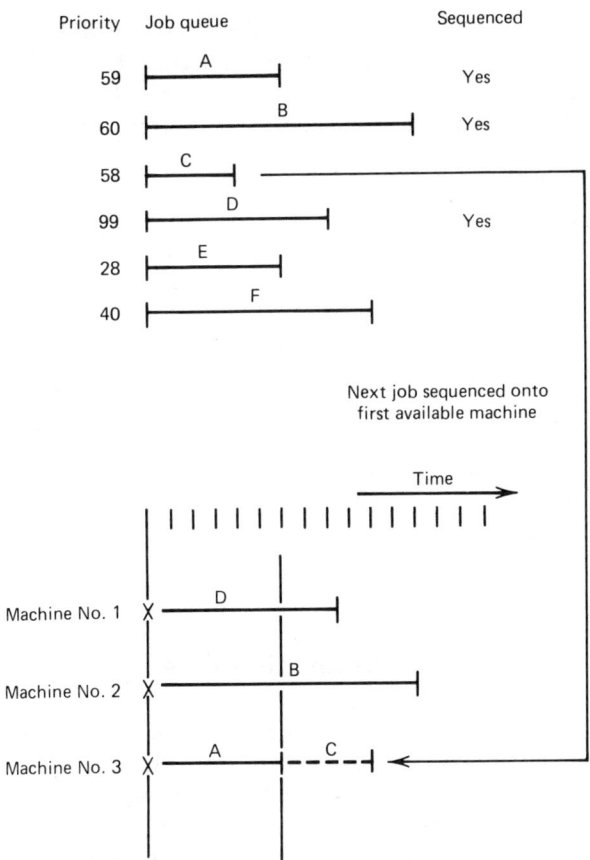

Figure 9.4 Scheduling by job priority.

After an operation has been sequenced onto a machine, the completed job is made available for sequencing at the next work center. The availability date and time are determined by using the interoperation time calculated earlier. The queue time is excluded because the actual queue is being simulated by this process.

The next work center is now loaded in a similar manner from the available jobs in its queue.

When all work centers have been sequenced, the clock is incremented and processing starts again at the first work center. Sequencing then goes on for each machine from the time it left off.

The next time around, the queue contains operations previously available but not sequenced, as well as new operations available as a result of just being completed at other work centers. The sequencing is continued to cover a period of one or more days, as required.

When work has been sequenced up to the end of a shift, the system sets the clocks to coincide with the start of the next shift. It also checks for any scheduled jobs that will be delayed beyond their start date; these jobs have not yet been sequenced, and their start date will now be later than that planned in the order release phase.

The system checks back through the shift to see what work could have been sequenced into alternate work centers. The operations that are running late are scheduled, if possible, for alternate work centers if capacity is available. Each work center will be looked at in turn, in a predetermined sequence, for work to offload.

Work centers having idle time above a specified figure can also be checked to see whether work can be offloaded from other work centers. Alternate operations or routings are checked for during the sequencing process.

The level of overtime necessary to meet the schedule has been determined by management as a result of capacity requirements planning and order release planning. Operations are sequenced up to the specified level of overtime above the normal working hours.

9.2 Dispatching Rules

Theoretically, capacity planning has scheduled the work to the last detail, and the foremen simply have to carry out this plan by assigning jobs to their operators. In practice, it never works in this way. In spite of the fact that the load was balanced and all the competition for capacity resolved, the foremen still face the problem of jobs competing for capacity. When a

9.2 Dispatching Rules

machine becomes free, a decision must be made about its next operation. The short-term capacity planning attempts to solve this problem by considering only those jobs that are ready for processing. The priority rating is used to sequence these jobs.

Another approach is to construct a simple practical rule—for example, first come first served—that the foremen will use in sequencing the jobs waiting in the queue of a machine. The rule must be simple, so that the foremen can use it without the need for elaborate computations. They must also possess all the relevant information. Following this line of thought, the question arises as to whether any priority decision rule (dispatching rule) works better than some other rule and whether any decision rule is significantly superior to another. This leads to another question: How do we measure and define good scheduling? What are we actually trying to accomplish? There are many criteria by which one can define the goals of scheduling, such as:

- Minimum level of work-in-process.
- Maximum number of processes completed.
- Maximum number of jobs sent out of the shop.
- Minimum number of processes completed late.
- Minimum average lateness (tardiness) of all jobs in the shop per period.
- Minimum queue wait time of jobs in shop.
- Minimum number of jobs waiting in shop.
- Maximum shop capacity utilization.
- Minimum number of jobs waiting in queue for more than one period.
- Minimum size of jobs waiting in queue for more than one period.

The topic of the dispatching rule has been studied by many researchers (see the references at the end of the chapter). Many simulation studies have been done, and many dispatching rules have been tested. There are various types of dispatching rules: local rules, where priority assignment is based only on information about the jobs represented in the individual machine queue; global rules, where information from other machines, in addition to the one at which the queue is formed, is utilized; static rules, where the relative assignment of priorities does not change over time; and dynamic rules, where the relative assignment of priorities does change with time.

The following is a list of some dispatching rules that were studied; the list contains the symbol representing the rule; the definition of the rule, that is, from among the jobs in queue, which one to choose; and a list of references referring to this rule and to results pertaining to comparisons of this rule with other rules:

DDATE (EDD)	Has the earliest due date (5, 13, 16, 18, 20, 23, 26, 28, 32, 38, 45).
FASFS	Arrived first in the shop (5, 8, 11, 13, 16, 17, 20, 26, 38).
FCFS	Entered first this queue (5, 9, 11, 13, 16, 17, 18, 21, 24, 26, 31, 32, 33, 38, 42, 44).
2-Class FCFS (p)	Entered first this queue, with priority to the subgroup of "short" jobs; when partitioning so that a fraction p of all waiting jobs, having the shortest imminent processing times, falls into the "short" group (13, 16).
FCFS; SPT (n) (RSPT)	Entered first this queue, unless there are n or more waiting jobs; in the latter case choose by SPT (13, 16, 17).
FOPNR	Has the fewest operations remaining to be performed (13, 16, 23).
LOST	Has the longest imminent operation time plus actual set-up time (45).
LPT (LOT)	Has the longest processing (imminent operation) time (5, 9, 13, 16, 20, 21, 26, 31, 36).
LPT/O	Has the largest value of longest processing time per operation (34).
LPTR (NWKR, LRT)	Has the largest remaining processing time (5, 9, 13, 16, 17, 20).
LTWK (TWORK)	Has the longest total processing time (5, 13, 16, 21).
LV	Has the largest value (26).
MOPNR (NOP)	Has the most operations remaining to be performed (5, 13, 23, 38).
NINQ	Will go for its next operation to the queue with the least number of jobs (23, 26).
OPNDD	Has the earliest operation due date, assuming the allowed flow time is divided equally among operations (4, 13, 16, 18).
OPNSD	Has the earliest operation start date, assuming the allowed flow time is divided equally among operations (43).

9.2 Dispatching Rules

PT + (a)SLACK/ OPNR	Has the smallest weighted sum of imminent operation time and slack per remaining operation (13, 16, 36).
PT + (a)WINQ	Has the smallest weighted sum of imminent operation time and work in the following queue (13, 16).
PT + (a)SLACK + (b)WING	Has the smallest weighted sum of imminent operation time, slack time, and work in the following queue (27).
PTR/OPNR	Has the smallest ratio of remaining processing time to number of remaining operations (21).
RANDOM	Is selected by a random process (5, 6, 13, 15, 16, 21, 24, 26, 30, 31, 33, 34, 45, 46).
R.CYT/PTR	Has the smallest ratio of cycle time (time from arrival to due date) to remaining processing time, multiplied by an urgency index R (8).
RPT	Is next on the list of jobs which have been listed in reverse of the LPT schedule (10).
SIMSET	Has the shortest actual set-up time (45).
SLACK (MST, SLK)	Has the least remaining slack time (5, 13, 16, 18, 24, 26, 28, 32, 33, 38).
SLACK/OPNR	Has the smallest ratio of slack time to number of remaining operations (1, 4, 5, 8, 11, 13, 16, 18, 32, 36, 38, 43).
SLACK/OPNR; SLACK (SLACK)	Has the smallest ratio of slack time to number of remaining operations, unless there are jobs with negative slack; in the latter case choose by SLACK (24).
SLACK/TR	Has the smallest ratio of slack time to the remaining time (33).
SLACK/TR; SLACK (SLACK)	Has the smallest ratio of slack time to the remaining time, unless there are jobs with negative slack; in the latter case choose by SLACK (24).
SPST (SOST)	Has the shortest processing (imminent operation) time plus actual set-up time (45).
SPT (SOT)	Has the shortest processing (imminent operation) time (1, 4, 5, 8, 9, 11, 12, 13, 15, 16,

2-Class SPT	17, 18, 20, 21, 22, 24, 26, 28, 30, 31, 32, 33, 34, 36, 42, 44, 45). Has the shortest imminent operation time, with priority to the subgroup of preferred jobs; when partitioning the waiting jobs according to some preference system (8, 13, 15, 16, 20).
SPT; FCFS (k) (TSPT)	Has the shortest imminent operation time, unless a job has been held in this queue longer than time k; in the latter case choose by FCFS (13, 15, 16, 20).
SPT; FCFS (p)	Has, alternatively (for periods of fixed length), the shortest processing time or entered first this queue, p being the proportion of the cycle during which the FCFS rule is used (15).
SPT; SLACK (SLACK) (SI^x)	Has the shortest processing time, unless there are jobs with negative slack; in the latter case choose by SLACK (20, 24).
SPT; (SPT + SLACK) (SLACK)	Has the shortest processing time, unless there are jobs with negative slack; in the latter case choose by smallest sum of imminent operation time and slack time (28).
SPT/O	Has the smallest value of shortest processing time per operation (13, 34).
SPT/(PTR)[a]	Has the smallest ratio of processing time to the "weighted" remaining processing time (13, 16).
SPT/TWK	Has the smallest ratio of processing time to the total processing time (all operations) (13, 16).
SPT*TOT	Has the smallest value of the product of the processing time by the total operation time (38).
SPT + (a)XWINQ	Has the smallest weighted sum of imminent operation time and work in the following queue (13, 16).
SPTR (SRPT, LWKR)	Has the smallest remaining processing time (5, 9, 13, 16, 20, 21, 31).
STWK	Has the shortest total processing time (5, 21).

9.2 Dispatching Rules

TR/(PTR + QTR)	Has the smallest ratio of remaining time to the sum of remaining processing time and expected queue time (move plus wait) (43).
TR–PT	Has the smallest difference between the remaining time and the imminent operation time (20).
W	By the end of its imminent operation, the total cumulative waiting time of the rest of the queue will be minimum (21).
WINQ	Will go for its next operation to the queue with the least work (5, 13, 16, 26, 32).
WSPT	Has the smallest ratio of imminent operation time to given weights (4, 37).
XNINQ	Will go for its next operation to the queue with the least number of jobs, both present and expected (23).
XWINQ (AWINQ)	Will go for its next operation to the queue with the least work, both present and expected (5, 13, 16, 17, 23).

It seems that it is not difficult to devise a plausible dispatching rule. In some situations the rationale for using a particular rule may be that it helps relieve congestion in the shop, while in other instances the motivating factor may be a need to meet order due dates. The major value of simulation research is that it has been able to examine a wide variety of alternative rules and to identify a few simple but effective rules for each situation.

Many results concerning the optimality achieved upon application of some of these priority rules to specific objective functions are known. The following is a list of these known results; they pertain to a single machine with nonpreemptable single-operation jobs, all of which are known initially:

- Mean flow time is minimized by SPT sequencing.
- Mean (and total) inventory (i.e., number in the system) is minimized by SPT sequencing.
- Mean (and total) waiting time (where the waiting time of a job is defined to be the time it spends in the system prior to the start of its processing) is minimized by SPT sequencing.
- Maximum waiting time is minimized by SPT sequencing.
- Mean (and total) lateness is minimized by SPT sequencing.

- Weighted mean flow time is minimized by WSPT sequencing.
- Maximum job lateness and maximum job tardiness are minimized by EDD sequencing.
- Minimum job lateness and minimum job tardiness are maximized by SLACK (MST) sequencing.

Although the above results apply only to simple problems, they are important as possible good heuristic procedures or parts of good algorithms to be devised for a particular problem.

The simulation studies vary widely in their objective functions and job-shop characterizations, constraints, and conditions. However, even under these varying conditions, the performance of the SPT dispatching rule, along with that of related rules, continues to be impressive.

Eden (19) tested 10 dispatching rules to be used for controlling the service operation to semiautomatic machines. Some of the objectives of this type of system are maximum labor utilization, maximum machine utilization, and minimum total queue time of the production system. The results showed that the SST (similar to SPT) rule performed best and is most likely to provide the highest total utilization for the system. In spite of the rule being static, it performed better than all others, including eight dynamic, more complex rules, which attempted to anticipate the future state of the system. The SPT rule performed best and the FCFS rule worst; even the RANDOM selection rule performed better than FCFS.

The mechanism by which SPT reduces the lead time of a job in the system (flow time) should not be difficult to understand. By giving priority to short tasks, it accelerates the progress of several short jobs at the expense of a few long ones; this was proved mathematically for a single-machine problem by Baker (5). Thus it is only natural to expect that SPT would perform well in the job-shop situation.

The effect of shop load on the relative performance of dispatching rules was studied by Conway (13–18). He found that the performance of SPT is preferable only when shop loads are quite heavy.

To keep the rule practical, the dispatching rule may be a combination of a few rules. For example, the normal dispatching rule is FCFS, but when the length of an individual queue grows too big, the dispatching rule switches over to SPT. In particular, when the length of any queue reaches a certain number (e.g., nine), priority within that queue is assigned according to SPT. However, once the queue length drops, the dispatching rule reverts to FCFS. Since FCFS is the normal rule, long jobs do not typically encounter excessive delay. However, long jobs are sometimes delayed temporarily, and SPT provides relief to individual machines facing severe congestion.

9.2 Dispatching Rules

Most of the dispatching rule studies were made under the assumption that each operation required a unique machine. In practice, machine groups and alternate machines are often specified in the routing.

Wayson (44) studied this case. In his model, certain operations could be performed by one of several machines, so that the imminent operation could avoid a long queue if another machine assignment was feasible. Wayson compared the effects of the FCFS and SPT dispatching rules for this model when the performance measure is mean number of jobs in the system. He found, for example, that the behavior of SPT under unique machine assignments was similar to that of FCFS when alternate machine selection could occur about 20% of the time. In this case, FCFS yields the same number of jobs in queue as SPT without alternate selection. He also showed that as the degree of potential alternate selection rises, the difference between the two rules becomes less significant.

It is generally accepted that added flexibility in grouping similar machines (thus allowing a flexible routing of jobs) is advantageous; it reduces machine congestion and queue lengths.

Several important themes emerge from the extensive job-shop simulation experiments that have been discussed. An encouraging theme in job-shop work is the efficacy of local rules that utilize only a limited information base. The studies point out that there is no unique definition of the goals of production planning. There are numerous objectives, and they both vary in nature and conflict with one another. Attention is also drawn to the nature of the common objective of meeting due dates. If due dates are not established by capacity planning, they are not engineering values. The simulation of rules with the objective of minimum tardiness (lateness) may depend critically on the manner in which due dates are set and on their tightness. This means that a certain amount of caution is necessary when attempting to apply the conclusions from one simulation study to a different type of system.

The aim of these studies was to come up with a simple dispatching rule that the foreman, with the information on hand, would be able to apply. Today we might broaden the scope of these studies and incorporate them into the IMS. We can use the computer with the system data bank to decide which job should be assigned next. Thus simplicity is no longer a necessity, and a combination dispatching rule can be used. For example, a combination of SLACK, FCFS, SPT, PTR, SIMSET, network priority, and so on could be easily handled by a computer.

In the scheduling studies, the routing is taken as given data that scheduling must cope with. If SPT is so impressive, the routing can be designed to be composed of several short processing time elements instead of a few long operations. If flexibility is desirable for scheduling,

9.3 Job Recording

The released jobs for production are under shop-floor control and are actual commitment to deliver the goods under its control. The requirement planning phase considered the released orders on the item level as on hand, or on order, or scheduled receipt. Shop-floor control is responsible for expediting these orders until job termination. The control is then transferred to inventory. The expediting of the orders is done on the job operation level. All released orders are recorded in the shop open order file. This file is used for:

- Daily scheduling of all work centers.
- Daily material handling planning.
- Daily release of material and tooling.

The same file is used as a source of information in the periodic requirement planning and capacity planning. For all these applications, the file must contain up-to-date information. The information comes from the shop floor, where the actual manufacturing takes place.

The main objective of job recording is to supply feedback on what is actually occurring on the shop floor to the open order file. The following information is required for this objective:

Where the job took place	Plant
	Department
	Work center–cost center
Job identification	Shop order number (product number)
	Part number
	Batch number
	Operation number
Quantity produced	Quantity completed
	Quantity scrapped
Time consumed	Elapsed time

One of the principles of data processing is to expand the use of a single event reported to as many applications as possible. Information is costly and therefore should not be wasted. If with a small additional effort in reporting many applications can exploit the same data, then the effort is worthwhile. This principle can be applied in our case by adding the

Before that: process planning can furnish this flexibility. This topic is covered by Hal-Technology and will be discussed in Part III.

9.3 Job Recording

operator number to the data reported. In this way, additional objectives can be set.

The objective of job recording is to supply data for the following applications:

- Job assignment.
- Capacity planning.
- Requirement planning.
- Preventive maintenance planning.
- Work-in-process.
- Employee salaries.
- Incentive systems and employee efficiency.
- Costing.
- Computing hourly rate of cost centers.
- Foreman rating system.
- Vacation balance for employees.
- Personnel sickness, absenteeism, and lateness.
- Accident reporting.
- Miscellaneous reporting on employees, machines, items, and so on.

The main application to which job recording supplies data—job assignment—utilizes the information only as feedback to jobs that are recorded in the shop open order file. Thus it requires the reporting of only selected activities. However, it is not good practice to report only selected activities, since this calls for operator judgment in deciding when and what to report and will usually result in vital information being omitted. The other applications on the list call for reporting all activities. Therefore, the job recording system should be designed to cover all operator activities.

The sum of time per day per operator in all job records must be equal to the operator attendance time. Thus the job recording system can be used as the data source for employee salary and as a validation check that no job record is missing. Furthermore, the job recording system in its extended meaning should enable operators to report any event and to account for every minute of their attendance time. For simplicity, it is advisable to adhere to a single set of rules and procedures to cover all events.

To achieve this, all events and activities should be regarded as manufacturing activities and reported as such. The manufacturing activities are reported by product-part-batch-operation, and the same form should be preserved for the other activities. To distinguish between the actual manufacturing activities, and other activities, it is recommended that the product field in the nonmanufacturing activities be kept blank. All events are recorded in an appropriate account in the bookkeeping system. It is therefore recommended that the part field be used as the bookkeeping account number for the event. The events might include:

- Electricity failure.
- Machine cleaning.
- Arrival late to work.
- Accidents and required first aid.
- Waiting for a job.
- Absence.
- Union business.
- Material handling.
- Overhead activity.

Thus one assigns waiting time to the "working on waiting" account number. This is a rough classification of events. Each plant may create as many groups as desired, corresponding to its bookkeeping system and the objectives set forth.

The rough classification might not be sufficiently accurate for some applications and objectives. Waiting for a job, for example, is one of the parameters for rating foremen. It might be doing them an injustice if the reasons for waiting were not revealed. Some of them are controlled by the foreman, but many others are not. The operation field can be used to code the reason. The reasons might include the following:

- Waiting for foreman.
- Waiting for materials.
- Waiting for tools.
- Waiting for blueprint.
- Waiting for planning changes.

Employee absence is an important source of information for the personnel department, for salary, and for sociological studies (effect of sex, age, skill, etc.). The reason for absenteeism can be specified in the operation field by code. Some reasons are:

- Paid vacation.
- Authorized unpaid vacation.
- Unauthorized absence.
- Illness.

9.3 Job Recording

- Studies.
- Holiday.
- Mourning periods.

The degree of refinement is up to the individual plant. Procedures must be specified for all events, such as:

- *One operator operates a group of machines.* Operator time should be counted only once for salary and costing purposes. For machine utilization and preventive maintenance each machine should accumulate its own working time. The group of machines might be of the same type working on the same job with start and finish times also being identical. Each machine, on the other hand, might be working on a different job with its own start time, finish time, and quantity produced. One should bear in mind that the job recording is operator-oriented and not machine-oriented. A code field must be used to resolve this situation. If an incentive system is employed, things might become more complicated, and an additional code might be required to designate the scheme by which the incentive is to be computed.

- *Team work.* The operation time is the sum of the individual operators' time. Hence, all operators should record their own start and finish times separately. The quantity is the result of the team effort and should be recorded only once. The code field must designate that it is a team work. Only one member of the team should report the quantity. Certain operators on the team might be shifted to other jobs, while others might join the team. The job identification and code will designate to the system the status of the team and total working hours for the task.

- *Set up and tear down.* In some plants the set up and tear down is performed by special operators and the manufacturing by another. The set up might include the manufacturing and inspection of a few pieces before the machine is turned over to the operator. The information that the machine is ready for manufacturing is important for job assignment. The set-up operation can be regarded as an overhead for costing purposes or as a direct job, and this decision has bearing on job recording procedures. In any case, a code indicating that the set up is complete must be incorporated into the job recording.

- *Interruptions.* A job can be interrupted for many reasons. Operators might wait at their machines during the interruption or the foreman might shift them to another job. In the latter case, there are no problems in reporting. If the operator returns to the first job later in the day, the quantity might be recorded only once. It is up to the company to decide how to treat interruptions. Probably, interruptions of up to a certain length of time should be incorporated into the job time, while longer interruptions should be recorded as a separate job, as discussed earlier in this section.

• *Job completion.* A job can be regarded as complete when the quantity produced is equal to the quantity ordered. However, seldom do the two quantities match absolutely. An automatic completion can be applied by assigning a tolerance for quantities. Otherwise, a separate completion reporting is required.

• *Incentive systems.* Incentive systems, if applied in the plant, might complicate the recording procedures. The job recording system must correspond to the system or systems used. A piece rate, a daily basis, a batch basis, group scheme systems, and so on might be used. Each one calls for different procedures.

Forming job recording procedures is tedious. General company regulations and application requirements for data should be carefully studied and analyzed. The job recording system is a key to many applications. Calculation of job elapsed time is computed by subtracting the start time from the finish time. However, this computation may prove erroneous if breakfast or lunch breaks fall within the start and finish time limits. It does not pose any problem if the whole plant has its breaks simultaneously. If, however, there is more than one working shift and different break times (e.g., in order to ease the pressure on the cafeteria), or some other special agreement as to working hours, the calculation of job elapsed time becomes a problem. An auxiliary file indicating planned shifts and break hours for each employee for each day of the month is required. This information is also required in order to check as to whether operators arrive on time and whether they report their late arrival and overtime on the same day. The operator arrival time before the authorized time is not counted as attendance time. Overtime authorization is also kept in this file.

If operators are working on a rush job during planned breaks, they must specify this with a code in order to arrive at the correct job elapsed time.

Job recording is a burden and operators do not do it gladly. The combination of job recording and attendance (salary) reporting is of some assistance, since operators are aware that their failure to report their jobs will result in their not receiving their full salary. Some companies employ a department clerk for the purpose of job recording in order to relieve the operator of this duty. In any case, it is advisable to design the system in such a manner as to keep reporting to a minimum. The planned shift file is of assistance. Planned vacation and illness can be reported in one record covering a long period. The file may also contain information on the last

9.3 Job Recording

job reported for each operator. If a job extends over a period of several days, operators must report only arrival and departure. The system will assume that they are continuing to work on the last reported job. Indirect jobs, such as machine cleaning, do not offset the above assumption.

Job recording in its extended meaning is the sole source of information on what is actually occurring on the shop floor. It activates and supplies data to many applications. The activities occurring with job completion reporting, for example, are demonstrated in Figure 9.5.

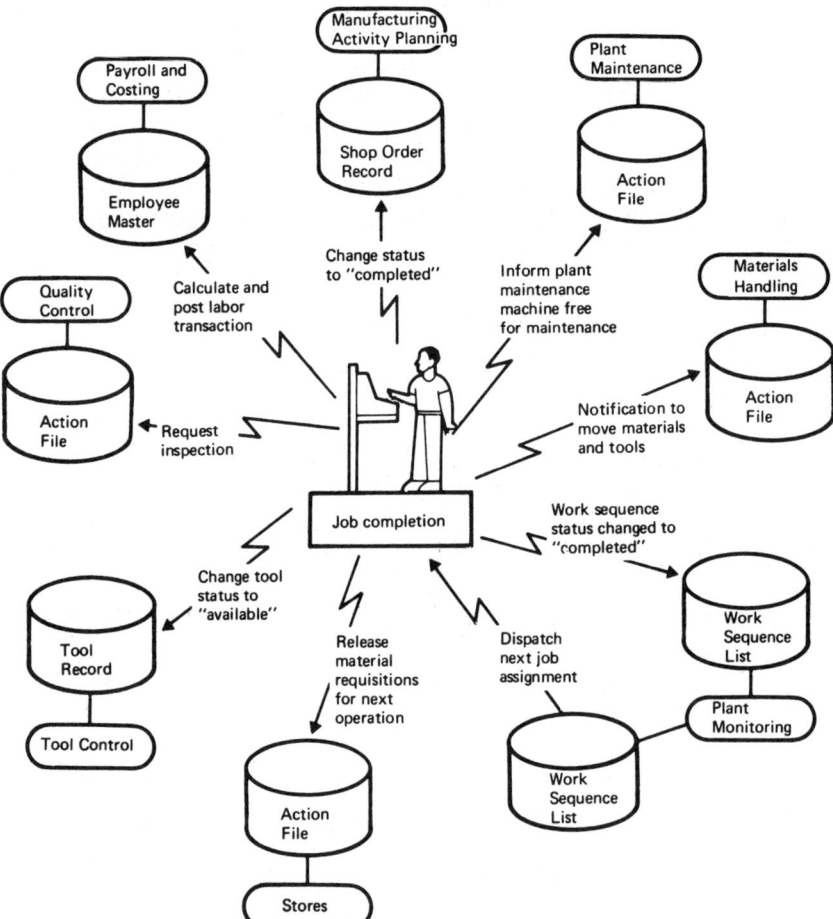

Figure 9.5 Activities occurring with Job Completion Reporting. (From *Copics*, Vol. VI, Chapter 8, G320-1979-0. Copyright 1972 by International Business Machines Corporation. Reproduced with permission.)

9.4 Job Recording Design for Reliability

Job recording must be reliable. If errors occur in reporting, irreversible damage will occur. Items not required will be produced, thereby occupying overloaded machines and wasting materials. Fortunately, manufacturing activities are planned and can be forecast. This feature can be used in validating job recording reports. No effort in validation is wasteful. The greater the number of checks, the more reliable the system will be.

The standard validation procedure as discussed in Section 2.5 should be applied. This type of validation checks each transaction on its own merits for being numeric, completeness, and valid codes used.

Special validation for job recording can be designed to check each transaction with respect to its logic and agreement with predicted activity. The "two-way" data processing concept will be used to eliminate the necessity of information reporting and to synchronize information. The files participating in the system are shown in Figure 9.6.

Some of the recommended validation checks are as follows.

Arrival Transaction

Operators can arrive at the shop only if they have previously departed or if they are new operators and shop employees.

Departure Transaction

Operators can depart from the shop only if they have previously entered.

Start-of-Job Transaction

Operators may start a job only if they have previously entered the shop. The job reported must be a planned job, and match a record in the shop open order file, or one of the indirect jobs as recorded in the standard job file.

For the planned jobs it is not sufficient that the job identification exists in the file. A check to determine if this operation is really ready for processing must be made. An operation is ready for processing if the previous operation has been completed and if materials (and tools) are ready.

It is good practice to assign the first operation of each item as the material issue. When material is issued, it is checked against the shop

9.4 Job Recording Design for Reliability 241

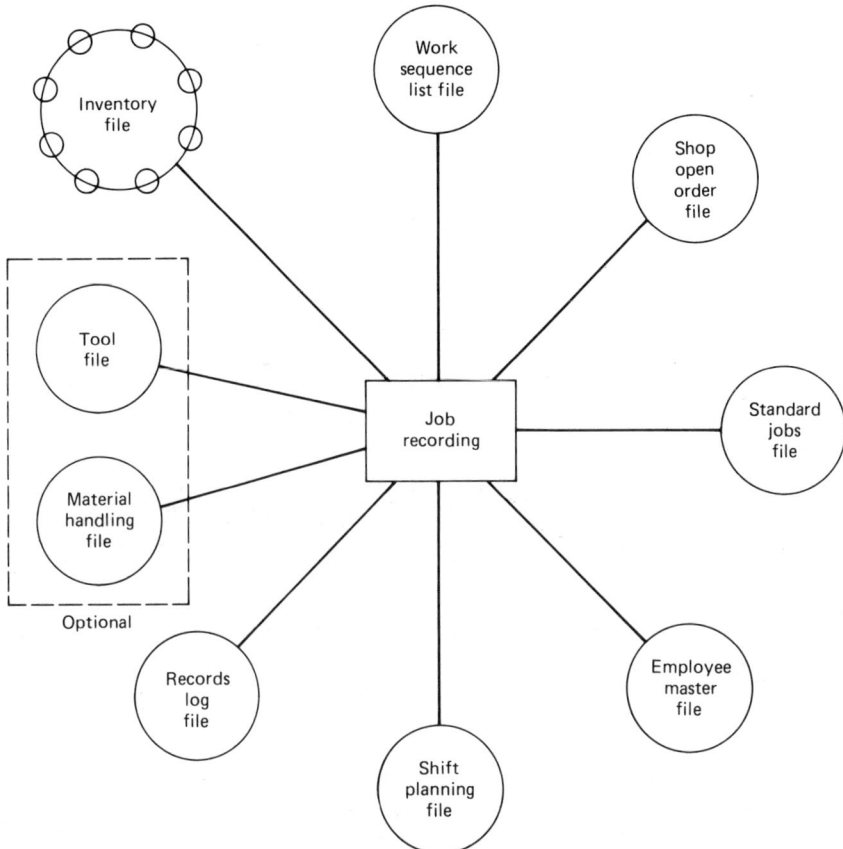

Figure 9.6 Files used in job recording validation. (See Figure 7.6 concerning the inventory file.)

open order file. If valid, the first operation is updated, and the quantity of items that can be produced from the quantity of material issued is recorded as quantity completed. The check for previous operation actually checks to ensure that the sequence of operations is preserved. It is possible that the sequence will be altered either by the foreman or by the process planner. If the sequence check does not correspond to that recorded in the file, a message is generated to the foreman. There are three possibilities in such cases:

1. It is an error. The job must be stopped.
2. It is a temporary change initiated by the foreman to solve an immediate

job assigning problem. The job should be allowed to continue and a sequence change for this batch alone should be entered into the open order file or work sequence list file, if used.
3. It is a permanent change. The job should be allowed to continue, and a change in the working files as described in case 2 should be made. A record to update the master routing file is generated. This change will thus be incorporated in all future planning.

The department and machine numbers specified in the open order file are compared to those reported in the transaction. If they do not match, a message is generated to the production center for approval. It is possible that an error in job identification has been made.

The job number is checked to ensure that it corresponds to the work sequence list. It is possible that although all previous conditions prevail, there are other jobs with higher priorities that should be processed first. A message to the production center indicating this situation is generated. A decision must be made as to whether to stop this job, if still possible, or to accept it as a fact.

Quantity Reporting Transaction

The job to which the quantity relates must be one about which a start-of-job transaction has been previously received. The quantity reported must be equal to or smaller than the quantity reported in the previous operation. By this method, follow-up of quantities from the material issued until the last operation is achieved. It sometimes occurs that the quantity tends to increase in production. There are three possible reasons for the occurrence of this phenomenon:

1. *Error in reported quantity.* This occurs mainly when piece rate incentive systems are applied.

2. *Error in job identification.* The quantity is correct, but relates to a different job number.

3. *Foreman decision to combine batches.* Sometimes a few batches of the same item are in production, each with offset toward the others. One batch can be delayed and another can arrive earlier than planned. It is possible that for a certain operation two or more batches meet. The foreman, for economic reasons, might decide to process the two batches with one setup. The reported quantity is correct, but the files and information system might be distorted. A message of such occurrence is sent, and a decision as to whether the two batches are going to be combined for the rest of the operations or whether they were specific for this operation

9.4 Job Recording Design for Reliability

only must be made. In correcting this situation, it should be remembered that these data are also used for costing.

Job Completion Transaction

A job can be completed only if the quantity of the last operation is within the tolerance of the ordered quantity. Moreover, all operations of this item, with the quantity recorded, must be completed. Sometimes an operation is unintentionally skipped. If an operation has been intentionally deleted, a check should be made to see if it is temporary for the above batch or permanent. If it is a permanent change, a record is generated to update the master routing file.

It is essential to keep the master routing file as well as the bill of material file updated for reliable planning. This calls for a discipline in reporting that is enforced only by procedure measurements. People sometimes like to perform rather than to report on activities. The reliability measures taken in the system could be used as an automatic updating tool for the crucial master files. Some data, such as operation time or standard scrap factor, are unknown or are theoretical values for those who have to report them to the master file. For reliability and ease of reporting, a feedback from the action files can be used to update master files with these data.

Similar logic should be applied in the validation of other reported transactions.

The depth of system control desired is up to the individual plant. Material handling, for example, can be left for manual treatment and regarded as an indirect operation; on the other hand, it can be treated as part of the system. In such cases, the completion transaction records it in a file, and the transfer of material is planned by the system. One should remember that any planned activity must have feedback. The reporting might, however, become too cumbersome and the possibility of error increased.

Inspection can be regarded as an overhead. It is recommended that in-line inspection be regarded as any other operation. It should be part of the routing and reported as such. This will establish a milestone in quantity checking throughout the manufacturing process.

The final quantity checking is made by the quantity received by inventory. This quantity is compared with the quantity reported in the last operation.

Any validated transaction updates all the relevant files. This is required for future validation, reducing individual reporting, synchronizing information in all applications, and eliminating reporting errors by eliminating the need to report.

The use of mechanical data collection equipment reduces the number of stages between the manual reporting on a form and its transfer into a computer-readable form. Since many errors can be introduced in these stages, data collection equipment contributes to the increase of reliability in the system.

Data collection through the use of terminals is a data reporting function that takes place in a non-EDP (electronic data processing) environment. Data collection is performed from the production floor or workshops by personnel untrained in the use of terminals.

Production and factory employees work far from the computer site. They are at a geographic distance from the EDP center and thus perhaps feel themselves on the periphery of the system. Production workers are required to perform their jobs according to certain production norms and production efficiencies; the work pressures they face are not similar in content to the work pressures placed on the EDP and data entry personnel who work at the computer center.

Data collection systems are used by production personnel who do not have EDP or terminal hands-on experience and whose work objectives and expectations may be at variance with total system objectives. These are the same employees upon whom the system depends for basic input, and who rely on that system output to help fulfill their production objectives.

Certain characteristics distinguish data collection from other modes of data acquisition.

The data processing system depends on accurate, timely input from remote production sites. Factory personnel must satisfy these system requirements (by reporting) and must satisfy their own work objectives at the same time. Simply put, the data reporting is a by-product of the worker's function. Ideally, the data collection system does not impact negatively on the employee's main function, which is production. The worker must be comfortable working with the data collection terminal, and the data collection system must improve the worker's situation from a production point of view.

Employees are provided with a unique identification badge. Upon reporting to and departing from work, they insert these badges into the badge reader. (They do the same during specified time periods of the work shift.) When employees insert the badge, the system will record their identification number, the terminal number, and will add such constant information as time of day, date, and transaction code.

9.4 Job Recording Design for Reliability

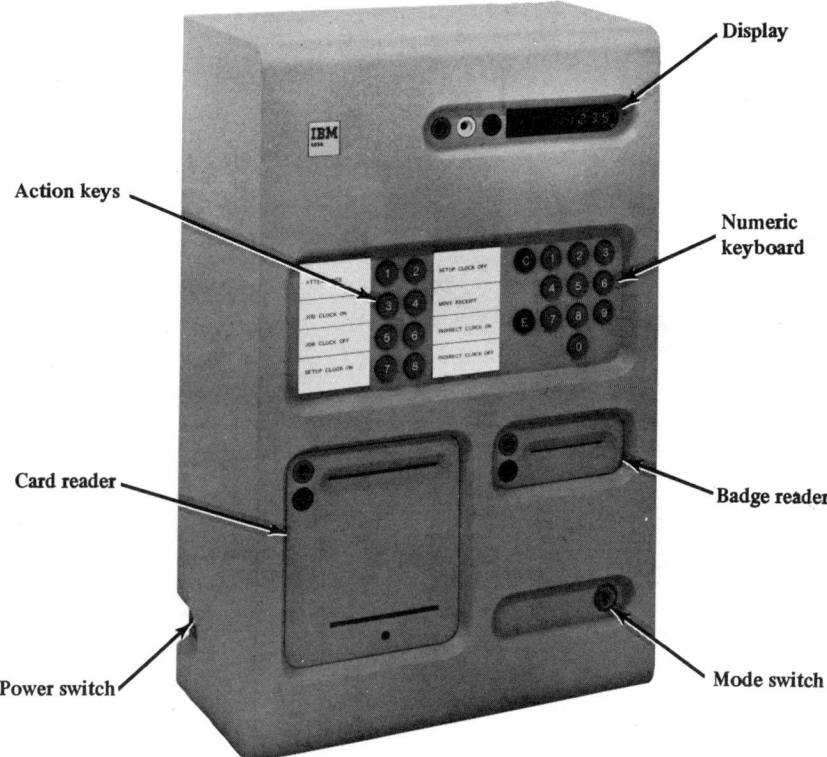

Figure 9.7 IBM 5235 Data Entry Station. (From *IBM 5230 User Guide*, GA34-0040-1. Copyright 1976 by International Business Machines Corporation. Reproduced with permission.)

Upon releasing a job to production, a prepunched card is prepared. This card is used to report the start of a job. To report this, operators insert their badges into the badge reader and the job-card into the card reader. Additional variable information, such as quantity and machine number, can be inserted by a numeric keyboard.

An action key indicates the nature of the transaction reported. The system inserts such information as hour, date, terminal number, and location.

Figure 9.7 shows an example of the data collection hardware available today—the IBM 5235 Data Entry Station. Figure 9.8 illustrates a plan communication system.

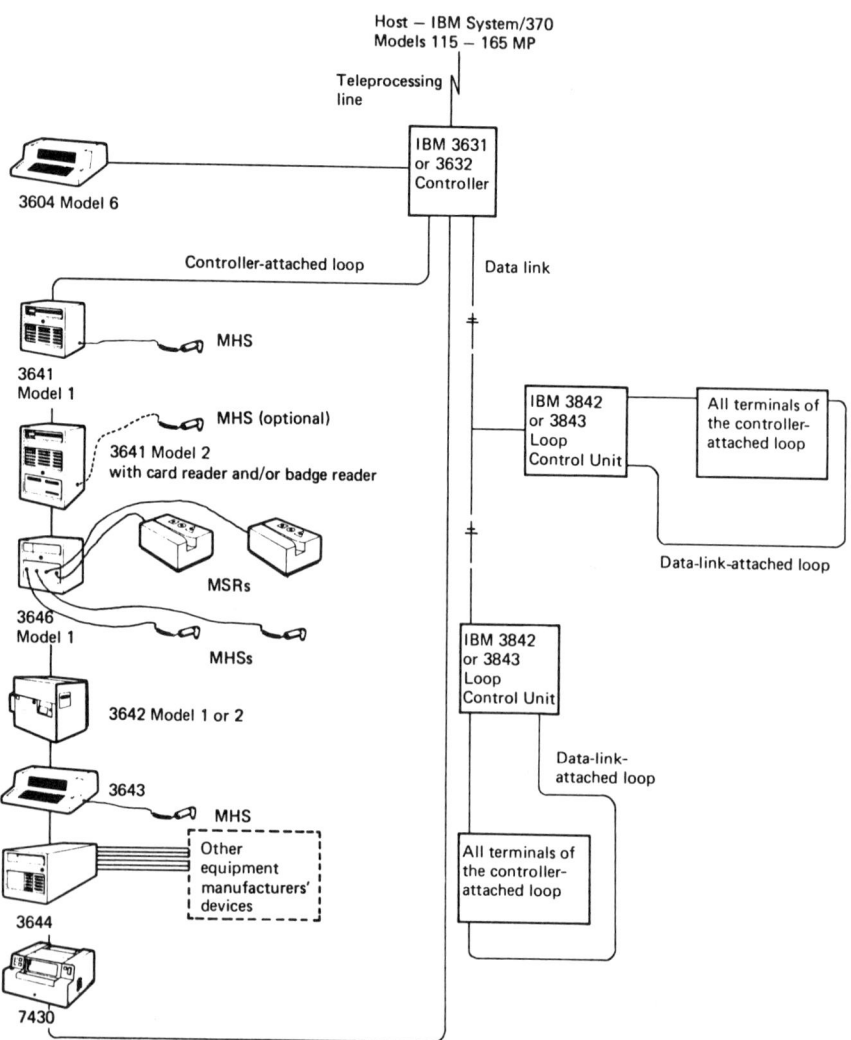

Figure 9.8 Plant communication system. (From *IBM 3630 Application Description,* GE19-5222-0. Copyright 1977 by International Business Machines Corporation. Reproduced with permission.)

9.5 Control Reports

The shop floor is controlled by foremen and production managers. The objective of the IMS (which is referred to by some as a production information and control system) is to supply data and information to those in actual control. Some of the common reports generated by the system are:

- *Daily release of jobs report.* This report is a work sequence list, as shown in Figure 9.2 and discussed in Section 9.1.
- *Daily performance report.* This report is a log of all activities performed on the previous day. It contains computed efficiency per operation and messages of such unusual occurrences as unplanned termination of job, operator efficiency above 150% or below 80%, quantity exceeds requirement, and operator coming late or leaving early. This report comes in two cross sections—by job and by employee. The first purpose of this report is to furnish the foremen with condensed information on the activities in their departments the previous day. Although the foremen are probably already aware of these occurrences, this report supplies quantitative data and emphasizes the unusual events that require action. (Some prefer the report to list only exceptions.) The second purpose of this report is to increase reliability. Despite all the validation and checks performed, errors might still exist. The more people working and using the information, the more likely the errors are to be discovered. Foremen or operators, merely by looking at the printed report, may discover errors. The data in the report will affect the salary paid to operators. Thus they are the most likely ones to detect errors—at least those that result in reducing their salaries.
- *Daily exception report.* This report results from comparing the planned daily work sequence list with actual performance. It lists those jobs that were planned and which, for some reason or another, were not done, as well as those jobs that were not planned but were actually carried out. This report is of interest to the production planning personnel. Its purpose is to serve as an early warning to the expediters and to point out any difficulty encountered in carrying out the work as planned.
- *Daily management report.* If management desires, a daily summary report can be processed. This report may include information on such key follow-up points as total number of employees, absentee ratio, direct to indirect hours ratio, average efficiency, direct to unplanned maintenance hours ratio, accidents, percent of jobs performed as planned, and available capacity to actual capacity ratio per period. The purpose of this report is to condense on one sheet of paper the key figures that designate,

in general terms, the plant performance. Managers must decide what they regard as vital information, and the system will probably be able to process it.

- *Periodic machine status report.* This report summarizes the machine breakage and machine loading per period. It provides a good indication of whether or not a machine should be replaced. If so desired, a "breakage as a function of operator" report can be prepared. The information is available in the files of the job recording system.
- *Periodic scrap summarized report.* This report is an exception report and indicates whether any individual operator is responsible for more than reasonable scrap. A study to correlate the scrap to a particular machine, day of the week, or time of day can be conducted by the information available on file.
- *Periodic efficiency report.* This report indicates the efficiency of each operator over a period. It might serve, together with the previous report, in operator evaluation. It may also indicate to the foreman what type of jobs the operator is more capable of performing.
- *Periodic direct to indirect ratio report.* This report indicates the distribution of working hours in each department between direct and indirect jobs. The report may also include a financial ratio, including indirect material used by the department. The figures in this report can be compared to the standard allowance and used for foreman evaluation.
- *Periodic management information report.* This report is similar to the daily management report with the exception that it covers condensed information over a longer period of time.

Data stored in the files of the IMS probably contain all the information that will be required by managers of different levels of the company. They should not hesitate to contact data processing personnel to discuss their needs. They will probably be amazed at the capabilities of the system. The job recording system keeps track of operator activity at any given moment, while inventory transactions keep track of any inventory activity. This is a powerful combination that can supply any need for information, simulation, and study.

References

1. Aggarwal, S. C. and B. A. McCarl, "The Development and Evaluation of a Cost-Based Composite Scheduling Rule," *Naval Research Logistics Quarterly*, Vol. 21, No. 1 (March 1974), pp. 155–169.
2. Baker, C. T. and B. P. Dzielinski, "Simulation of a Simplified Job Shop," *Management Science*, Vol. 6, No. 3 (April 1960).

References

3. Baker, K. R., "Priority Dispatching in the Single Channel Queue with Sequence-Dependent Setups," *Journal of Industrial Engineering*, Vol. 19, No. 4 (April 1968).
4. Baker, K. R., "Control Policies for an Integrated Production Inventory System," Ph. D. dissertation, Department of Operations Research, Cornell University, Ithaca, NY (September 1969).
5. Baker, K. R., *Introduction to Sequencing and Scheduling*, John Wiley & Sons, Inc., New York (1974).
6. Bakhru, A. N. and M. R. Rao, "An Experimental Investigation of Job-Shop Scheduling," Research Report, Department of Industrial Engineering, Cornell University, Ithaca, NY (1964).
7. Barash, M. M., et al., "The Optimal Planning of Computerized Manufacturing Sysitems," NSF GRANT No. APR74 15256, Report No. 1 (November 1975).
8. Berry, W. L., "Priority Scheduling and Inventory Control in Job Lot Manufacturing Systems," *AIIE Transactions*, Vol. 4, No. 4 (December 1972), pp. 267-276.
9. Blick, R. G., "Heuristics for Scheduling the General n/m Job Shop Problem," Master's thesis, Union College, Schenectady, NY (April 1969), 100 pages.
10. Bruno, J. L., E. G. Coffman, Jr., and R. Sethi, "Scheduling Independent Tasks to Reduce Mean Finishing Time," *Communications of the ACM*, Vol. 17, No. 7 (1974), pp. 382-387.
11. Carroll, D. C., "Heuristic Sequencing of Single and Multiple Component Jobs," Ph. D. dissertation, Sloan School of Management, Massachusetts Institute of Technology, Cambridge, MA (1965).
12. Coffman, E. G., Jr., J. L. Bruno, R. L. Graham, W. H. Kohler, R. Sethi, K. Steiglitz, and J. D. Ullman, *Computer and Job-Shop Scheduling Theory*, John Wiley & Sons, Inc., New York (1976).
13. Conway, R. W., "An Experimental Investigation of Priority Assignment in a Job Shop," Rand Corporation Memorandum, RM-3789-PR (February 1964).
14. Conway, R. W., B. M. Johnson, and W. L. Maxwell, "An Experimental Investigation of Priority Dispatching," *Journal of Industrial Engineering*, Vol. 11, No. 3 (1960), pp. 221-229.
15. Conway, R. W., and W. L. Maxwell, "Network Dispatching by the Shortest Operation Discipline," *Operations Research*, Vol. 10, No. 1 (1962), pp. 51-73.
16. Conway, R. W., W. L. Maxwell, and L. W. Miller, *Theory of Scheduling*, Addison-Wesley, Reading, MA (1967).
17. Conway, R. W., "Priority Dispatching and Work In Process Inventory in a Job Shop," *Journal of Industrial Engineering*, Vol. 16, No. 2 (March 1965).
18. Conway, R. W., "Priority Dispatching and Job Lateness in a Job Shop," *Journal of Industrial Engineering*, Vol. 16, No. 4 (July 1965).
19. Eden, C. L., "Rules for Scheduling Semi-Automatic Machines with Deterministic Cycle Times," *International Journal of Production Research*, Vol. 13, No. 1 (January, 1975), pp. 41-56.
20. Eilon, S., and D. J. Cotterill, "A Modified SI Rule in Job Shop Scheduling," *International Journal of Production Research*, Vol. 7, No. 2 (1968), pp. 135-146.
21. Eilon, S., and A. J. Pace, "Job Shop Scheduling with Regular Batch Arrivals," *Proceedings, Institution of Mechanical Engineers*, Vol. 184, No. 1, Pt. 1 (1970), pp. 301-310.
22. Eilon, S., I. G. Chowdhury, and S. S. Serghiou, "Experiments with the SI Rule in Job-Shop Scheduling," *Simulation*, Vol. 24, No. 2 (February 1975), pp. 45-48.

23. Gerac, L. P., "An Investigation of Priority Dispatching Rules and Job Release Schedules in the Airsearch Job Shop," Arizona State University, Tempe, AZ (1969).
24. Gere, W. S., "Heuristics in Job Shop Scheduling," *Management Science*, Vol. 13, No. 3 (1966), pp. 167–190.
25. Giffler, B., G. L. Thompson, and V. Van Ness, "Numerical Experience with the Linear and Monte Carlo Algorithms for Solving Production Scheduling Problems," in Muth and Thompson, eds., *Industrial Scheduling*, Prentice-Hall, Englewood Cliffs, NJ (1963).
26. Ginsberg, A. S., H. M. Markowitz, and P. M. Oldfeather, "Programming by Questionnaire," Rand Corporation Memorandum, RM-4460-PR (1965).
27. Harding, J., D. Gentry, and J. Parker, "Job Shop Scheduling against Due Dates," *Industrial Engineering*, Vol. 1, No. 6 (1969), pp. 17–29.
28. Holloway, C. A., and R. T. Nelson, "Job Shop Scheduling with Due Dates and Variable Processing Times," *Management Science*, Vol. 20, No. 9 (May 1974), pp. 1264–1275.
29. Horn, W. A., "Single-Machine Job Sequencing with Treelike Precedence Ordering and Linear Delay Penalties," *SIAM Journal on Applied Mathematics*, Vol. 23 (1972), pp. 189–202.
30. Hottenstein, M. P., "Expediting in Job-Order-Control Systems: A Simulation Study," *AIIE Transactions*, Vol. 2, No. 1 (March 1970), pp. 46–54.
31. Jeremiah, B., A. Lalchandani, and L. Schrage, "Heuristic Rules toward Optimum Scheduling," Research Report, Department of Industrial Engineering, Cornell University, Ithaca, NY (1964).
32. Jones, C. H., "An Economic Evaluation of Job Shop Dispatching Rules," *Management Science*, Vol. 20, No. 3 (November 1973), pp. 293–307.
33. Mellor, P., "A Review of Job Shop Scheduling," *Operational Research Quarterly*, Vol. 17, No. 2 (June 1966), pp. 161–171.
34. Muhlemann, A. P., and A. G. Lockett, "A Scheduling Problem with Machine Flexibility," *International Journal of Production Research*, Vol. 13, No. 1 (January 1975), pp. 57–73.
35. Neimeier, H. A., "An Investigation of Alternative Routing in a Job Shop," Master's thesis, Cornell University, Ithaca, NY (June 1967).
36. Oral, M., and J. L. Malouin, "Evaluation of the Shortest Processing Time Scheduling Rule with Truncation Process," *AIIE Transactions*, Vol. 5, No. 4 (1973), pp. 357–365.
37. Rau, J. G., "Selected Comments Concerning Optimization Theory of Functions of Permutations," in S. E. Elmaghraby, ed., *Symposium on the Theory of Scheduling and Its Applications*, Springer-Verlag, New York (1973).
38. Rochette, R., and R. P. Sadowski, "A Statistical Comparison of the Performance of Simple Dispatching Rules for a Particular Set of Job Shops," *International Journal of Production Research*, Vol. 14, No. 1 (January 1976), pp. 63–75.
39. Rochette, R., "A Statistical Analysis of a Job Shop Scheduling Problem," unpublished Ph. D. dissertation, Department of Industrial Engineering and Operations Research, University of Massachusetts, Amherst, MA (1975).
40. Rowe, A. G., "Sequential Decision Rules in Production Scheduling," Ph.D. dissertation, University of California, Los Angeles, CA (1958).
41. Rowe, A. G. "Towards a Theory of Scheduling," *Journal of Industrial Engineering*, Vol. 11 (1960), p. 125.
42. Serghiou, S. S., "Robustness of Results in Job Shop Simulation," Master's thesis, Imperial College, London (1973).

References

43. Steinhoff, H. W., "Two Recent Developments in Scheduling Applications," in *Symposium on the Theory of Scheduling and Its Applications, Lecture Notes in Economics and Mathematical Systems*, Springer, Berlin (1973), pp. 92–108.
44. Wayson, R. D., "The Effect of Alternate Machines on Two Priority Dispatching Disciplines in the General Job Shop," Master's thesis, Department of Operations Research, Cornell University, Ithaca, NY (1965).
45. Wilbrecht, J. K., and W. B. Prescott, "The Influence of Set Up Time on Job Shop Performance," *Management Science: Application*, Vol. 16, No. 4 (December 1969), pp. B-247–B-280.
46. Yamamoto, M., "Project Network to deno Jinin-Kikai Haibun Keikakuho (Dai I vu)" ["Resource Allocation in a Project Network (Part I)"], *Keiei-Kagaku*, Vol. 10, No. 3 (April 1967), pp. 160–174.

Chapter Ten
Cost Planning and Control

The Master production schedule is a statement of the company's manufacturing objectives stated in engineering form. For management planning and control purposes, it is converted into a common denominator—money. It serves as a basis for preparation of the company budget. The budget is a company operating plan, expressed in money over period. It forecasts sales and profits. On the basis of the planned income, it plans expenses for different purposes—for direct manufacturing, department overheads, plant overheads, research and development, and expansion and profit distribution. Expense must equal income. Most of the income is from the sale of the manufactured products. This amount is arrived at by the multiplication of sales quantity by standard cost. Standard cost is computed by the cost planning system. It is based on the engineering data and shop-floor historical data.

A budget cannot be prepared without a costing system. The purpose of cost planning is to forecast the cost of each of the company's products. This becomes the standard cost. The budget and the standard cost are the operating plan. They also serve as control tools. They are the yardstick by which management measures the actual performance of company activities. Any deviation from the plan calls for management interference. The reasons for deviation should be analyzed and either corrective measures taken or the plan reexamined and altered to match reality. The control phase is concerned with the actual occurrences in the plant. Some of the company activities are fixed, while others are variable. It is advisable, for control's sake, to separate them. The overall budget can then be divided into several operating budgets, each one controlled separately. These budgets might include the production budget, which is heavily dependent on orders received and is thus variable, and the investment, research and development, and finance budgets, which are more fixed in nature.

Control is concerned with variances and should have the ability to ascertain reasons. Variances in the production budget can be caused by a change in the volume ordered or by the actual cost being higher than the standard cost. Control of each one of the variables should be performed separately.

The objective of cost control is to compute the cost variance of each of the company's products. It should be broken down into the same cost elements as cost planning and be given as closely as possible to the occurrence period.

Cost planning and control obtains its basic information directly from production planning and control. Both systems work with the same basic data, although there is a conversion of dimensions—from time and material to money. If production planning and control is available, it can without any difficulty be extended to serve as a cost planning and control system. If so done, the costing system benefits in the following ways:

- Data have an improved level of accuracy.
- Data duplication is eliminated.
- New types of measurements for management are available.
- New techniques, such as cost simulation, can be applied to production data.
- Up-to-date data are available at any desired moment.
- Cost improvement technique can be applied.

Figure 10.1 shows the relationship of the IMS to cost planning and control.

10.1 Conversion Cost

The product is defined in engineering terms by the bill of material and routing. The former specifies the items and material required, while the latter specifies the operations required. Cost planning and control is concerned with:

- What it should cost to make and sell each of the company's products.
- What it actually costs to make and sell each of the company's products.
- Where variances occur and how they can be controlled.

Cost planning and control is equivalent to production planning and control, with the exception of dimensions. Production planning and control is concerned with:

- What quantity of material and items should be purchased in order to manufacture the company's products in the quantity forecasted.

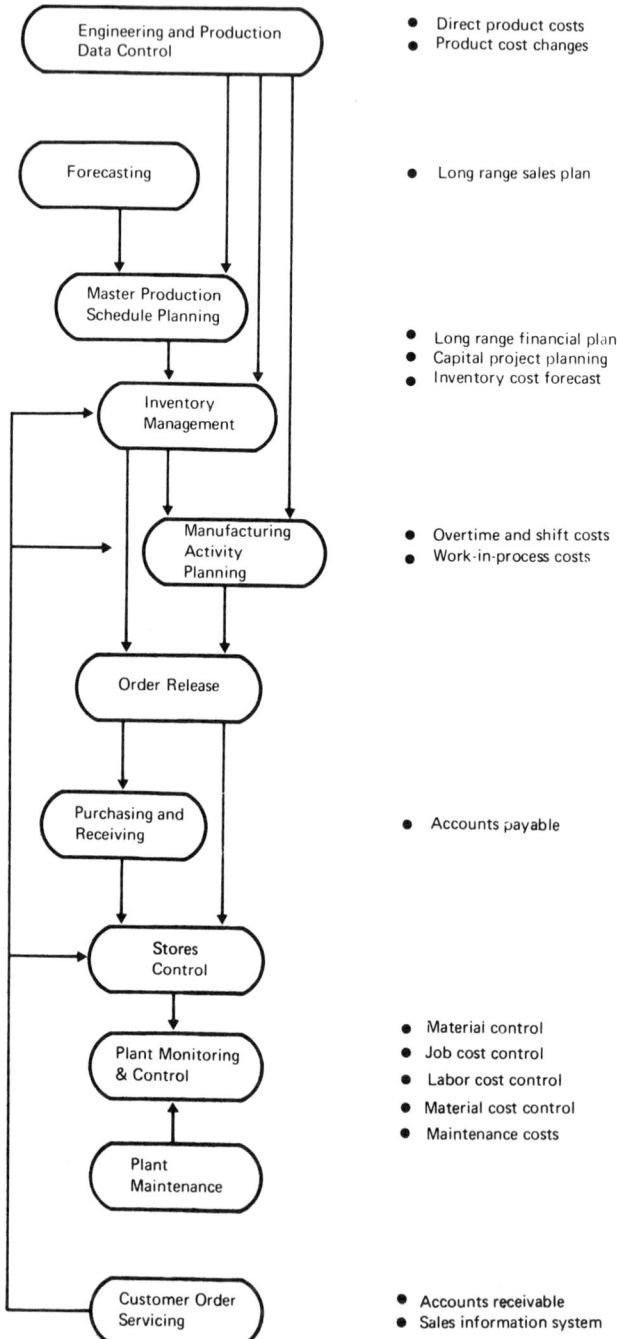

Figure 10.1 The relationship of the IMS to cost planning and control. (From *Copics,* Vol. VII, Chapter 12, G320-1980-0. Copyright 1972 by International Business Machines Corporation. Reproduced with permission.)

10.1 Conversion Cost

- What the scheduling should be in order to meet the delivery dates of the company's products.
- What it takes in material and labor to make the company's products.
- Where variances occur and how they may be controlled.

To convert the production planning and control system to management dimensions—money—a conversion factor must be used. This conversion factor should have the quality of being able to equalize the total manufacturing costs to the total amount of company expenses.

The main expenses in manufacturing are time and material. For conversion purposes, two factors are needed—one to convert time into money and another to convert material quantity into money.

Standard Direct Hourly Rate

The purpose of this factor is to convert direct labor time into cost in such a manner that company expenses are covered. The reasoning is that in order to give the operator a chance to work direct hours, there are many expenses, such as supervision, machines, power, and engineering. These expenses must be absorbed into the direct labor hourly rate. A list of expense items is given in Figure 10.2. Each one of these expense items must be analyzed and assigned a value. This is usually done by taking the previous year's figures, compensation being allowed for the present year's anticipated changes. The sum of all these expenses is the annual (or period) amount covered by the predicted number of direct hours for the year (or period). Thus the hourly rate is computed by

$$\text{Hourly rate} = \frac{\text{total company expenses (dollars)}}{\text{total direct hours}}$$

Most of the required data are available in the IMS files. The above equation is mathematically correct; it assigns a single hourly rate to be used throughout the plant. However, this is not always satisfactory, since, on the one hand, there are machining departments that require expensive machines, while, on the other hand, there are, also assembly departments that only require a screwdriver. To refine the calculation, a department hourly rate is used. It is computed in exactly the same way, the total expenses and total direct hours relating individually to each department. The weighted average of the hourly rates of all departments is equal to the plant hourly rate. In some plants this situation is not acceptable. The argument is that in the same department there are NC machines and drill presses. It is not fair to charge each one the same

Labor	Charges
Operating labor:	Trucking
Direct labor	Power
General foremen	Operation supplies
Clerks etc.	Light
Material handling	Heat
Trucking	Fuel
Maintenance labor:	Water
Maintenance buildings	Compensation insurance
Maintenance machinery	Maintenance:
Maintenance tools	Maintenance buildings
General labor:	Maintenance machinery
Engineers	Maintenance tools
Accounting	General:
Planning	Engineer
Payroll	Laboratory
Time-study	Office supplies
Watchmen	Traveling
Personnel	Telephone and telegraph
Laboratory	Recreation and welfare
Receiving	Miscellaneous
Stores	

Fixed
  Depreciation buildings
  Depreciation machinery
  Depreciation tools
  Taxes, buildings
  Taxes, machinery
  Insurance, buildings
  Insurance, machinery

Figure 10.2 List of expenses.

hourly rate. Thus a further refinement is made by computing a cost center hourly rate within the department in the same way. If one wishes to carry it to an extreme, an hourly rate is assigned to each individual machine. This raises the question of operator salary. Not all operators receive the same wage. Why work on an average wage when it is possible to assign hourly rates to each operator?

Another approach to the hourly rate is from the controlling standpoint. A general philosophy is to establish responsibility for controllable costs. People can only be held responsible for variations if the elements that constitute the cost are controlled by them.

The supervisor and foreman have control over such cost elements as idle time, assignment of jobs to workers and machines, efficiency, and amount of scrap produced; they do not have any control at all over such elements as amount of depreciation, property taxes and administrative expenses. Therefore, for purposes of control, these elements should be withdrawn from the hourly rate and treated separately.

The total number of direct hours used in the equation that computes hourly rate is in direct proportion to the hourly rate. However, not all expense elements vary linearly. General supervision, administration, and engineering will probably not be increased substantially by the added direct hours; depreciation on buildings and machinery, as well as taxes, will probably not be affected at all. Thus the hourly rate is not in absolute agreement with reality. It should be borne in mind that the conversion of time to cost serves two purposes: cost planning and cost control and improvement. This conversion must take place for cost planning and hence for budget preparation. For control purposes, one might prefer to use the time elements instead of cost and to control each expense element separately. One can find fault with any proposed costing system, since there will always be elements that are not treated with absolute justice. A working system is needed, and any decisions and rules that meet the planning and controlling purposes will suffice. The tendency to set up standard hourly rates in great detail sometimes defeats the purpose. For example, even the setting of hourly rates for each machine and each operator is not sufficiently refined. Overtime hourly rate is not the same as regular hourly rate. Efficiency and incentive paid by hours saved should have a different hourly rate value. All these refinements bring us closer to true cost. However, rather than doing their jobs, the foremen will be busy computing cost effectiveness, from the mathematical standpoint, in an effort to appear efficient. The following case demonstrates this situation.

The foreman receives a rush order. The operator and machine that would normally perform the job are already occupied. If the foreman substitutes the rush job on the required machine, extra time and cost for tear down and set up of the incompleted job will result. Another alternative is to use an available idle operator and machine. However, using a refined cost standard, the hourly rates of the alternate operator and machine are higher than the original values. The foreman will choose the alternative that looks best on the records, although, in fact, using the idle operator and machine requires almost no net incremental expenditure by the company. The defect in this control concept is in the way the foreman is evaluated. If the hourly rate was not so refined, the foreman's decision would have been correct from both aspects. The standards are built for control purposes and should leave some freedom in order to minimize the total costs under control.

Material Conversion Cost

The factor for conversion of material quantity into cost is individual to each item stored in inventory. Usually, the unit price of each item is part of the inventory item record. Basically, the price is the purchasing price. However, during the year, several purchase orders will be issued with different quantities, different delivery lead times, and, probably, to different vendors. For these reasons the price paid for each order might be different. In Chapter 7, several pricing methods that could be applied in inventory accounting were discussed. The costing is, no doubt, affected by the pricing method used. For cost planning, the market current price should be used; therefore, this price is also kept as part of the item inventory record. For control, the actual cost is used; thus the current unit price of each item is employed. The variance total value in this computation might be misleading. Figure 10.3 demonstrates this situation. The variance might be in quantity used or in unit price. The control should be aware of this situation and compensate for it. The problem becomes even more critical as more than one inventory item is used to manufacture the product.

There are many strong arguments that this cost conversion factor does not result in absolute justice. The floor space, the storekeepers' salaries, lift truck depreciation, and so on are inventory and material handling expenses and should be covered by inventory items cost. Thus they should not be projected on to the direct hourly rate, but rather added to the purchase price.

One should remember that costing is not a science. Its purpose is to estimate future cost and serve as a management control tool. Too much

		Material		Variance		
Case	Quantity	Unit price	Total value	Total value	Quantity	Unit price
Standard material cost	1.0	1.0	1.0	—	—	—
Case 1	1.0	1.0	1.0	0	0	0
Case 2	0.8	1.25	1.0	0	−0.2	+0.25
Case 3	1.25	0.8	1.0	0	+0.25	−0.2
Case 4	1.2	0.7	0.84	−0.16	+0.2	−0.3
Case 5	0.7	1.2	0.84	−0.16	−0.3	+0.2
Case 6	1.2	1.2	1.44	+0.44	+0.2	+0.2

Figure 10.3 Variances in material cost.

refinement usually defeats the purpose. Any costing system, however detailed, can be justifiably criticized. Valid arguments can be made to show that there are cases where injustice has been committed. A costing system is nevertheless required. Any system that answers the necessary requirements and is practical for implementation will be satisfactory as long as clear and unique procedures are set and the meaning of the figures is understood.

10.2 Cost Planning (Standard Cost)

Cost planning is the assignment of a standard cost to each of the company's products. The standard cost is used as a basis for the company's budget and as a yardstick by which to measure and control operations. The standard (planned) cost is computed from the engineering and manufacturing data base. In order to compute the cost of a product, the items that constitute it and the quantity of each item in a single product unit must be known. This information is stored in the bill of material file. In addition, the operations required to produce each item must also be known. The operations should give details about the machine, length of operation, and set-up time planned. This information is available in the routing file. These two files contain time and quantity information. To transform time into cost, the conversion factors must be used. The material unit price is in the inventory file, while the hourly rate is in the machine file (both may be kept as constants in the program). Figure 10.4 shows the files used in cost planning. The cost planning system can be prepared either as a simple integrated program or as a set of independent programs, working in series, each performing one task. In the second type, the individual programs will prepare data and insert them into the bill of material file. Then a summing-up program is used by working only with the bill of material file.

Figure 10.5 shows the logic of a program (or of a module in a program) used in computing the labor cost of an item. It utilizes the operation specified in the routing file as time elements (direct and set up) and a machine file or a table as a conversion factor. The multiplication of one by the other together with summing up of all operation costs result in the standard cost of item labor.

The standard material usage per item is specified in the bill of material file. The cost conversion factor for each item is in the inventory file. The multiplication of one by the other results in the standard cost of item material.

Computing the standard product cost can be done by two methods,

Cost Planning and Control

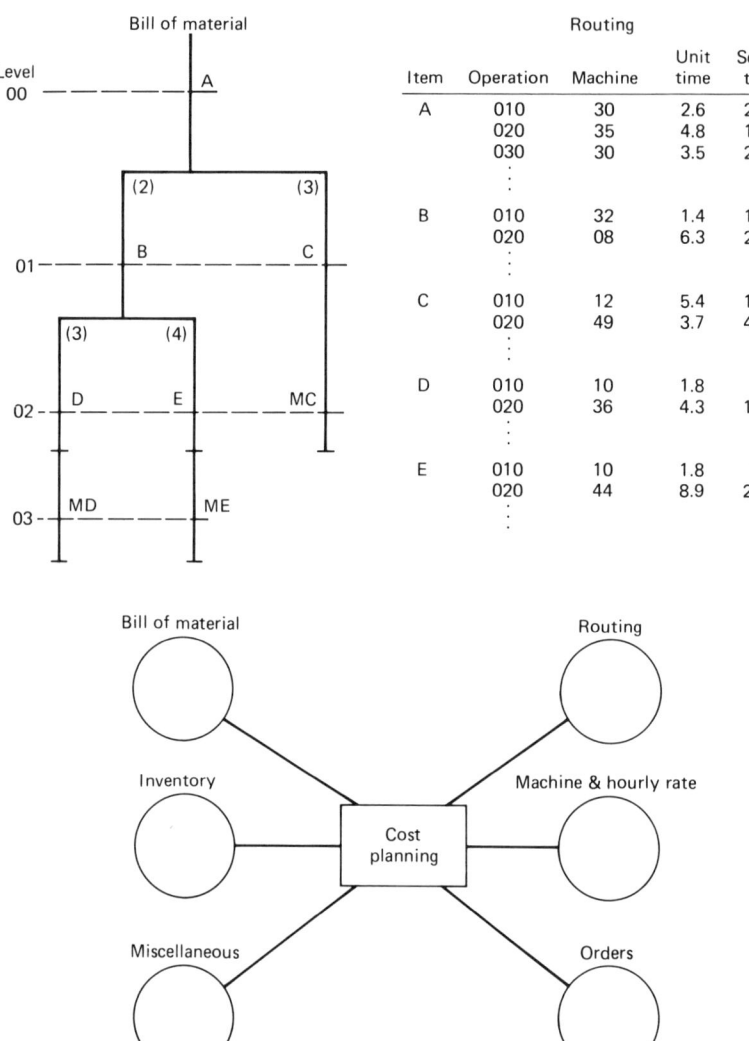

Figure 10.4 Files used in cost planning.

both of which utilize the bill of material file. The first is a downward method. In this case the computation starts from the product record. Explosion of the product results in a list of all items and their quantity per unit product. Each retrieved item record carries information on its added labor standard cost and/or material standard cost. The sum of all these values is the product standard cost.

10.2 Cost Planning (Standard Cost)

Example Figure 10.4 defines product A by the product tree method. The item A record is retrieved. It contains the standard labor cost for assembly items B and C. This value should have been prepared and inserted into this record in a previous program. If this has not been done, the program branches into the routing file (by a pointer in the item A record) and retrieves the first operation record of item A. It sums up all operation costs as demonstrated in Figure 10.5 and branches back (through a pointer in the routing records). This standard labor cost represents the added cost of this assembly operation. This cost is stored in the summary cost counter. Then by the use of the product structure record, the item B record is retrieved. It contains the labor standard cost for assembly items D and E. This cost is multiplied by the quantity of items B in one unit of product A (data in the product structure record). This cost is added to the summary cost counter. The item D record is then retrieved. The standard labor cost of processing this item is multiplied by the quantity per assembly B and added to the summary counter. The material MD record is retrieved. It contains the standard material cost of item D. This cost is added to the summary counter. Items E, ME, C, and MC are treated in a similar fashion. This is the end of product explosion, and the value in the summary cost counter is the standard cost of product A.

This top-down method gives the value desired; however, it does not give the standard cost of each item and subassembly that compose the product. Naturally, a request can be made to compute the cost of any item or subassembly present in the bill of material file. However, it will be costly and time-consuming.

The second method is an upward one. It computes the cost from the bottom of the product tree upward. It works on all products defined in the bill of material file simultaneously and gives the standard cost of each item, subassembly, and product. It uses the low-level code of the bill of material file.

In each item record in the bill of material, two fields are assigned. One is the added cost of the particular item, while the other is the accumulated cost of the item. Initially, the accumulated field is set to zero. The added cost is either inserted into the record as a precomputed value or is computed in the main computation program. In both cases the computation is the same and, as shown in Figure 10.5, uses the routing file.

Cost computation starts by handling the items having the lowest-level code. Each item retrieves all the where-used items and adds its cost to the accumulated cost field. When all items having the lowest-level code have been processed, items with low-level code minus one are treated. The added cost field is added to the accumulated cost field, and the result is

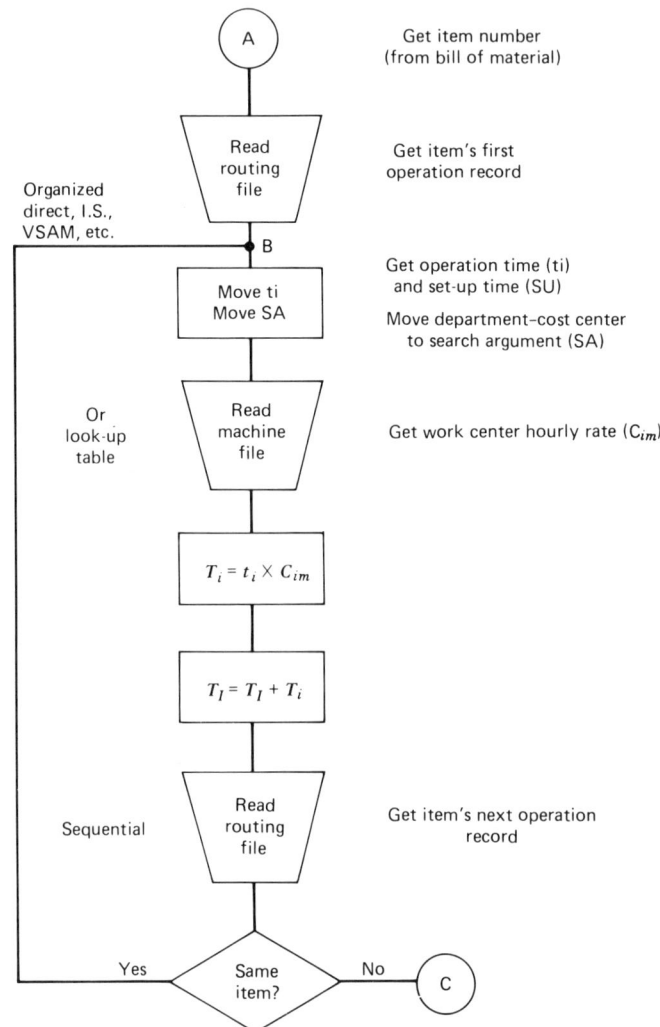

Figure 10.5 Logic in computing the labor cost of an item.

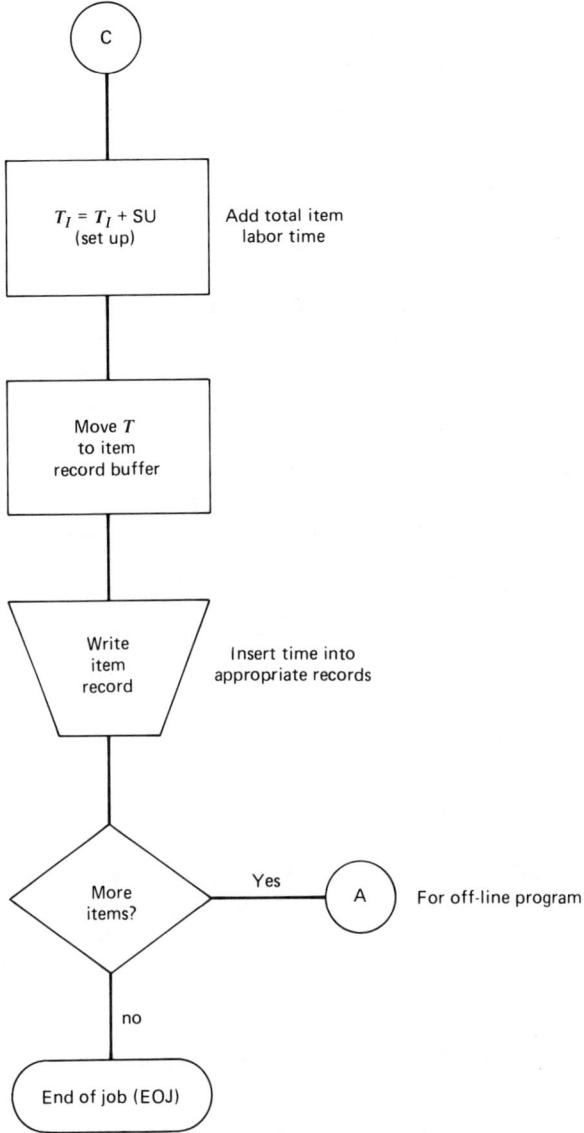

Figure 10.5 *(Continued)*

added to all where-used items. This process continues until the low-level code to be treated is zero. Each item in the bill of material file now carries its standard cost in the accumulated cost field. If we follow product A as shown in Figure 10.4, for example, the computation starts with level 03, item MD adding its cost to item D followed by item ME adding its cost to item E. Since no more items at low level 03 exist, we start treating items with low-level code 02. Item D adds its cost to item B, item E adds its cost to item B, and item MC adds its cost to item C. Since there are no more items with low-level code 02, we start treating items with low-level code 01. Item B adds its cost to item A and item C adds its cost to item A—up to low-level code 00. Since no more items exist at low-level code 01 the computation terminates. On the record of each item there are two fields—added cost and accumulated cost. The added cost field designates the cost contribution of the item itself, including processing or assembly cost. The accumulated cost field represents the standard cost, including the cost of all items and materials used from its level down to materials. The standard cost will be the same regardless of the method of computation.

The standard cost is composed of two elements:

1. *Cost conversion factor.* The methods by which the conversion factor is established as well as its accuracy and validity were discussed in Section 10.1.
2. *Time standards for each operation.* Time standards can be established in two ways: by past performance data or by time study.

By definition, a time standard should permit a qualified operator to work at a normal pace indefinitely without undue fatigue. In fact, the time standard is usually set at such a level that the average operator can readily do 20–30% more work than the standard requires.

The actual time can vary widely from the standard time. Under a straight hourly wage system, it is expected that actual time will be quite variable and usually longer than standard time. Under the incentive pay system, the actual time will usually be more stable and shorter than the standard time. A correction factor can be used in computing the standard cost.

A standard cost based on the above methods provides detailed information on the cost of each item and each task. Therefore, it is a good measuring and variance analysis tool.

10.3 Actual Cost

The actual cost is affected by shop-floor activities. Seldom do things work out as planned. The operator may either work efficiently and cut down the

10.3 Actual Cost

standard time, or, on a bad day, perform with poor efficiency, the job hence taking longer. Parts may be scrapped and extra material and time invested in the batch. Rework operations might be added or some operations skipped or modified in a certain batch. All of these affect the actual cost. Just as reality varies from initial planning in production scheduling, so does it in costing, the actual cost varying from standard cost.

The actual cost is computed by adding the actual cost of labor and material used on the shop floor. It does not rely on the bill of material and routing.

Gathering data and computing actual cost are performed by exploiting existing systems. Each cost-controllable job carries a unique name—a code number. This name can be the order number or product-part-batch-operation. Since this information is also used for production planning, the costing system should adopt the same detail level, although it does not necessarily require the same degree of detail level. The two systems must be in agreement with one another, using the same code names and refinements. Costing is a by-product of production planning. In designing the production planning and control system, one should bear in mind costing requirements. Production planning requires a tremendous amount of information. Information is costly. If not used by as many applications as possible, including costing, it constitutes a waste of money.

The shop open order file contains information on all open orders. It is used for capacity scheduling and for validation purposes in the inventory system, as discussed in Section 7.4. It is recommended that the record be extended to include cost information as well. A record detail was shown in Figure 7.7. The cost portion contains accumulation fields for material issued for that job, labor direct hours, and labor cost. Each open job order has a job record where concentrated information concerning the job is kept, as well as a record for each operation required to perform the job.

Material and items used will be accumulated in the job record, while labor will update the operations records. The total of all operations will be kept in the job record.

Material and items issue documents, including information on the item identification number and job identification code. The item identification number is used for updating inventory records, while the job identification code is used to update the shop open order file. The same identification code is used in the job recording system.

The actual cost of each inventory item is kept in the inventory file.

All inventory issued for production transactions (identified by a transaction code; see Section 7.2) will add their costs to the material accumulation field of the job record in the shop open order file. This transaction has passed a validation check, and at this stage the system does not bother to check if the quantity and item code number correspond to the data in the bill of material file. It relies on the inventory validation check to do this. Thus this field contains the actual cost of all material and items issued from stock for that particular job.

Subcontracted operations and items are transferred to production through the inventory system, even if they are under the control of the production system and not the inventory system. Both the price transfer and the actual transfer are carried out through the inventory system, designated by the work status code. The price is transferred and taken care of through the inventory system. The job recording system reports the activities of all employees, at any moment of their attendance time. By applying the two-way data processing concept, each job reported record will add its operation elapsed time to the accumulated hours field in the appropriate operation and job record of the shop open order file. Each job record contains information on the department and cost center number. Thus a conversion from time to cost can take place. The operation records following a job record need not necessarily be in agreement with the master routing file. A routing may possibly be modified for a certain batch, and this may not affect the master file. Thus the accumulated labor direct hours and cost are not dependent on the routing; they are rather the accumulation of what has actually taken place on the shop floor.

Set-up time, if regarded as direct labor, is reported by the job recording system. It carries the job and operation identification plus an indication that it is a set-up job. Thus time and cost can be accumulated in the general counter or, if desired, by a special accumulation field.

The sum of the accumulation fields in the shop open order file is the actual cost. It gives actual cost per operation and actual cost per job. Job is defined here as an order to the shop. If the order is to produce a part, the material is the raw material cost, while the labor is the machining time and cost. If the order is for an assembly, the material is the items required for the assembly plus auxiliary material, while the labor is the assembly operations time and cost. If the order is for a product, the material is the subassemblies, items, and auxiliary material, while labor is the assembly operations time and cost.

Thus the actual cost is built as a pyramid, as shown in Figure 10.6. The cost of each level is computed and transferred to inventory. When the upper-level job starts, the cost of the previous level is transferred from the inventory system to the production system. The added cost of this level is accumulated, and the total is again transferred to the inventory system. This continues up to the finished product.

Storing the accumulated costs in the extended portion of the shop open order file is not essential to the costing system as described. It is convenient from the data processing and reliability points of view. The inven-

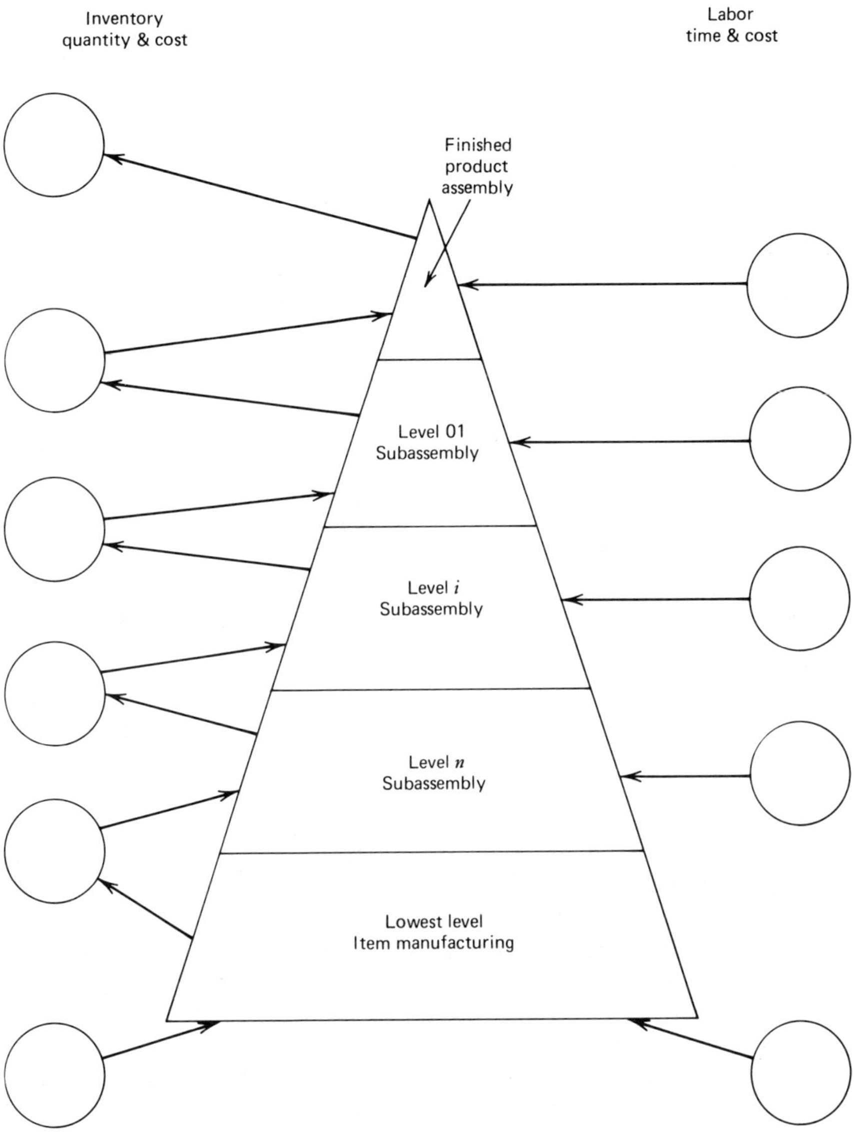

Inventory
quantity & cost

Labor
time & cost

Finished product assembly

Level 01 Subassembly

Level *i* Subassembly

Level *n* Subassembly

Lowest level Item manufacturing

Figure 10.6 Actual cost built as a pyramid.

tory issue transactions and the job reporting records can be stored for a period in separate files. A special off-line batch processing program can be prepared to calculate the actual cost by following the above logic.

10.4 Cost Control

In the previous sections the establishment of standard cost (estimated cost) and actual cost was discussed. Cost control is the phase in which comparison of the two takes place and variances are computed.

Cost control is a management tool. A controlling system should provide the following:

- An indication that something is wrong.
- An indication of what is wrong.
- An indication of why it is wrong (if possible).
- An adequate trail of information so as to be able to determine the what and why and possibly remedy the situation.

Each level of management is interested in a different level of control information. The highest-level manager may initially be interested only in if the order is profitable or not. A superintendent might be interested in knowing which portion of the job is making or losing money, while low-level management might be interested in pinpointing the operations in which the variance lies.

A control system must provide timely information. Merely knowing that an operation is over budget is not meaningful for control purposes, since nothing can be done to remedy the situation. Knowledge that continuation of the operation at its present rate will cause over budgeting is required. Corrective action can thus take place.

The costing system based on the IMS can provide such information. The system carries up-to-date information on all the company's manufacturing activities. The cost accumulation fields in the records of the shop open order file contain information on the quantity produced in each operation and the actual cost. For purposes of convenience, the standard cost can also be kept in this file. (At any rate, it is available in the system.) At any desired time, a variance and anticipated completion cost report can be prepared for any item on order. Figure 10.7 demonstrates such a report.

The report is divided into two sections: labor cost and material and items cost. Each section is further divided into two parts: actual completion and anticipated completion cost. The actual completion cost is based on data available in the shop open order file records.

Product	Part	Batch	Quantity Ordered	Description							08/08/78
12345	4321	13	1,000	Bracket							Percent completion = 33.8%

		Actual completion cost							Anticipated completion cost		
		Standard		Actual		Variance		Unit actual time			
	Quantity	Hours	Cost	Hours	Cost	Hours	Cost		Standard	Actual	Variance

Labor

Operation											
010	1,000	267	8,277	216	6,480	+51 (19%)	1,797	0.216	8,277	6,480	1,797 (+22%)
020	800	200	6,000	160	5,280	+40 (20%)	720	0.200	7,320	6,450	870 (+12%)
030	500	208	6,656	166	5,312	+42 (20%)	1,344	0.332	13,250	10,484	2,766 (+20%)
040	400	200	5,200	207	5,796	−7 (3%)	596	0.517	12,800	14,295	−1,495 (−11%)
050	300	90	3,330	75	2,400	+15 (17%)	930	0.250	10,980	7,870	3,110 (+28%)
060	200	33	964	47	1,488	−14 (42%)	524	0.233	4,750	7,320	−2,570 (−54%)
070	100	33	1,320	38	1,520	−5 (15%)	200	0.380	12,905	15,050	−2,145 (−17%)
080	—								23,135	23,135	0
Totals		1,031	31,747	909	28,276		3,741 (+12%)		93,417	91,084	2,333 (+2.5%)

Material and items

Code name											
123456789			21,000		23,154		2,154 (−10%)		21,000	23,154	2,154 (−10.2%)
Item total			52,747		51,430		1,317 (+2.5%)		114,417	114,238	179 (0%)

Figure 10.7 Variance and anticipated completion cost report.

A record for each planned operation exists. This record contains the standard time, the actual finished quantity, and the actual hours and cost of the finished quantity. The actual completion part of the report converts the quantity finished to standard hours and cost. Thus a variance can be computed at any stage of production. In order to anticipate completion cost, it is assumed that the rate and efficiency encountered in the finished quantity of an operation will prevail until the whole quantity is completed.

Operations that have not yet been started are assumed to be at standard pace. With these assumptions we can project the actual time and cost to the ordered quantity and anticipate total labor variance for the item. The projection is not absolutely proportional, since set-up time and other preparatory activities are performed at the starting time of an operation and last for the whole quantity. On the other hand, tear down is done when the whole quantity is finished. An algorithm can be worked out to fit the level of refinement desired.

Material cost is treated by a similar method. The actual issues using the actual unit price from inventory thus results in actual cost. In the case of partial issues, it is assumed that the rest of the quantity will be at the same unit price. Unissued material can use standard quantity and standard unit price. However, if the present inventory unit price is used instead, it will result in a more real anticipated completion cost.

An indication of how far the anticipated completion cost is based on facts, and how much on assumption, is the percent completion value. This value is based on time rather than on cost.

One should bear in mind that this report is a dynamic one and varies as a function of the date it is prepared. For example, the report in Figure 10.7 anticipates no cost variance, in spite of the fact that material actually costs 10.2% higher than standard and large variances in the individual operations occur. However, if this report was prepared at a later date, after operation 080 had started, the anticipated completion might be quite different.

Figure 10.8 shows a similar report, in which actual completion and anticipated completion cost for a product are given. The labor section is computed exactly as in the previous example. The material and items section lists all the required components for the assembly. This list is taken from the assembly demand file (discussed in Section 6.5), which is one of the inventory reference files (see Figure 7.6), or from the bill of material file. The first is preferred, since it includes working data and not master data. The actual completion cost is for the quantity per product issued. This cost is projected, following the previous assumptions, to give the anticipated completion cost.

A report of the type shown in Figure 10.8 can contain as much detailed

Product	Part	Batch	Quantity	Description						08/08/78
2345	0000	14	1,000	Gear Box						Percent completion = 23%

		Actual completion cost			Anticipated completion cost		
	Quantity	Standard	Actual	Variance	Standard	Actual	Variance

Labor

Operation							
010	400	3,240	3,693	−453 (−14%)	8,020	9,120	−1,100 (−14%)
020	400	1,325	1,192	+133 (+10%)	3,175	2,870	+305 (+10%)
030	300	2,480	3,125	−645 (−28%)	7,945	10,180	−2,235 (−28%)
040	200	1,120	1,057	+63 (+6%)	5,380	5,055	325 (6%)
050	100	870	790	+80 (+9%)	8,430	7,670	760 (+9%)
060	0	—	—	—	6,350	6,350	0
070	0	—	—	—	2,980	2,980	0
080	0	—	—	—	4,125	4,125	0
Labor total		9,035	9,857	−822 (−9%)	46,405	48,350	−1,945 (−6%)

Material and items

Code							
A	500	26,000	29,000	−3,000 (−12%)	52,000	58,000	−6,000 (−12%)
B	600	40,800	37,800	+3,000 (+7%)	68,000	63,000	+5,000 (+7%)
C	400	13,000	14,560	−1,560 (−12%)	32,500	36,400	−3,900 (−12%)
D	200	16,880	18,020	−1,140 (−7%)	84,400	90,100	−5,700 (−7%)
E	—	—	—	—	42,300	42,300	0
123	1,000	15,400	14,200	+1,200 (+8%)	15,400	14,200	+1,200 (+8%)
456	500	11,350	12,150	−800 (−7%)	22,700	24,300	−1,600 (−7%)
Total		123,430	125,730	−2,300	317,300	328,300	−11,000 (−3.5%)
Product total		132,465	135,587	−3,122 (−2.3%)	363,705	376,650	−12,945 (−3.5%)

Figure 10.8 Product variance and anticipated completion cost report.

information as desired. It is restricted only by the desired ease of reading and the number of characters that can be printed on one line.

For working purposes, it is recommended that a series of reports be made, each one complementing the other. For top management, an exception report is probably most suitable. It should contain only those orders in which the anticipated completion variance exceeds a predetermined value and include only the product total line.

The report demonstrated in Figure 10.8 can be used for general variance analysis. This report shows if the variance is caused by material or labor, and in what item or material. A report of the type demonstrated in Figure 10.7 can be used to analyze each item separately. It shows if the variance is in time per operation or in hourly rate (working in a different department or cost center from that planned).

Further reports may be used to analyze material with respect to quantity or unit price variance (and so on). The information is available in the files, and any desired report can probably be prepared. The problem is that if too much detailed information is given, one might not see the overall picture. The reports are prepared for use by people. Users must specify what they would like to receive and in what form.

Cost control, as described above, relates only to direct labor and material expenses. It is based on the standard and actual hourly rates and the inventory pricing system. The hourly rate depends on total company expenses—indirect and overhead costs, investments, development, depreciation, and so on—and on anticipated total direct labor hours, that is, on sales volume.

There might be variances in any of the above activities and expenses. Thus cost control alone does not guarantee profit control. Controls over all other expenses must be set and carried out by the use of the data available in the IMS. Some of these controls were discussed in Sections 5.4 and 9.5.

Some of the variances in cost control are operational, while others are caused by changes. Product design might be changed, manufacturing methods varied, new union contracts might cause wage increase, material price changes, and so on. All these changes increase the variance of actual cost from standard cost. The operational variances can be controlled by the manufacturing supervisors, but the changes are beyond their control and responsibility.

This problem can be solved by working with dynamic standard costs. As discussed before, a dynamic budget can be prepared whenever sales volume changes. The work-in-process value is immediately available in the shop open order file; new material prices are available in the inventory file; and the production plan is also available. A monthly balance sheet, variance reports of indirect labor and expenses, can be prepared with no great effort. Thus a new standard cost that incorporates changes can be prepared whenever desired. In this manner the cost control report will be meaningful and easy to work with. An extra report can be prepared to show the original standard cost and the current standard cost. This report isolates variances due to changes and can also serve as a cost improvement tool, while the cost control report will serve its original purpose of control over direct manufacturing activities.

The description of cost control has been somewhat simplified here. There are many other situations and expenses that management must decide how to handle and set procedures. One case, for example, is when the number of reject items in a batch falls below prediction. This results in an actual completed quantity greater than ordered (required).

The question of what quantity should be used in computing the unit cost is actually a finance problem and as such is beyond the scope of this book. Our main interest was to demonstrate that the tools and data for cost control are available in the IMS as a by-product of the production control system.

PART II
GENERATIVE PROCESS PLANNING (GPP)

Chapter Eleven
Generative Process Planning— Prerequisite

Process planning is an engineering task that dictates the manufacturing cost and the efficiency of plant operations. This holds true for all types of manufacturing. Engineering tasks are generally based on human skill and intuition. In mass-production or chemical processes a lengthy study is made in order to reach an optimum process. Such studies are not practical in batch-type manufacturing, since the cost of the study will probably be greater than the possible savings in manufacturing. It is, therefore, doubtful whether optimum processes are employed in batch-type manufacturing. This situation is particularly critical in the metal-cutting field, where about 75% of all items are produced in batches of 50 or less and where approximately 50 billion dollars are spent each year in the United States.

In order to ensure an optimum manufacturing process in batch-type production, a practical and economical method of establishing optimum process planning is needed, that is, generative process planning. This part of the book is devoted to such programs.

Generative process planning should follow some general concepts, such as separation of computational and noncomputational elements. However, each specific manufacturing field should be studied in great detail with respect to its parameters and their effect on the process.

A full and in-depth understanding of the manufacturing field is a prerequisite for the construction of a generative process planning program for it.

Generative process planning programs are being developed for different manufacturing fields, such as sheet metal work, welding, injection molding, textiles, building construction, electronics, hydraulics, and chemical processes. Since each manufacturing field has its own specific problems, the discussion here will be focused on the steps and studies undertaken to construct a generative process planning program for one specific field—metal cutting–lathe work. By means of this example, we will illustrate how generative process planning programs should be constructed. Many of the concepts—the mathematical techniques—used are process-independent and can thus be applied to other manufacturing fields.

11.1 State-of-the-Art

As in any other process, the economics of metal-cutting call for the increase of productivity or the lowering of manufacturing cost. These objectives can be achieved by:

- Technological improvements.
- Effective utilization of available technologies.

Technological Improvements

Research work has contributed immensely to the improvement of metal-cutting technology. In turning, for example, 40 years ago 2-hp machines were used to cut steel, whereas today 25–35-hp machines are available; 40 years ago the nominal volume of chip removal rate was about 2 in.^{3} per minute, whereas today a rate of 20–30 in.^{3} per minute is possible.

This does not mean that steel has become easier to cut; the required horsepower per cubic inch is unchanged. However, the research work has resulted in a better understanding of such aspects of the metal-removal process as the forces, chatter, temperature, and tool wear mechanism. This knowledge is used in developing new machines and tools that contribute to increased productivity in metal cutting.

The high-speed tool has given way to the plain tungsten carbide tip. In recent days, plain tungsten carbide material has rapidly been superseded by coated tungsten carbide and a homogeneous ceramic material. The basic need is to increase the high-temperature hardness of the cutting edge while at the same time retaining sufficient toughness to resist edge chipping. The homogeneous ceramic cutting tip admirably fulfills the first requirement, but suffers from brittleness and, consequently, chipping under fluctuating or discontinuous cutting conditions. Hence, its use tends to be restricted to wrought materials and premachined or precision castings. However, in those applications that provide stable cutting conditions, the increase in cutting speed and reduction in machining time are considerable when compared to those achievable with plain tungsten tips. For example, to obtain a tool life of about 20 minutes with the homogeneous ceramic tip, cutting speeds of 300–400 m per minute are utilized, whereas cutting speeds of 150–170 m per minute are required to obtain the same tool life with a plain tungsten carbide tip. In order to obtain the property of high-temperature cutting-edge hardness combined with toughness, the plain tungsten carbide material has been provided with a thin (several micrometers) coating of an extremely hard and wear-resistant material, such as titanium carbide or ceramic. The titanium carbide-coated tips used on medium-carbon and nickel chrome alloy steels will give a tool life of approximately 20 minutes at a cutting speed of around 200 m per minute; this represents a 50% increase over the cutting speed that could be employed with plain tungsten carbide tips to give a similar tool life. The increase in cutting speed in machining ferrous materials with ceramic-coated tips approaches that obtainable with the homogeneous ceramic tip tool, giving a 70% increase in cutting speed compared to the plain tungsten carbide tip for equivalent tool life. The coated carbides and ceramics also show marked increases in attainable cutting speeds when used on high-duty alloys. Further increases in the cutting speeds on these very difficult materials have been achieved by the development of a hot machining technique that utilizes a plasma arc to locally heat the cutting zone. The basis of this technique is to heat the material being cut to a temperature at which its mechanical properties are altered to such an extent that it is more readily machinable. For most difficult-to-machine materials, increases in cutting speed (compared to conventional machining) of between 5–10 times have been achieved.

The development of cutting-tool materials capable of adequate durability at considerably higher cutting speeds leads to the demand for increased spindle speeds. At the same time, there is still the need for slow spindle speed with high spindle stiffness for some operations such as threading and machining exotic materials. The need has therefore arisen for spindle drives capable of providing spindle speeds from virtually 0 to 5,000 RPM. In order to obtain the property of high-speed running combined with adequate spindle stiffness, a new type of bearing design is required. Recently, a variable-preload bearing has been introduced. At the higher spindle speeds of 4,000–5,000 RPM now being introduced on turning machines, the problem of heat generation in the spindle drive and the resulting possibility of clutch or gear failure or loss of accuracy due to thermal distortion arise. A new type of driving mechanism is required. The tendency is to considerably simplify the drive by utilizing high-power DC motors that drive directly onto the machine spindle or through a two-speed gearbox. This type of drive is also easier to control. Increased spindle speed also aggravates the problem of ensuring adequate gripping force in the item-holding equipment and also the danger of serious injury to operators should the workpiece be released at high speed. A modified chucking design is required. Modern power-operated chucks using high mechanical advantages, usually through wedge-type devices, ensure gripping forces of several thousand pounds per jaw at speeds of up to 3,000 RPM. However, the effect of centrifugal force is to reduce the gripping force in direct relation to approximately the square of the speed. The wedge-type chucks commonly in use are of the self-locking type, and, consequently, their built-in friction tends to reduce the loss of grip due to centrifugal force. In order to counteract and indeed harness the effect of centrifugal force, compensated chucks have been introduced. This involves linking a free mass to each jaw in such a manner that the centrifugal force acting on the free mass is transmitted to the chuck jaws. The effect of the compensation is to provide virtually constant gripping force with increasing rotational speed. In addition, high-speed turning discharges cuttings from the workpiece in large quantities, at high velocity, and often at an unacceptably high noise level. When using large-diameter chucks there is a danger that a spindle speed will be selected that will cause the chuck to burst due to the stresses set up in the chuck body. This has led to the total enclosure of the working area with heavily ribbed steel guards that can be lined with sound-absorbing material. Developments and technologies from electronics and computer science are also utilized. The NC (numerical control), CNC (computer numerical control), and, lately, DNC (direct numerical control) machines have been introduced. These enable greater flexibility in production and transfer most of the setup from the shop floor to the office. Consequently, a change in machine design is called for. The operator is kept further away from the machine and has to rely on instruments and electronics.

The measurements are made on the tool-holder movement rather than on the workpiece itself. Thus the machine has to be more accurately constructed in order to maintain the precision obtained on universal machines. Additional functions, such as automatic changing of cutting-tool tip, automatic loading and unloading, and automatic in-process gauging and adjustment, are under development. The machine builders are really doing a great job of incorporating new techniques into the machine design. The machines are becoming more and more sophisticated (and costly), and the operator is being kept away from the machine. In fact, a new kind of operator, possessing a different kind of skill, is now required. In the old days, operators were told to machine a part and they did. The operators were craftsmen who probably understood the metal-cutting process better than the young engineer at the office. They were near the work—they heard the noise, saw the chips coming out, touched the part and tool, and felt their temperature. Today the operator must know something about electronics, computers, programming, instrumentation, and metal cutting. It has been said that in the old days there were smart operators and dumb machines; today it is the other way around.

The part programmers are in a location remote from the machine. They must consider computer capabilities, programming language, software, and metal cutting. I wonder which of these plays the leading role in their work. Many times one is carried away by tools and forgets the main purpose of their existence.

The utilization of technological improvement has no doubt contributed to increased productivity. The improvement is based mainly on more powerful machines and increased cutting speed.

Utilization of Available Technologies

The other means to increase productivity is the effective utilization of available technologies. This course of action can, with less expense and effort, contribute more than the last 40 years of technological improvements have done. It is realized that power and cutting speed are of secondary importance in the economic consideration of metal cutting. When choosing the optimum sequence of operations, machine selection is the prime consideration. No one can expect a machine to be efficient over an entire range of diameters, operations, speeds, and powers. To demonstrate this point, let us examine a part that requires four machining operations at a particular power (25 hp, 10 hp, 5 hp, and 2 hp). One possibility is to manufacture the part on four different machines, each with the appropriate power. Thus the technological possibilities will be fully utilized; however, the overall part manufacturing efficiency will suffer. Extra time and cost must be added to cover gripping the part four times, physical handling of the parts between operations, set-up time, and production planning expenses. It is possible that these extra expenses will offset the economical utilization of equipment.

Another possibility is to machine all four operations on one machine, thus reducing the in-between operator expenses to a minimum; however, the machine will then not be effectively utilized. To select a 2-hp machine means that the first operation (25 hp) does not benefit from the technological improvement. To select a 25-hp machine means that the last operation (2 hp) does not utilize machine capabilities.

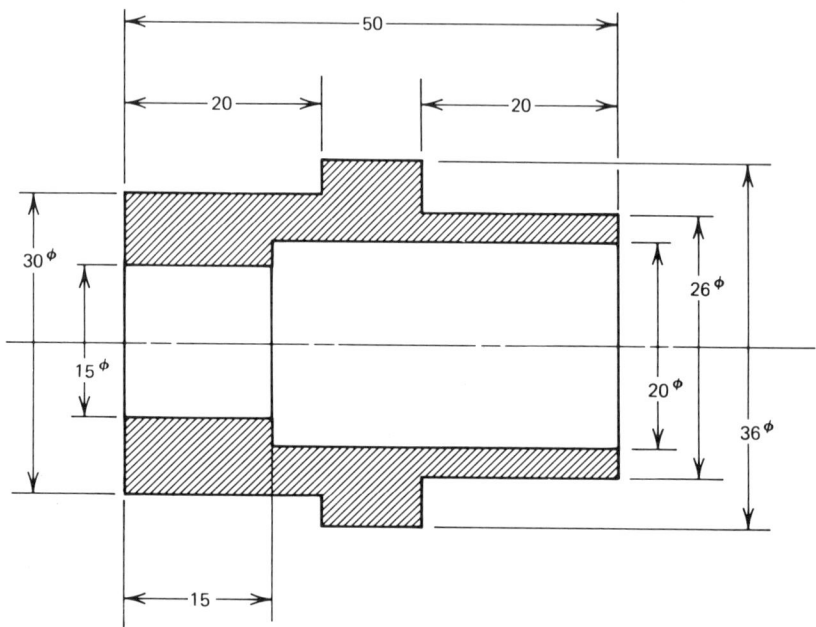

Figure 11.1 Part number 120.

The hourly rate for this operation is that of a 25-hp machine, not that of a 2-hp machine. As usual, the more powerful the machine, the lower the maximum spindle speed (RPM) of the machine. It is, therefore, possible that the 2-hp operation will be processed at a lower cutting speed than if it were performed on a 2-hp machine.

It is clearly seen that machine selection is one of the dominant factors in the economic considerations. To stress this point, the machine time and cost of machining part number 120 (see Figure 11.1) on 11 different machines are given in Figure 11.2. The time and cost shown are for direct machining. It is clearly seen that the time varies from 2.11 to 5.24 minutes, while the cost varies from $27 to $176.

Different operations can be used to machine the part. Figure 11.3 shows two of the many different sequences of operations for machining the same part. Naturally, both cannot be optimal, that is, one is better than the other. Unnecessary operations, even when performed with the economical cutting condition, are still a waste.

It is surprising to learn that the economic problem in metal cutting starts and ends with the solution to the economical cutting speed of a simple operation (although there is agreement that feed rate is more cost influen-

Machine number	Power (hp)	Maximum RPM	Hourly rate ($)	Production time (Minutes)	Production cost ($)
220	1	1,600	16	5.24	139.68
100	3	1,800	25	3.31	138.51
250	5	1,700	20	2.57	85.66
110	7.5	1,600	15	2.66	111.68
260	10	900	36	2.55	153.96
120	10	1,400	30	2.67	134.25
300	15	710	9	2.62	39.21
200	15	710	19	2.62	82.77
130	15	900	35	2.82	165.96
270	18	2,000	50	2.11	175.98
280	25	700	6	2.69	26.91

Figure 11.2 Machining time and cost of part number 120 (see Figure 11.1).

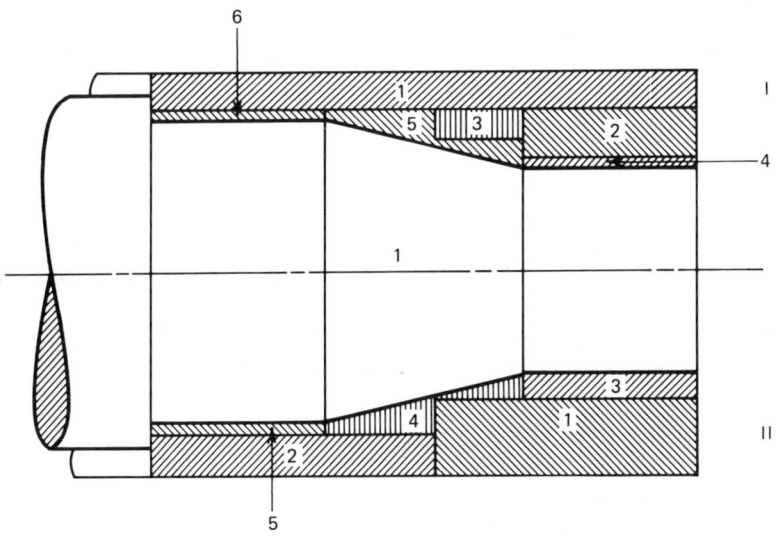

Figure 11.3 Two of the many possible sequences of operations for machining a part.

283

tial than speed). It is assumed that the operation, depth of cut, and machine were optimally selected. As will be shown in Section 11.2, these assumptions are not valid.

11.2 Economics of the Basic Turning Operation

Most of the research work on the economics of metal cutting has considered only a single operation. These works assumed three cost factors:

1. Indirect operation: machine set up, chucking, tool adjustment, and so on.
2. Direct operation: actual cutting time.
3. Tooling.

Thus the cost of an operation can be expressed as

$$C = C_1 T_p + C_2 T + C_3 \frac{T}{T_L} \tag{11.1}$$

where C = total cost of the operation (dollars)
 C_1 = indirect labor rate (dollars per minute)
 T_p = indirect operation time (minutes)
 C_2 = direct labor rate (dollars per minute)
 T = machining time (minutes)
 C_3 = cost of cutting tool (one corner)(dollars)
 T_L = tool life (minutes).

Machining time can be expressed in terms of the operation being performed as

$$T = \frac{L}{N \cdot S} = \frac{\pi DL}{V \cdot S} \tag{11.2}$$

where L = length of the cutting operation (mm)
 N = spindle speed (RPM)
 S = feed rate (mm per revolution)
 D = part diameter (mm)
 V = cutting speed (mm per minute).

The equation that relates cutting-tool life to cutting conditions is the extended Taylor equation:

$$VT_L^n S^p A^q = K \rightarrow T_L = \frac{K^{1/n}}{V^{1/n} S^{p/n} A^{q/n}} \tag{11.3}$$

11.2 Economics of the Basic Turning Operation

where A = depth of cut (mm)
n, p, q = exponents
K = constant (equivalent to cutting speed of 1-minute tool life).

Equation 11.1 can be rewritten using the values of eqs. 11.2 and 11.3:

$$C = C_1 T_p + C_2 \frac{\pi DL}{VS} + C_3 \frac{\pi DL}{VS} \cdot \frac{V^{1/n} S^{p/n} A^{q/n}}{K^{1/n}} \quad (11.4)$$

In order to arrive at minimum cost, the derivative of eq. 11.4 with respect to cutting speed (V) is taken and set equal to zero:

$$\frac{dC}{dV} = -C_2 \frac{\pi DL}{V^2 S} + C_3 \frac{\pi DL}{SK^{1/n}} \cdot S^{p/n} \cdot A^{q/n} \cdot \left(\frac{1}{n} - 1\right) V^{(1/n - 2)} = 0$$

$$-C_2 + C_3 \frac{V^{1/n} S^{p/n} A^{q/n}}{K^{1/n}} \left(\frac{1}{n} - 1\right) = 0 \quad (11.5)$$

Equation 11.3 is used to simplify eq. 11.5:

$$-C_2 + C_3 \frac{1}{T_L}\left(\frac{1}{n} - 1\right) = 0$$

$$\frac{C_3}{C_2} \cdot \frac{1}{T_L}\left(\frac{1}{n} - 1\right) = 1 \rightarrow T_L = \frac{C_3}{C_2}\left(\frac{1}{n} - 1\right) \quad (11.6)$$

The surprising meaning of the results is that the optimal tool life has nothing to do with the cutting conditions. It depends on the hourly labor rate and the cost of the tool. In a way, the results stand to reason. The tool life of an expensive tool should be longer than that of a less expensive one. If the hourly labor rate is high, it will be the controlling factor, since labor time will be utilized more extensively; thus tool life should be shorter. However, these controlling factors are subjective and depend on company costing policy, inflation, taxes, and labor unions. These factors are dynamic and independently variable. What this actually means is that if one wants to work at optimum cutting speed, they should continuously change the routing sheet, adjusting it to market price changes and union contracts.

At optimum cutting speed, the ratio between tool cost and labor cost is

$$\text{Ratio} = \frac{C_3 T/T_L}{C_2 T} = \frac{C_3}{C_2} \cdot \frac{C_2}{C_3(1/n - 1)} = \frac{1}{1/n - 1}$$

For a value of $n = 0.15$, the ratio is 0.176. This ratio diminishes when working at lower than optimum cutting speeds. When the optimum cutting speed is reduced by approximately 10%, the effect of tool cost becomes negligible. When set-up cost is proportionally higher than direct machining cost (by a factor of three), a 10% change in cutting speed has very little effect on total part cost. These facts probably explain why the differences between various economic modules are very small. The above results stand to reason, since a change in cutting speed of 10% results in

$$\frac{T_{L_1}}{T_{L_2}} = \left(\frac{V_2}{V_1}\right)^{1/n} = 1.1^{(1/0.15)} = 1.8 \rightarrow 80\% \text{ of tool life}$$

With all its faults, this type of economic module is the only one available today. To use this module, it is necessary to know the value of the parameters in the equation. In order to know the value of C_2, one must select the machine for the operation before knowing the cutting speed to be used. However, the cutting speed is the factor that determines machine power requirements. Most of the available economic modules find that the power is the limiting factor in using the optimum cutting speed. To lessen power requirements, the cutting forces can be reduced by lowering the feed rate or depth of cut. As will be shown later, neither remedy is efficient. Reducing cutting speed is much more economical. An interesting attempt to overcome this difficulty was made by Bartalucci (1). He assumed a linear relationship between machine power and hourly rate:

$$M = K + HQ \tag{11.7}$$

where M = machine tool cost (dollars)
K = constant (dollars)
H = cost coefficient (dollars per hp)
Q = machine power (hp).

The power requirements for an operation can be expressed by the following equation:

$$Q = \frac{PS^x A^y V}{4,500} \tag{11.8}$$

where P = cutting force coefficient (N per mm²)
S = feed rate (mm per revolution)
A = depth of cut (mm)
V = cutting speed (mm per minute)
x, y = exponents.

The hourly rate for a machine can be expressed as

$$C_2 = C_0 + K_1 + \frac{H_1 PS^x A^y V}{4,500} \tag{11.9}$$

11.2 Economics of the Basic Turning Operation

where C_0 = labor direct cost per hour (dollars per minute)
K_1 = fixed hourly cost of the machine (dollars per minute)
H_1 = variable (by power) hourly rate of the machine (collars per minute).

The value of C_2 in eq. 11.9 is substituted for the value of C_2 in eq. 11.4; thereby the use of the module does not require preselection of machine. The results of this module are shown in Figure 11.4. It is interesting to note that the power required to work at the optimum cutting speed is

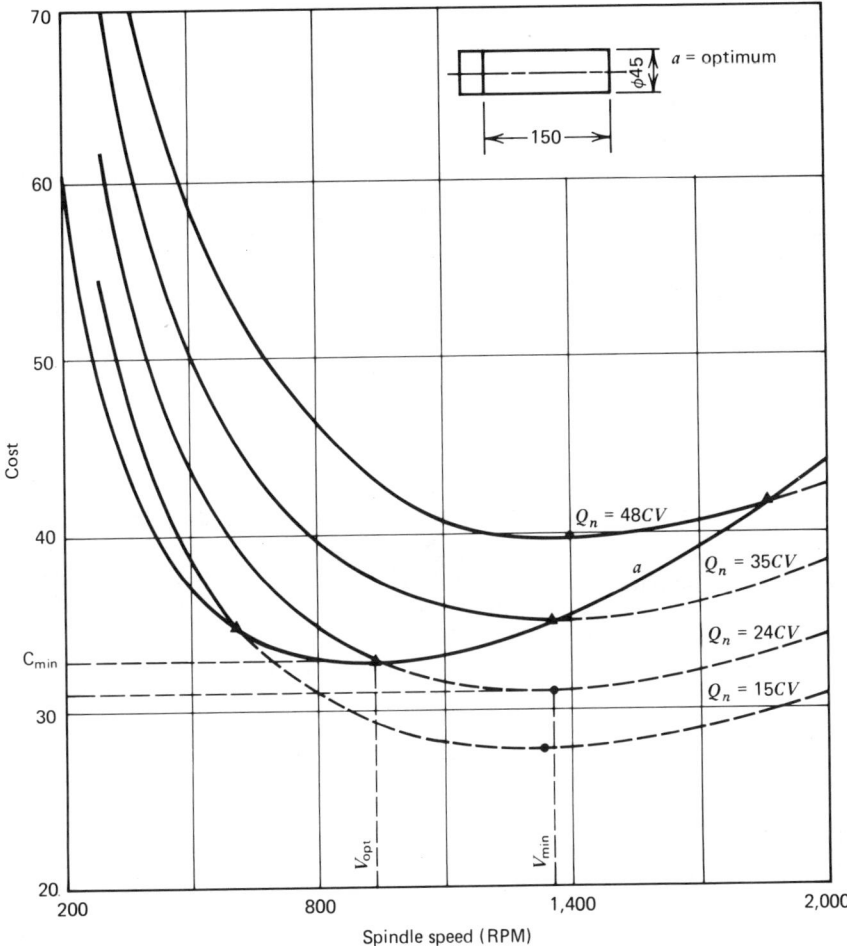

Figure 11.4 Bartalucci module machining cost for four machines having different powers.

greater than that available. The curves for machines of different powers are parallel, but offset by cost. The higher the power, the higher the cost. It is, therefore, necessary to draw a curve that passes through the maximum power points of the different machines and to select a machine according to the minimum cost value of this curve. It is clearly seen that the optimum cutting speed is not the minimum-cost cutting speed.

In addition to the power constraint, there is the speed constraint, that is, the required optimum speed is not available on the machine. With most universal machines there is a finite number of speeds available. They are usually arranged according to a geometric series distribution with a coefficient of 1.25. This means that there might be a deviation of up to 20% between the required and the available speeds. It has been shown earlier that a 10% deviation from the optimum cutting speed results in an 80% change of tool life and that the tool cost effect becomes negligible. Thus the whole module loses its mathematical basis. It is very difficult to carry out a theoretical analysis of this effect, since the speed requirements depend on the item and other cutting conditions. Even machines with similar powers might have a different number of speeds and different speed distributions.

To study this effect, the cutting speeds that are required to machine a particular part were matched with the available speeds on six different machines. Figure 11.5 shows the deviations between the required and available speeds. The deviations are divided into three groups:

1. The required speed is within the speed range of the machine, but does not exist.
2. The required speed is above the maximum available speed on the machine.
3. Average deviation.

Machine number	Lowest speed (RPM)	Geometric coefficient	Number of speeds	Deviation in RPM		
				Within range	Above maximum	Overall
1	9	1.25	25	44.76	571.80	274.91
2	32	1.25	19	46.14	622.21	305.24
3	50	1.25	19	63.68	758.63	154.72
4	9	1.27	19	20.69	1,140.86	910.82
5	9	1.12	25	9.28	1,170.65	1,013.44
6	45	1.21	19	34.86	878.32	469.48

Figure 11.5 Deviation between the required and available speeds.

It is possible that for a different part the values in Figure 11.5 will be different. However, some conclusions can be drawn.

The minimum deviation within the range of available speeds occurs when the geometric coefficient is small. This is not surprising, and we would expect such a result even without a study. However, there are some surprising results. The second best choice is the machine with the highest coefficient. This fact leads us to the conclusion that besides the geometrical coefficient of speed distribution, the minimum available speed is of utmost importance in machine selection.

The average deviation column shows that none of the machines has a high enough speed to fit requirements. In this case the minimum speed has a major effect, even greater than that of the number of available speeds on the machine.

The above analysis was not carried out according to scientific methods. However, it points out that in constructing an economic module, machine selection should include a consideration of speed distribution in addition to hourly rate and power.

11.3 Process Planning

An economic machining process should take into consideration the following factors:

- Geometric shape of the part.
- Tolerances and surface finish required.
- Raw material.
- Quantity required.
- Optimization criteria.
- Available facilities.

The first three factors are given in the part drawing, and they serve only the process planning phase. The last three factors are dominant factors in the process planning and production planning phases.

Process planning based on design specifications finally dictates the cost of the product. The efficiency of processing is directly related to the quality of the process planning. Systematic and consistent process planning not only affects the routine functions, such as capacity planning, machine and tooling utilization, and incentive pay system, but is also the key to future facility planning. Process planning plays a vital role in the successful exploitation of such new manufacturing methods as the cellular and multimachine manufacturing systems.

The process planning is, however, dependent on engineering. Comprehensive standards as well as consistency in adherence to these standards are vital in suppressing diversity in process planning.

Designing an economic process calls for thousands of computations. One has to examine all the possible combinations of operations, machines, and cutting conditions in order to be sure that the advocated process is the most economic one. It is a huge job and almost impossible without the aid of a computer. That is why we tend to rely on our past achievements and usually employ the same process for similar parts. However, each process planner has a different background, that is, will determine a different process for the same part.

The role of the Process Planning Department—converting design data to work instructions—is increasing in importance and complexity. Process planning includes many complex disciplines of a very precise nature, including sequencing, machine selection, time study, and programming, just to name a few. The many, many variables involved makes it very difficult to maintain consistency and discipline.

A study was made in order to evaluate the consistency and uniqueness of the process planning in our shops today. Drawings of eight parts of different complexities were selected and given to process planners in four of our metal-cutting plants in order to have them specify the machining process. The number of advocated methods was the same as the number of participants.

Figure 11.6 shows a simple part drawing and the processes proposed by the different process planners. It is amazing that in order to produce the 18-mm hole, the following operations were recommended:

1. Drill 15 mm + drill 18 mm.
2. Drill 16 mm + bore to 18 mm.
3. Drill 17.5 mm + bore to 18 mm.
4. Drill 16 mm + bore to 18 mm.

From these recommendations one can guess the past experience of each of the participants.

Figure 11.7 shows another part and the processes recommended by the four process planners. Again, each one submitted a different process. Naturally, they cannot all be economical processes. It is of interest to note that even the raw material dimensions vary from one planner to the next.

Dr. Bootwalla from Cincinnati Milacron conducted a similar study. In his paper to CAM-I (2) he writes:

> Let us review the results of one study conducted on a family of spur gears in Cincinnati. A sample of 425 gears, relatively simple in nature, revealed 377

11.3 Process Planning

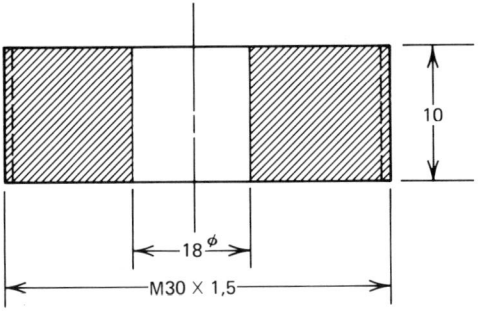

Operation number	Process planner			
	One	Two	Three	Four
1	Drill 15$^\phi$	Face cut	Turn 30	Face cut
2	Drill 18$^\phi$	Turn 30	Countersink	Turn 30
3	Turn 30	Thread cut	Drill 17.5$^\phi$	Countersink
4	Thread cut	Drill 16$^\phi$	Bore 18$^\phi$	Drill 16$^\phi$
5	Face cut	Bore 18$^\phi$	Thread cut	Bore 18$^\phi$
6	Cut off	Cut off	Face cut	Chamfer
7	Second face cut	Second face cut	Cut off	Cut off
8			Second face cut	Second face cut
9				Chamfer
10				Thread cut

Figure 11.6 Processes proposed by four process planners.

different process plans (operation sequence and machine groups), 54 different types of machine requirements and 15 different material types.

Such a wide deviation in process planning has a major effect on all activities with flow of work through the shop. In this particular example, there were 252 unique combinations of machine tools for successive operations. One does not need much imagination to visualize the maze of problems such a situation creates in the shop.

Let us now focus in on the situation at individual operation level within a Process Plan. The following examples are based on a turning operation for a part family of 93 parts. There is a high degree of similarity of geometrical shape within the part family and all the parts lie within a defined size range.

For this operation, 15 unique machine types are specified in the current Process Plans. This group of machines represents all types of lathes from a simple manually operated to a full automatic NC turret lathe of various vintage.

Operation Number	Process planner			
	One	Two	Three	Four
Raw material	$130^\phi \times 130$	$130^\phi \times 130$	$130^\phi \times 128$	$127^\phi \times 127$
10			Chuck on drill press	
20			Drill 75^ϕ	
30	Chuck on lathe	Chuck on lathe	Chuck on lathe	Chuck on lathe
40	Face cut	Face cut	Face cut	Drill 30^ϕ
50	Ext. cut 114^ϕ	Drill 20^ϕ	Ext. cut 110^ϕ	Bore 60^ϕ
60	Ext. cut 128^ϕ	Drill 38^ϕ	Ext. cut 100^ϕ	Cut 110^ϕ
70	Ext. cut 105^ϕ	Drill 55^ϕ	Thread cut	Cut 100^ϕ
80	Drill 25^ϕ	Bore 63^ϕ	Bore 80^ϕ	Face cut
90	Drill 40^ϕ	Bore 80^ϕ		Bore 80^ϕ
100	Drill 60^ϕ	Cut 110^ϕ		Thread cut
110	Bore 80^ϕ	Cut 100^ϕ		
120	Cut 125^ϕ	Thread cut		
130	Cut 110^ϕ			
140	Cut 100^ϕ			
150	Thread cut			
160	Chuck	Chuck	Chuck	Chuck
170	Face cut	Face cut	Face cut	Cut 125^ϕ
180	Bore cone	Cut 125^ϕ	Cut 125^ϕ	Cut 100^ϕ
190	Cut 100^ϕ	Cut 100^ϕ	Cut 100^ϕ	Face cut
200	Thread cut	Thread cut	Thread cut	Thread cut
210		Bore cone	Bore cone	Bore cone
220		Burr		Burr

Figure 11.7 Processes proposed by four process planners.

This diversity, of course, has its effect on the set-up times. The set-up times do not vary from machine type to machine type, as would be expected, but also within each machine type which should not be the case for a part family with a very high degree of similarity and a given size range.

The piece time analysis also shows a similar trend. In this case, some variation in the piece time per machine type is to be expected due to variations in part characteristics within the family. At this time, we do not know how much of the total variations are due to Part Family characteristics and how much goes back to the nature of manual time study.

Further analysis was done to identify if any decision rules were being used in the selection of the machines. There appears to be no real trend with considerable overlapping of the dimension range between machine types. The same is true of the relationship between machine type and the average batch size.

I am almost sure that anyone, anywhere, who conducts such a study will get similar results.

11.3 Process Planning

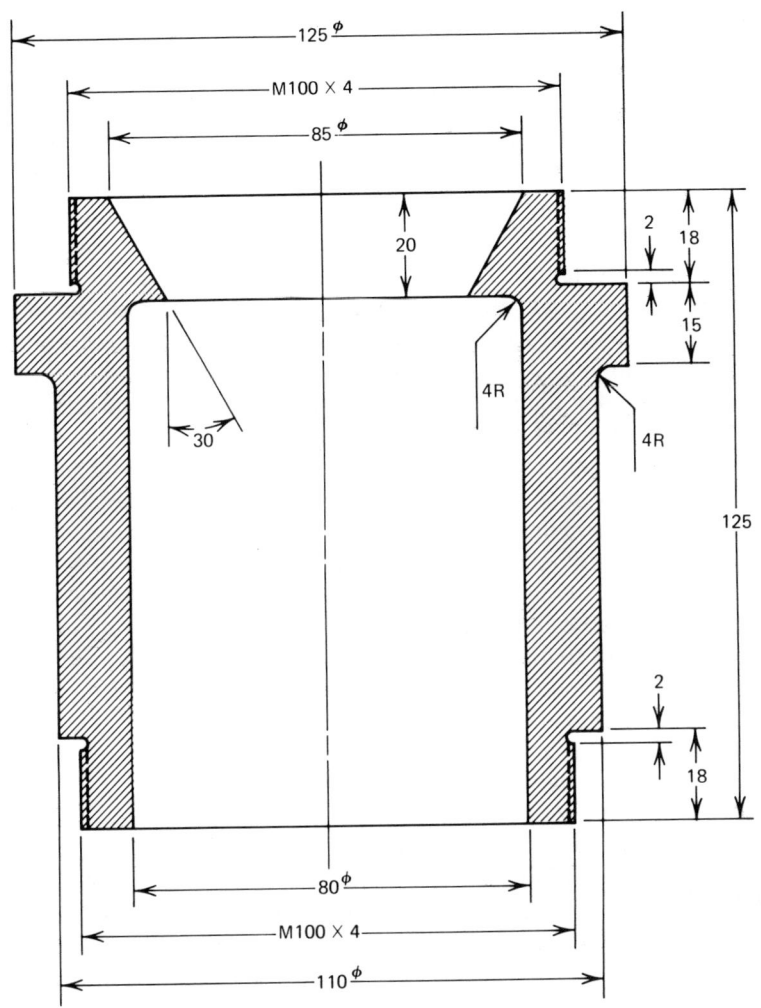

Figure 11.7 *(Continued)*

The situation in process planning as shown is neither unique nor surprising, since at present it is not a science, but strictly a human-oriented activity, highly dependent on individual skills, human memory and mood, and a mass of reference manuals.

The above must not be interpreted to mean that we should not use our accumulated experience in process planning. There is nothing as valuable as experience. Industry has been operating in such a manner for years and, in a way, quite successfully.

Process planning is the input to production planning and scheduling. A certain process plan might cause machine overloading whereas another would not. An overloading condition is resolved by shifting jobs forward or backward and thus increasing work-in-process or causing a delay in delivery. These conditions are not a must and can actually be caused by using a nonoptimum process plan. However, existing technology does not have the tools to handle or even be aware of this situation.

One of the major handicaps encountered in analysis for a cellular manufacturing system is the inconsistencies in the existing process plans. Accurate analysis and, more important, the efficient implementation of the cellular manufacturing system necessitate accurate and consistent process planning.

The process plan dictates the manufacturing cost and the efficiency of the shop operation. Hence, the extreme diversity and inconsistency inherent in our present methods can no longer be tolerated. Besides affecting the efficiency of our current operation, it is a roadblock in moving toward advanced manufacturing concepts. Today's methods of process planning, which remain a sort of black magic, lacking proper planning tools, must be systematized and made part of a manufacturing science.

A highly desirable solution would be a fully automatic process planning system. Such a system would take the description of the part in the form of a mathematical model, production- and shop-dependent data as inputs, and produce a complete process plan automatically. This type of system is generally known as a generative system. The development of such a system is an extremely complex problem.

The need to do something in the process planning area is, however, very pressing. An ad hoc improvement is required and can be achieved by a retrieval-type program.

These two types of process planning programs—retrieval and generative—are discussed in Sections 11.4 and 11.5, respectively.

11.4 Retrieval-Type Process Planning Programs

A retrieval-type system logic is derived from the GT methods of classifying and coding machined parts for the purpose of segregating them into family groups. Each part family will be composed of "like" parts having a sufficient number of common attributes to prescribe a common manufacturing method.

The "sameness" of a group of parts will be determined by analysis of the classification codes of the encoded part spectrum. Sorting by discrete values or sets of ranges of values for individual attributes embedded in the

11.4 Retrieval-Type Process Planning Programs

part codes will reduce the encoded part spectrum to increasingly numerous, homogeneous groups. The final reduction will result in part families, each with a membership of parts naturally susceptible to fabrication by a basically common method. Refinement and/or subdivision of these groups will then probably be necessary to accommodate the constraints, capabilities, and general characteristics of the production facility.

The two entities—part families and standard plans—are products of the user's classification and coding activities and of the effort required in analyzing the product mix and manufacturing facilities. Part families and standard plans are stored in the system files.

Primary operation of the system is investigated by invoking the part family search function. The user keys in the individual part classification code of the engineering part to be process planned. The part family file is then searched in an attempt to establish the existence of a family to which, by user definition, the input class code should below. If a family concurrence exists, the standard plan for the part family is retrieved and displayed on the CRT graphic display terminal, so that the user can modify or extend the same to suit the specific part being planned. If part family refinement and standard planning are of quality, the adaptation of the standard plan will be straightforward, with the user's decision-making reduced to a minimum. The system provides for permanent on-line storage of complete part-specific process plans. The user effects this storage via a simple system command. Hard copies of individual plans can be made, when appropriate, for shop distribution, reference, and other purposes. Computer Aided Manufacturing—International, Inc., has developed such a program, called CAPP (3), which is now in the public domain.

The retrieval-type system is basically a file organization system in which data can be stored and retrieved efficiently in many cross sections. The data themselves are the responsibility of the user. Process planning information is kept in a three-level structure, as follows.

Level 1 Header data and general manufacturing data. User-specified header menu built from system-recognized header elements. This level includes such general data as:
- Part number.
- Part name.
- Classification code.
- Engineering change.
- Material type and form.
- Quantity.
- Heat treatment.

Level 2 Standard plan information structure. Each part family will retrieve a standard plan, such as:

Part family 026
VMILL	Machine on vertical mill
HMILL	Machine on horizontal mill
ELATH	Turn on engine lathe
JBORE	Bore on jigbore
DRILR	Drill on radial drill
INSP	Inspect

and so on.

Level 3 Each of the above basic operations is detailed with respect to:
- Department
- Machine number
- Set-up data and time
- Work elements or work instructions
- Work element parameters

The detailed records are retrieved on request and can be modified if so desired.

The above is, of course, an oversimplification of the system, but does serve to demonstrate the concept. The system cannot be operated without a great deal of initial effort by the users. The use of a classification and coding system is a must.

There are three types of classification and coding systems: design-, process-, and resource-oriented (see Part I, Section 3.6). Each one has its advantages and disadvantages. In the present case a process-oriented classification and coding system is required. However, due to the extreme diversity and inconsistency inherent in present-day process planning, it is very difficult to construct the desired classification and coding system. Hence a design-oriented classification and coding system is often used.

Classification and coding are tedious jobs. One advantage in making the initial effort to implement the retrieval-type process planning system is that it focuses attention on the state of process planning today. When it is found that the same part family has a few hundred process plans, they must be analyzed and standardized according to the economical process plan. It is possible that this, by itself, is sufficient justification for the preparatory work.

Economic process planning is sensitive to minute changes in part geometry, relationship between adjoining segments of the part, quantity, and so on.

The retrieval-type program is based on the standard process plan for similar parts, that is, a family of parts. There is no doubt that it introduces consistency into process planning, but does it also result in an economical

11.4 Retrieval-Type Process Planning Programs

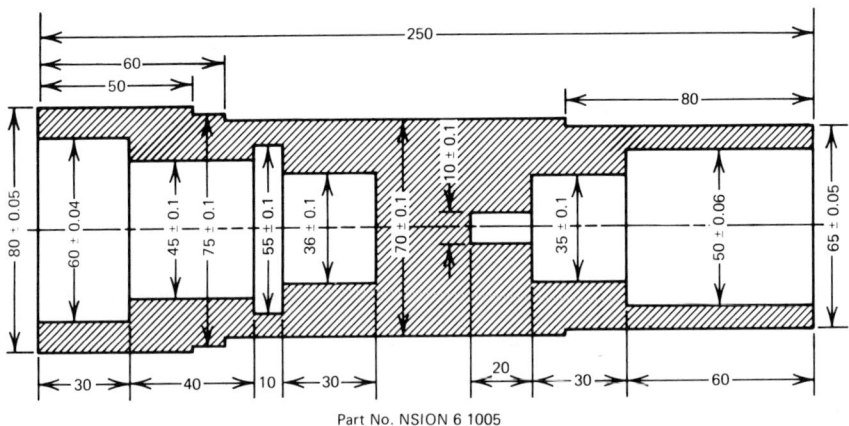

Part No. NSION 6 1005

Figure 11.8 An experimental part.

process plan? To examine this question, economical process plans were computed for the experimental part shown in Figure 11.8. While maintaining its geometric shape, the following were changed:

- Tolerances.
- Raw material hardness (BHN).
- Quantity.
- Optimization criteria.

Machine number	Power (hp)	Tolerance	Maximum spindle speed (RPM)	Hourly rate
100	3.0	0.1	1,800	10.0
110	7.5	0.1	1,600	15.0
120	10.0	0.02	1,400	30.0
121	10.0	0.1	1,400	10.0
130	15.0	0.04	900	35.0
200	15.0	0.06	710	19.0
220	1.0	0.02	1,700	16.0
250	5.0	0.02	1,700	20.0
260	10.0	0.02	900	36.0
270	18.0	0.02	2,000	50.0
280	25.0	0.02	700	40.0
281	25.0	0.1	700	9.0
300	15.0	0.1	710	9.0

Figure 11.9 Machines used to produce the experimental part shown in Figure 11.8.

500 pieces—maximum production

Minimum tolerance ≥ 0.1 mm				Minimum tolerance < 0.1 mm			
170 BHN		400 BHN		170 BHN		400 BHN	
Side 1	Side 2	Side 1	Side 2	Side 1	Side 2	Side 1	Side 2
121	281	281	281	270	281	260	281
200	270		270		270		270
			200				200
21.80 minutes		36.25 minutes		21.83 minutes		36.25 minutes	
$11.09		$17.05		$16.78		$23.86	

500 pieces—minimum cost

300	281	281	281	281	281	281	281
			300	250	300	250	250
				300	250		300
23.97 minutes		39.08 minutes		25.76 minutes		41.83 minutes	
$3.66		$5.96		$5.08		$8.02	

50 pieces—maximum production

270	270	281	281	270	270	260	280
22.86 minutes		37.48 minutes		22.86 minutes		37.48 minutes	
$19.73		$6.30		$19.73		$24.63	

50 pieces—minimum cost

300	281	281	281	250	300	250	281
				300	250	281	250
							300
24.87 minutes		40.30 minutes		28.28 minutes		44.59 minutes	
$4.41		$6.72		$8.13		$11.40	

Figure 11.10 Economical processes for the experimental part shown in Figure 11.8. The numbers in the upper part of each box of the table are the machine numbers.

To produce this part, the machines listed in Figure 11.9 were used. Figure 11.10 shows the economical process for each of the above variations. For 16 combinations of the same part, the process repeated itself in only three cases, giving 13 different economical process plans. One may argue (and rightfully) that by using the retrieval-type process plan, the user is allowed to modify any standard process plan stored in the files. It is true, however, that this is not the original intention of the program.

There is no doubt that a retrieval-type process plan is one step forward. However, it is only one intermediate step toward a generative-type process plan.

11.5 Generative Process Planning

Process planning is the task that defines how the part is to be manufactured in the most practical, economical way. It is basically a decision-making task. Decisions have to be made on the following, interdependent elements:

1. Which operations.
2. On which machine.
3. With which tools.
4. Under what cutting conditions.

Like in any decision-making process, three elements are necessary: goals, alternatives, and selection of the best alternatives. This method of decision-making can be applied to process planning.

The task of selecting cutting conditions was discussed in Section 11.2. There are many mathematical modules that can be used in selecting the optimal cutting conditions. The problem, as has been pointed out, is that in using these modules the machine and tool selected must be known in order to choose the parameter values for the economical cutting speed equation. From a decision point of view, this does not create any problem. There are many alternatives—each existing machine in the shop is an alternative. One should compute the economical cutting speed, subject to the constraints, for each machine. Thus the operation time and cost can be computed. All that is left to be done is select the best alternative. This decision is actually a purely mathematical computation. One does not need to posses any technical know-how. One just has to multiply, divide, add, and subtract numbers, compare the results, and decide which number is larger than the other. In this way, the best decision will be arrived at. It does require a lot of computations, but the best decision will be reached for the optimum single operation.

It has been pointed out that the economics of a single operation will not necessarily result in the economic process of the part. If each operation is performed on a different machine, transfer expenses of the part between machines, gripping the part separately for each operation, set-up time, and other expenses must be allowed for. It is highly probable that these expenses will offset the savings gained by performing each operation by the most economical method. Therefore, the number of alternative processes increases. The possibility of combining several single operations on one machine should be considered, but which operations should be combined? Again, from a decision-making standpoint, the answer is simple. All possible combinations are alternatives. All the process planner has to do is to compute the economical cutting speed for all possible combinations of operations on all the machines available at the shop, compute the time and cost of all combinations, and sum up all operations costs, thus arriving at the process time and cost of the part. The decision is thus straightforward, that is, choose the best result for the particular goal. This is a simple comparison of numbers. It is a purely computable process, and no technical know-how is required to make the computations and to arrive at the best possible decisions. It takes a lot of computations, many more than you suspect. The number of possible combinations is

$$P = N! \times M^N$$

where P = number of combinations
N = number of single operations per part
M = number of available machines.

To illustrate this huge number, let us consider a simple part with only 10 operations and a small machine shop with only 10 machines. Thus

$$P = 10! \times 10^{10} = 3.6288 \times 10^{16}$$

If we assume that it takes 1 second to compute an alternative, the task will last

$$\text{Time} = \frac{3.6288 \times 10^{16} \text{ (number of alternatives)}}{3.1536 \times 10^7 \text{ (seconds per year)}} \approx 1{,}000{,}000{,}000 \text{ years}$$

Naturally, this is impractical; however, it does not mean that the logic of the decision-making process is wrong. Theoretically, the logic is correct. It is a mathematical problem, not a metal-cutting problem, the logic of which is based on computations.

Since the best decision cannot be reached, we tend to rely on intuition. Process planners eliminate all those alternatives they feel will probably not have a chance of being the best. Actually, they reduce the number of alternatives to less than five (if not to one) in order to make their decisions. They have no choice, since they cannot spend years in order to arrive at the most economical process plan. They would rather have a process plan than wait for a better one.

The selection of what machine and cutting conditions to use is a computational—not engineering—decision and as such is much better made by a computer, since the latter is much faster and more accurate than a human.

Generative process planning will examine and compute all of the above-mentioned combinations, a task too big even for a computer, considering the huge number of alternatives. If we follow the previous example and assume that the computer will evaluate each alternative in

11.5 Generative Process Planning

one-millionth of a second, it will still take 1,000 years, which is impractical even for a computer. Mathematical methods must be developed in order to solve this combinatorial problem in a reasonable computer time.

The problem of deciding which operations are required is of a different nature: It is noncomputable. This decision requires technical know-how in the field of metal cutting. The decision-making process remains; however, now the problem is how to generate alternatives rather than how to evaluate them. The number of alternatives is finite. The finished operations are dictated by part geometry. However, if it is impossible to machine the part to its final form in one pass (as is usually the case), rough cuts are taken. There are several alternate operations for such cuts; a simple example of this is given in Figure 11.3.

The general equation that gives part machining time can be expressed as

$$T_p = \sum_{i=1}^{n} \left(\frac{L_i}{N_i S_i} + T_{si} \right) + \sum_{i=1}^{q} T_{Di} + \sum_{i=1}^{p} \frac{T_{Hi}}{M} \quad (11.10)$$

where T_p = Total part processing time (minutes)
 L_i = Length of cut in operation i (mm)
 N_i = Revolutions per minute in operation i (RPM)
 S_i = Feed rate in operation i (mm per revolution)
 T_{si} = Handling time (adjust tool, engage feed, etc.) in operation i (minutes)
 T_{Di} = Chucking time (sum of number of grippings per part) (minutes)
 T_{Hi} = Set-up time of operation i (minutes)
 M = Quantity of parts per setup.

By examining eq. 11.10 some rough conclusions concerning the optimization of machining time can be assumed:

- The number of chuckings should be kept to a minimum.
- If possible, the chucking method should be such that the gripping time is minimal.
- The number of operations should be kept to a minimum.
- Economical cutting conditions should be used.

The direct machine time is computed by the simple equation

$$T_i = \frac{L_i}{N_i S_i} \quad (11.11)$$

At first sight, it looks easy to handle. However, if the factors that affect each one of these variables are carefully examined, it is found that:

  L Length of cut
    Part geometry
    Part specifications
    Depth of cut
    Chucking location
  N Cutting speed
    Tool
    Machine power
    Available speed
    Cutting forces
    Part material
    Feed rate
    Depth of cut
  S Feed rate
    Surface finish
    Feed tool constraints
    Cutting forces
     Torsion stress in the part
     Gripping forces
     Part deflection
      Tolerance
      Chucking type
      Chucking location
      Depth of cut
    Machine moment
    Depth of cut
     Chatter
     Part length
     Chucking location
     Chucking type
     Part specifications

A large number of factors are all interrelated. The depth of cut, for example, affects each of the parameters—and each one differently. Conventional mathematical optimization techniques cannot be employed in this case: There are too many variables, too few auxiliary equations, and the variables are neither linear nor continuous functions.

 The way to handle this type of problem is by an algorithm based on metal-cutting technology. The main problem in constructing the algorithm is determining with which of the many parameters to begin. As may be concluded from the previous list of parameters, they are all interdependent. Once a parameter is set, it limits the others. This constraint is artificial: If a different value is assigned to the parameter, it will result in a different value of the constraint. This type of constraint prevents the possibility of reaching optimum operations. The true constraint to be considered is the part strength. This is the only given parameter of the task. It is clear that if the forces acting in the metal-removal process break the part, the task of producing the part will not be possible. If the forces deflect the part to such a degree that it becomes impossible to keep the required tolerances, it is a real constraint. However, if the forces exceed the part gripping forces, it is an artificial constraint. It only means that our previous decision about chuck selection was not good and must be altered. As a general rule, no artificial constraints are being considered at all.

Any decision—except those concerning part geometry, specifications, and material—can be changed in order to overcome a constraint that arises due to a previous decision. This may result in an endless loop of computing, decisions, and change of decisions, since one decision affects many parameters, each one in a different manner. A careful study should be made in order to establish the best sequence of decisions, that is, the one that results in a minimum number of changes and a short decision loop. A study and understanding of the metal-cutting parameters can give the desired algorithm.

11.6 Summary

The above discussion has established the basic concepts of a generative process planning program; they are as follows:

1. There must be a separation between the computable and the noncomputable elements of process planning. The noncomputable elements are the definitions of the operations required. The computable elements are machine and sequence of operations selection.
2. In order to arrive at optimum process planning, no artificial constraints are to be considered. The only constraint to be considered is the part to be produced.
3. Concepts 1 and 2 lead to the necessity of having two separate phases in the program:
   a. *Theoretical process plan.* This is composed of the most economical operations and cutting conditions. It is theoretical in that it disregards the available facilities. It is practical from the standpoints of metal-cutting technology and part constraints.
   b. *Practical process plan.* This is a mathematical phase. It considers

the available facilities and their constraints. It adjusts and modifies the theoretical operations and cutting conditions to suit the machine constraints. It thus creates the infinite process alternatives and then computes—and selects—the best alternative.

To construct the generative process plan along the above concepts, knowledge in the following fields is required:

- *Metal-cutting process.* A thorough understanding of the metal-cutting process is necessary (see Chapter 12).
- *Mathematical techniques for solving the combinatorial problem.* Chapter 13 demonstrates the type of research and mathematical technique used in solving this problem. The solution obtained will hold true for any type of machining (not only turning). Furthermore, it is the link to capacity planning, as will be discussed in Part III of this book.
- *Computer science.* A computer must be employed to solve the huge amount of computations required. To arrive at an economical generative process planning program, the process time, memory size, and disk space must be kept to a minimum. This calls for unconventional programming. A good understanding of computer capabilities is required.
- *Part description system.* In order to arrive at a generative process plan, one with no human interference, the computer must "see" the finished part geometry and specifications. This is required in order to compute the required operations. This system will be referred to in Chapter 14 and will be fully detailed in Part III of this book.

References

1. Bartalucci, B., R. Bedini, and G. G. Lisini, "On the Optimum Selection of Machine Tool in Turning Operation," CIRP General Assembly, Nottingham, England (September 6–13, 1968).
2. Bootwalla, M., "Planning for Group Technology," *Proceedings of CAM-I's Coding and Classification Workshop*, P-75-PPP-01, Arlington, TX (June 23–24, 1975), pp. 183–194.
3. Barnes, R. D., "CAM-I Automated Process Planning—The CAPP system," Proceedings of CAM-I's International Seminar Development for a Decade, Step 1—Plans and Procedures, P-76-MM-02, Atlanta, GA (April 21–23, 1976) pp. 284–302.

Chapter Twelve
Generative Process Planning— Engineering

The proposed generative process planning program is divided, for reasons that were discussed in Chapter 11, into two main modules. The first one establishes the operations that are required to produce a part. It is theoretical, since it does not consider available facilities and their constraints. This is done in order to guarantee optimum operation. In the second module practical considerations are put into effect, and hence the most economical process planning is reached.

Specifying which operation should be carried out in order to produce the part is an engineering problem. There are many parameters to consider, such as chucking type and location, depth of cut, feed rate, length of cut, chatter, deflection of part, tolerances, and surface finish.

All of these variables are interdependent and affect each other. The effect on one parameter might be in conflict with the effect of another parameter. Values must be assigned and decisions made concerning all parameters. It is almost impossible to solve the problem of specifying the required operations in a conventional, mathematical way. It has, therefore, been decided to use both a logical block diagram approach and an algorithm to solve this problem. In order to arrive at the most economical process, all alternative operations will be generated and evaluated. This calls for thousands of computations; hence, a computer is used to perform this task. The main problem in constructing the algorithm is selection of the sequence of handling parameters value. It is possible that a certain sequence will result in an endless loop of decisions, since each previous decision compels the decision of the following parameters. These types of constraints are artificial and exist only because of the sequence of decisions used. Such artificial constraints are not allowed on our program. Therefore, when limiting conditions are encountered, the previous decision is altered in order to remove and overcome the constraint.

This chapter is devoted to metal-cutting technology for turning. Its aim is to understand both the process and the interdependence of parameters. This knowledge is used in constructing the algorithm in such a way that it does not result in an endless loop. Furthermore, in order to arrive at an economical computer program, certain decisions that can be computed only once, based on mathematics and metal-cutting know-how, and the results used in the algorithm of the computer program, saves computer time in recomputing the same element over and over again when the result may be foreseen.

12.1 Basic Cutting Conditions Equation

The metal-removal process is very complicated. Until the present, researchers in this field had been unable to reach an agreement and develop a mathematical module that would be of use to science and practical application.

The generative process planning program is concerned with optimization and arriving at economical processes. It must have an optimization module that both ties together and gives the interrelationship between the parameters involved.

Most theoretical metal-cutting researchers are in agreement that the main criterion in the selection of cutting speed is tool life. However, the term tool life is not explicitly defined. There are several definitions of tool life:

- Tool breakage.
- Tool wear to a predetermined value.
- Surface finish arrived at by a worn tool.
- Predetermined dimension deviation as a result of tool wear.

There is no doubt that tool wear is one of the most influential parameters. It is known that there are three mechanisms of tool wear:

1. Wear by friction (e.g., grinding of the tool).
2. Wear by chipping of the crystals on the cutting corner.
3. Wear by diffusion between tool and material.

F. W. Taylor dedicated his life to investigating the relationship between cutting speed and tool life. For 26 years, he conducted experiments and

12.1 Basic Cutting Conditions Equation

gathered data, and in 1907 he published his famous equation (1):

$$VT_L^n = C \tag{12.1}$$

where V = cutting speed (m per minute)
 T_L = tool life (minutes)
 n = exponent that depends on tool and part material
 C = constant; the equivalent of cutting speed that will result in 1-minute tool life.

This famous equation is used by most scientists today to compute the economical cutting speed.

Many papers were published on the subject of the reliability of this equation. In his discussion of R. C. Brewer's (2) paper, "On the Economics of the Basic Turning Operation," B. N. Colding points out that the Taylor exponent is not of a fixed value, but changes with feed rate and depth of cut.

B. K. Lambert (3), in his doctoral dissertation at the Texas Technological College, investigated the results of work duration on tool wear: Do short intermittent operations result in the same total tool life that one continuous cutting operation would? He experimented with eight tools and reached the conclusion that there is a difference in tool life. If a cutting speed is chosen on the basis of intermittent experiments, an error of up to 100% in tool life may result when a continuous cut is taken. A. Bhattacharyya (4) proves that the Taylor exponent n has a turning point that is a function of temperature. Actually, there are two wear zones: one in which the wear is insensitive to temperature and another that starts at a critical temperature and is sensitive to it. From that temperature on, the wear rate increases. Above the critical temperature, the wear mechanism is that of diffusion between tool and material. This phenomenon is also recognized by H. Opitz (5). He claims that at high temperatures there are diffusion and chemical reactions between the tool and the part material. To prove his point, he carried out an experiment in which a carbide tool and steel were brought into contact and both inserted into an oven. The temperature was raised to 1200°C. When the items were removed from the oven, they seemed to be welded to one another.

The value of the exponent n is also dependent on the predetermined value of wear (see Figure 12.1). If this value is changed, the value of the exponent will be changed. The predetermination of this value is arbitrary and depends on general agreement between scientists. In spite of the fact that many attack Taylor's equation (eq. 12.1), very few have a better one to offer.

M. Kronenberg (6) shows that by dimensional analysis of heat balance

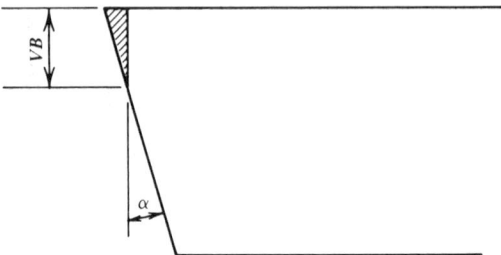

Figure 12.1 Predetermined value (VB) of tool wear.

one can arrive at Taylor's equation. His deduction is, therefore, that Taylor's equation holds true only for constant temperatures. Kronenberg introduces a new set of equations for computing cutting speed and forces. His logic is that tool life can be defined by the volume of chips removed. The volume is a function of the depth of cut, feed rate, and cutting speed. Hence, cut surface, that is, depth of cut multiplied by feed rate, must be considered as a primary parameter. He defines specific cutting speed as a 60-minute tool life when the cutting area is equal to 1 mm² (0.001 in.²). As a correction factor, he uses the slenderness ratio (depth of cut divided by feed rate). The specific cutting speed also depends on the combination of material and cutting tool. He proves that the primary parameter of the material is its hardness.

Kronenberg's equation is as follows:

$$V = \frac{C_v (G/5)^f}{(1{,}000R)^z (T_L/60)^y} \tag{12.2}$$

where
V = cutting speed (in. per minute)
C_v = specific cutting speed (in. per minute); see text
G = depth of cut to feed rate ratio (slenderness ratio)
R = cutting area: depth of cut multiplied by feed rate (in.²)
T_L = tool life (minutes)
f, z, y = exponents.

The values of the exponents given in Kronenberg's book are a result of extensive study and the use of experimental data found by many researchers, such as Slezinger, Taylor, the AWF Institute in Germany, and the American Society of Mechanical Engineers.

Waxen (7) and some other researchers recommend using the following equation:

$$V T_L^n B^m = C \tag{12.3}$$

12.1 Basic Cutting Conditions Equation

where V = cutting speed (m per minute)
        T_L = tool life (minutes)
        C = constant
  n, m = exponents
        B = chip equivalent thickness (CET).

The CET is expressed as

$$B = \frac{A \cdot S}{\left(\frac{A - R(1 - \sin Sc)}{\cos Sc}\right) + R\left(\frac{T}{2} - Sc + \sin^{-1}\frac{S}{2R}\right)} \quad (12.4)$$

where A = depth of cut (mm)
        S = feed rate (mm per revolution)
        R = tool nose radius (mm)
      Sc = cutting-tool angle.

(The denominator is merely the length of the active part of the cutting tool.)

The beauty of this equation is that tool geometry is introduced into the cutting speed equation. It is very helpful in laboratory tests where the parameters are known or can be measured, and it enables us to test the effect of each of the parameters separately. However, from a user standpoint, it increases the number of variables that have to be determined before the cutting speed can be computed. Even without this addition, there are too many degrees of freedom. This addition may make the solution almost impossible. Kronenberg proves that it is possible, by mathematical manipulations, to bring the CET equation to the form

$$B = CA^x S^y \quad (12.5)$$

Hence, his equation takes the CET into consideration. In my study, the results show that for a ratio of depth of cut to tool nose radius within the limits of 10:1 to 2.5:1 and a cutting angle of 30 deg, the difference between Kronenberg's (12.2) and Waxen's (12.3) equations is negligible. A completely different approach to the selection of economical cutting speed is taken by the U.S. Air Force and General Electric. They claim that today it is impossible to construct an equation that will connect cutting speed and tool life. Therefore, the use of tables is recommended. (It is of interest to note that the U.S. Navy is using Taylor's equation).

In a preliminary generative process planning study, all of the data published in a table form were keypunched. A computer program was written to compute, from the given data, the tool life exponent, the ratio of depth of cut to feed rate, the ratio of depth of cut to tool nose radius, and to compare the given data with data computed for the same cutting conditions by using the Kronenberg equation. It was found that 75% of the given cutting speed data in tables are within 0.7–1.3 of the computed cutting speed values. In metal-cutting research work, a correlation of 30% is considered good and acceptable.

Using the Kronenberg equations and parameters results in the following advantages:

- Using equations instead of tables simplifies programming and reduces computer execution time.
- Kronenberg's equations are easy to use; the number of variables is minimal and must be known anyway.
- The values of the exponents are all from one source.
- The values of the exponents have been verified by comparison with many other sources.

Therefore, we shall use Kronenberg's equations as the basic cutting conditions equations in our generative process planning programs.

The use of these equations is not essential to the process planning program. If one wishes (or when better equations are available) these equations can be replaced by others without affecting the logic of the program.

12.2 Parameters Analysis

The Product of Cutting Speed by Feed Rate

Machining time in turning is computed by eq. 11.2:

$$T = \frac{L}{N \cdot S}$$

It is clear that in order to minimize T, the product $N \cdot S$ (spindle speed multiplied by feed rate) should be maximized. However, the equation does not indicate which values should be assigned for the spindle speed, and which for the feed rate, and what are the interdependence of these two. This problem has been investigated by some researchers, and their results will be used.

Lentz and Koren (8) analyze this problem from the tool standpoint. Their conclusion is that the optimal feed rate, for minimum cost, is a function of the feed rate (p) and tool life (n) exponents in the expanded Taylor equation (see eq. 11.3). They examine three options:

1. Working at minimum feed rate.

2. Working at maximum feed rate.
3. The exponents' values are equal.

They conclude that in the common case, where tool life is controlled by the width of tool wear, maximum feed rate will result in minimum part cost. Brewer (2) reaches the same conclusion by a different approach. He develops an optimization model, as shown in eq. 11.1. He then looks for the minimum cost first with a constant feed rate and varied cutting speed, and then the cutting speed is kept constant and the feed rate varied.

The results are compared by a diagram, where one axis is cutting speed, the other feed rate, and the curves represent cost. The conclusion is that the feed rate has a more profound effect on cost than cutting speed. Therefore, to optimize cutting conditions, feed rate must be as large as possible.

Other researchers agree with the above conclusion. The generative process planning programs adapt this conclusion. It means that the feed rate should be selected prior to cutting speed and that an attempt should be made to work at maximum feed rate.

Maximum Feed Rate

The expression maximum feed rate is used freely by most researchers. However, few define what maximum feed rate is and what controls its value. Depiereux (9) investigated this problem in his doctoral dissertation. He proves that at high feed rates the width of tool wear is no longer the parameter that controls tool life. At high feed rates, plastic deformation of the cutting edge and oxidation of the secondary cutting plane exist. The maximum feed rate is a function of cutting speed. At cutting speeds of 120–150 m per minute, the higher the feed rate, the higher the deformation and oxidation rate. He concludes that maximum feed rate should be 0.8 mm per revolution. Peters (10) recommends using the value of 0.5 mm per revolution as the upper limit feed rate. Lentz and Koren (8) recommend 0.6 mm per revolution as the maximum feed rate.

On the basis of this study, 0.6 mm per revolution is used as the maximum feed rate in the generative process planning.

Feed Rate and Cutting Forces

In their research on forces and stress, Crockall (11) and Peters (10) assume a beam firmly chucked on one side and free on the other side. We will use the same assumption in our analysis, that is, the controlling factor is part deflection. The cutting forces can be computed by the equation

$$P = CA^x S^y \tag{12.6}$$

where P = force (N)
 A = depth of cut (mm)
 S = feed rate (mm per revolution)
 C = coefficient that considers the force element that controls deflection (N per mm²)
 x, y = exponents.

The force (P_1) that will deflect the part to a known dimension (D) can be computed from beam deflection theory:

$$P_1 = \frac{3EID}{L^3} \tag{12.7}$$

where E = modulus of elasticity of part material (N per cm²)
 I = moment of inertia of part cross section (cm⁴)
 L = beam free length (cm)
 D = deflection (cm).

Substitution of the force from eq. 12.6 into eq. 12.7 results in

$$CA^x S^y = \frac{3EID}{L^3}$$

Solving for feed rate gives

$$S = \left(\frac{3EID}{CA^x L^3}\right)^{1/y} \tag{12.8}$$

The parameters in eq. 12.8 are known. However, for a rough cut, where the length and depth of cut are not dictated by the part, these two parameters are variables. It is possible that for the first trial of depth and length of cut a feed rate below the maximum value will result from eq. 12.8. Conventional economical modules will regard length and depth of cut as a constraint and accept the computed feed rate. Our approach is that an attempt will be made to overcome the constraint. In the present case, the depth of cut, chucking location and type, and length of cut can be changed. These changes affect the feed rate and might result in more economical conditions. The next section investigates these conditions.

The Effect of Depth of Cut on Feed Rate

For this investigation, all parameters but depth of cut are assumed to have a constant value. Therefore, eq. 12.8 can be rewritten as

$$S = \frac{K}{A^{x/y}} \tag{12.9}$$

12.2 Parameters Analysis

According to Kronenberg, numerical values of x are always greater than those of y. Therefore, $x/y > 1$. A decrease in the depth of cut will not result in a proportional increase of feed rate.

Machining time is given by eq. 11.2:

$$T = \frac{L}{N \cdot S} = \frac{\pi DL}{V \cdot S}$$

If the extended Taylor equation (eq. 11.3)

$$VT_L^n S^p A^q = K$$

is substituted for V in eq. 11.2, the time can be expressed as

$$T = \frac{\pi DL}{S} \cdot \frac{T_L^n S^p A^q}{K} = \frac{K_2 A^q}{S^{1-p}} \quad (12.10)$$

To insulate the effect of depth of cut on machining time, eq. 12.9 is used as follows:

$$T = \frac{K_2 A^q}{(K)^{1-p}} A^{(x/y)(1-p)} = K_3 A^{(x/y)(1-p)+q} = K_3 A^r \quad (12.11)$$

where $r = (x/y)(1 - p) + q$.

Decreasing the depth of cut actually means adding a rough cut operation. Therefore, we may initially assume that in order to overcome the constraint, integer numbers (H) of cuts are used, all with the same depth of cut. In such a case, additional time T_H should be allowed for idle tool travel and tool adjustment. Thus the economic value of H can be computed by the following equation:

$$T - H \cdot T_2 \geq (H - 1) \cdot T_H \quad (12.12)$$

where T is the machining time for the case of one cut, and T_2 is machining time of the individual decreased depth of cut pass. Introducing eq. 12.11 into eq. 12.12 results in

$$K_3 A^r - HK_3\left(\frac{A}{H}\right)^r \geq (H - 1)T_H$$

or

$$K_3 A^r\left(1 - \frac{H}{H^r}\right) = T(1 - H^{1-r}) \geq (H - 1)T_H$$

or

$$\frac{1 - H^{1-r}}{H - 1} \geq \frac{T_H}{T} \quad (12.13)$$

The value of r can be:

- $r < 1$. In such a case, H^{1-r} is greater than one, and the left side of eq. 12.13 is negative. This means that there is no possibility of reducing machining time by increasing the number of operations.
- $r = 1$. In such a case, the left side of eq. 12.13 becomes zero, and the previous conclusion holds.
- $r > 1$. In such a case, the left side of eq. 12.13 is positive. This means that a possibility of reducing machining time by increasing the number of operations exists.

To demonstrate the above conclusions, let us examine the case of cutting steel with a carbide tool. The exponents are $x = 0.963$, $y = 0.643$, and $n = 0.26$; assume $T_H = 0.1$ minute. In such a case, it is economical to increase the number of operations (cuts) if the single operation machining time is greater than 1.4 minutes. However, eq. 12.13 does not consider the maximum feed rate limit. Therefore, it might be more practical to use the computed feed rate value rather than the machining time limit in computing the economics of an increasing number of operations.

This can be done by using eq. 12.9. Let S_1 designate the feed rate computed for one cut and S_2 the feed rate for several cuts. Limiting S_2 to its maximum value,

$$S_2 = \frac{K}{A_2^{x/y}} \leq S_{max} = 0.6 \text{ (mm per RPM)}$$

$$S_1 = \frac{K}{A_1^{x/y}}$$

the feed rate ratio becomes

$$\frac{S_1}{S_2} = \frac{K}{A_1^{x/y}} \cdot \frac{A_2^{x/y}}{K} = \frac{A_2^{x/y}}{A_1}$$

or

$$S_1 = S_{max}\left(\frac{A_2}{A_1}\right)^{x/y} = 0.6\left(\frac{A_2}{A_1}\right)^{x/y} \tag{12.14}$$

For example, it is advantageous to split the cut into two passes when cutting steel with a carbide tool if the feed rate (S) is below 0.25 (this value is computed by eq. 12.14).

The same technique can be used to arrive at a borderline feed rate for three or more splits.

Splitting the cut into equal passes means that

$$A_2 = \frac{A_1}{H} \rightarrow H = \frac{A_1}{A_2}$$

12.2 Parameters Analysis

By using equation 12.14 for the ratio of depth of cut, the following general equation is derived:

$$H = \frac{A_1}{A_2} \rightarrow H = \left(\frac{0.6}{S_1}\right)^{y/x} \tag{12.15}$$

The value of H computed by eq. 12.15 is introduced into eq. 12.13 to compute the economics of splitting. The economical considerations in deciding which depth of cut to select are not the only parameters to be considered. Surface finish and chatter, for example, can limit the selection of depth of cut.

Nonsymmetrical Split of Depth of Cut

The conclusions previously derived assume that the cut is divided into an integer number of equivalent cuts. Let us examine the case when a nonsymmetrical split of depth of cut takes place.

The initial depth of cut is A and is divided into two cuts: JA and $(1 - J)A$. Computing cutting time with eq. 12.11 results in a total time of

$$T = K_3(JA)^r + K_3[(1 - J)A]^r \tag{12.16}$$

or

$$T = K_3 J^r A^r + K_3(1 - J)^r A^r = K_3 A^r [J^r + (1 - J)^r] \tag{12.17}$$

However, $K_3 A^r$ is the machining time (T_1) for one cut with the total depth of cut. Hence,

$$T = T_1[J^r + (1 - J)^r] \tag{12.18}$$

To arrive at minimum machining time as a function of J, we take the derivative of eq. 12.18 with respect to J and set it equal to zero:

$$\frac{dT}{dJ} = T_1 r[J^{r-1} - (1 - J)^{r-1}] = 0$$

It will be equal to zero only if $J = 1 - J \rightarrow J = 0.5$, that is, symmetrical division.

The second derivative indicates whether this value is minimum or maximum. Thus

$$\frac{d^2T}{dJ^2} = T_1 r(r - 1)[J^{r-2} + (1 - J)^{r-2}]$$

The meaning of this equation is that if $r > 1$, the result is minimum, while if $r \leq 1$, the result is maximum. This is in agreement with our previous conclusions.

Effect of Cutting Speed on Surface Finish

One of the factors that controls surface finish is the primary cutting edge. Its effect is explained by the fact that at low cutting speeds material builds up on the tool edge (built-up edge—BUE). This BUE scratches the surface. As the cutting speed increases, the temperature rises, and the BUE separates from the tool. The periodic build up and removal of the hard edge ruins the tool; it produces vibrations by lifting the tool so that it snaps back when the BUE fractures, which is detrimental to the surface finish.

Additional increase of cutting speed results in a good surface finish. When additional increase in cutting speed takes place, burn marks might appear on the machined surface.

The above conclusions have been reached by experiments in many laboratories and are only a qualitative description of the process. We could not find equations to express the above conclusions quantitatively, probably due to the fact that it is very difficult to isolate the cutting speed effect on surface finish from other more influential factors, such as feed marks and tool geometry.

The generative process planning program (GPPP) adopts this conclusion: Minimum and maximum cutting speed limits should be observed in order to arrive at a good surface finish. The values of these limits for steel are chosen according to the recommendations of Peters (10): 100 m per minute as the lower limit and 400 m per minute as the upper limit.

Effect of Feed Rate and Tool Nose Radius on Surface Finish

Tool geometry has an effect on the machined surface finish. Increasing the true rake angle and side cutting-edge angle, for example, improves the surface finish. However, it is very difficult to investigate this effect. Research, therefore, turns to the investigation of the effect of feed tool marks on the surface (see Figure 12.2). These marks depend mainly on the feed rate and tool nose radius. The commonly used equation for this function is

$$R_T = \frac{1}{8} \cdot \frac{S^2}{R} \tag{12.19}$$

where R_T = surface finish (RMS)
S = feed rate (in. per revolution)
R = tool nose radius (in.).

Shaw (12) claims that the results obtained at different research centers do not coincide with one another. He claims that different tool grinding

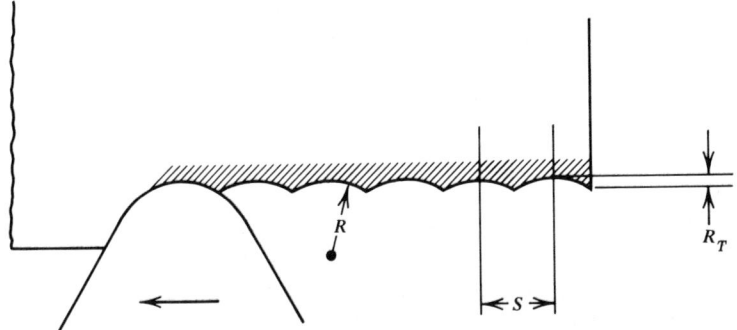

Figure 12.2 Surface finish—feed tool marks.

techniques will result in a different tool nose radius. Tool wear increases gradually during the cut and causes changes in surface finish as well as in dimension. He suggests defining tool wear by part tolerance and surface finish. This actually means that for each required tolerance a different tool life will be used.

Besides the classic surface finish equation (12.19) there are some other approaches that should be mentioned. Peters (10) suggests using K. V. Olsen's equation:

$$R_A = 10 \cdot \left(\frac{2V}{25}\right)^{(2S/R^{0.25}-1)} \qquad (12.20)$$

where R_A = surface finish (CLA—Center Line Average)
V = cutting speed (m per minute)
S = feed rate (mm per revolution)
R = tool nose radius (mm).

This equation is difficult to use, since it contains three parameters, which is probably why even Peters does not recommend using it as a working equation, but rather as an experimental tool to compare cutting conditions.

F. O. Rasch and A. Rolstadas (13) suggest using the following equation:

$$R_A = 2.95 S^{0.7} R^{-0.4} T^{0.3} \qquad (12.21)$$

where R_A = surface finish (CLA)
S = feed rate (mm per revolution)
R = tool nose radius (mm)
T = cutting time (minutes).

The extended equation includes the cutting speed as one of the parameters. However, they found that its exponent is very small and that within a deviation of 99.9% the cutting speed effect is negligible. Within a deviation of 95%, the values are somewhat larger, but, even so, have no real effect.

It has been decided to use the classic equation (12.19) in the GPPP.

Effect of Depth of Cut on Surface Finish

Equation 12.19 connects cutting conditions with surface finish. The depth of cut has no effect whatsoever on surface finish. Moreover, one may choose any feed rate desired as long as the tool nose radius is appropriately selected. These scientific conclusions contradict practical metal-cutting procedures.

Surface finish–depth of cut–feed rate do not appear as parameters in any single equation published. There are, however, some remarks and recommendations that sometimes contradict one another. Shaw (14) points out that at a tool nose radius below 0.005 in., the cutting edge might frequently break. It is customary to increase tool nose radius with increase of depth of cut. It is usually recommended that the radius be 10% of the depth of cut. One way to arrive at a good surface finish is to use low feed rates (see eq. 12.19). However, at low feed rates the energy required to remove a unit of volume of chips increases. Thus the temperature at the cutting tip increases. Shaw claims that the depth of cut should be 10 times greater than the feed rate. These two recommendations result in a 1:1 ratio of feed rate to tool nose radius. However, eq. 12.19 holds true only for cases where the radius is "much bigger" than the feed. No one defines what "much bigger" actually means. In the demonstrations, a ratio of 4:1 is used. (By the way, in the same example, a finish cut with 2.5 mm depth of cut is used, a value that is regarded high for a finish cut.) From practical experience it is known that there is a minimum value for the depth of cut. The theory explains that below this value the metal is not cut by the tool, but is pressed into the machined surface by it. Figure 12.3 illustrates this occurrence. R. Connolly and C. Rubenstein (15) prove this theory and investigate its effect on the cutting forces. Pekelharing claims that working with a depth of cut below its minimum value damages the surface finish and causes excessive tool wear. In studying eq. 12.19, Bhattacharyya (16) states that although it seems that the greater the tool nose radius, the better the surface finish, there should be limits to its value, since a large radius might cause chatter. He recommends limiting the tool nose radius of a carbide tool to 0.02–0.04 in. This limit is a function of the depth of cut; the radius may increase with increase of depth of cut. (It actually means that in order to arrive at good surface finish, the depth of cut of the finish cut should be large.)

12.2 Parameters Analysis

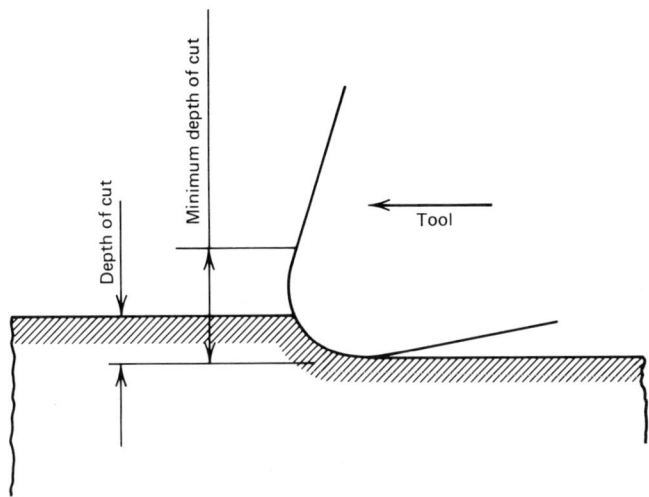

Figure 12.3 Minimum depth of cut.

The U.S. Air Force does not refer directly to this problem; however, in studying its cutting condition recommendations, as published in booklet number AD-843-994, we find constant ratios. For a carbide tool and depth of cut between 1.5–2.5 mm, they use a depth of cut to tool nose radius ratio of 2.5:1 and tool nose radius to feed rate ratio of 4:1.

No immediate conclusions can be reached from this study, and we will return to it later on to define how the GPPP determines the value of the depth of cut.

Surface Integrity

Surface finish is commonly defined by its geometric dimensions. However, in addition, it should consider strain and stresses in the machined surface—that is, work hardening. This field has not been thoroughly investigated, and quantitative equations are not available.

W. Graham and C. Rubenstein (17) conducted experiments in which a series of thin-wall disks were pressed, clamped, and turned under various cutting conditions. Later, the disks were separated and stress penetration measured. Their conclusions from this experiment are:

1. Stress penetration increases with:
   a. Increase of feed rate.
   b. Decrease of tool nose radius.
   c. Insignificantly with increase of cutting speed.

2. Stress level rate increases with:
   a. Increase of feed rate.
   b. Cutting speed did not have any effect (within the speed range used, i.e., 332–664 ft per minute).

It seems that the parameters that control work hardening are identical with those that control geometric surface finish. In addition, they found that at a very low feed rate (0.04 mm per revolution), the effect of work hardening increases. From the above, we may conclude that if the feed rate is kept within the constraints imposed by the tool and geometric surface finish, it will cover the work hardening constraints as well.

Chatter

Vibrations are associated with any chip formation process. Vibrations affect surface finish as well as dimension accuracy. If the vibration frequency is not synchronized with the spindle revolutions, a wavy surface will be generated instead of a cylindrical one, since the relative locations of the tool and the work piece are changing. Vibration frequency varies between 100–2,000 cycles per second when using carbide tools in machining steel. In some cases it is possible to count the chatter marks on the workpiece (or, rather, their distance e in inches). The following formula can be used to calculate the frequency of vibrations from the chatter marks:

$$f_r = \frac{V}{5e} \text{ (cycles per second)}$$

where V is cutting speed in ft per minute.

Chatter Effect on Tool Wear

It is customary to assume that vibration in the chipping process reduces tool life. Experiments conducted by Opitz and Salje definitely support this assumption. In some cases tool life was reduced by 60%, which indicates a 200% increase in tool wear. These conditions occurred due to instability of the machine and the tool, thereby causing radial and tangential deflections as a result of the cutting forces. Under such cutting conditions, the depth of cut and cutting speed change dynamically and are not constants, as Taylor's equation assures. Recently, we have witnessed experiments and research where controlled vibrations are introduced into the tool. These experiments were mainly conducted by the Russians. They claim that by this method tool life increases when machining hard metals.

12.2 Parameters Analysis

However, they do not reveal the cutting conditions used in such experiments. Weill reports that when the vibrations are in the feed direction and the depth of cut varies, there is an improvement in tool life for certain tool materials. Simonet reports that in his experiments with a vibration frequency of 0–300 cycles per second, he could not detect any significant change in tool life.

It seems that vibrations might have a good or bad effect on tool life, depending on their direction, amplitude, and frequency.

B. W. Rooks (18) wrote his doctoral dissertation on this subject, and his conclusions for cutting mild steel with a carbide tool are:

- Over a vibration frequency range of 0–150 cycles per second in the feed direction, the tool wear rate increases. The minimum increase of tool wear was at a frequency of 75 cycles per second.
- At a vibration amplitude of 0–0.010 in., a gradual increase of flank wear occurs, while crater wear has a maximum point that is a function of vibration frequency.
- Variations in vibration amplitude have more severe effects, on both sides, than frequency variations.
- Vibrations in the feed direction cause changes in the rake and clearance angles. Theoretical tool wear analysis under steady conditions with different angles gave good results.
- Torsional vibrations and vibrations in the chipping direction might improve tool life.
- In cutting Ni–Cr–Mo steel, vibrations improve chip breaking shapes and may thereby prevent cracks from developing in hard tool material, such as Ardoloy Ak. However, vibrations may cause cutting tip breakage in Ardoloy 5.8 tools.

In the GPPP we will try to eliminate chatter conditions in any case.

Chatter Causes and Prevention Methods

S. A. Tobias (19) devotes the first part of his book to general vibration theories and the second part to vibrations in metal-cutting processes (one chapter for each type of machine). The practical part of the book is a summary of test results achieved by many laboratories.

In the lathe chapter, he points out that in spite of the fact that chatter in lathe work was of prime interest in the research, there is no thorough understanding of the vibration mechanism.

It is of interest to note that the cutting conditions recommended for chatter prevention are similar to those that improve tool wear, stresses, and economics. This means that there is no conflict of parameters and that the economic cutting conditions will also prevent chatter.

Experiments conducted by Doi and Hahn indicate that cuts of very low depth of cut cause instability. They explain it by friction forces between tool and material. Tobias agrees with the findings, and attributes them to the fact that with a low depth of cut even low-amplitude vibrations cause separation of contact between the tool and material. This separation increases the amplitude, and an unstable condition results. Arnold found that vibration amplitude increases with tool wear. He claims that when a sharp tool is used, there will be no vibrations. As tool wear reaches a significant level, vibration will appear.

Hahn (20) claims that when the depth of cut is low, there is instability in cutting forces, and as a result vibrations will build up. He notes 0.05 mm as the lowest limit for the depth of cut.

I would suggest adding a safety factor to his lowest limit, since the machined surface might be wavy (e.g., as a result of the previous cut) and the depth of cut is not constant. Thus, in practice, a critical depth of cut may result, and vibration amplitude will increase. Doi (21) states that vibration tends to increase with low feed rate. From his diagrams it seems that the critical minimum feed rate value is 0.04 mm per revolution.

Tlusty (22) studied the vibration problem mainly for cases of wavy surfaces resulting from previous cuts on the same segment. He reached the conclusion that there is a maximum value of depth of cut such that above it vibrations will develop. In his study he refers to a depth of cut of 6 mm.

Tobias (19) reached the same conclusion. He states that the tendency to vibrate increases with depth of cut. The critical maximum depth of cut varies from machine to machine. It is a function of machine rigidity as well as of tool and material rigidity. It seems that there is a general agreement to this conclusion, and it has become the standard procedure in machine testing.

Kronenberg (6) found that the cutting force decreases when cutting speed increases up to a certain limit. From that point on, cutting force and speed are independent. The limit for carbides is at about 350 ft per minute. At deeper cuts, the low-speed limit can occur already at 175 ft per minute. From his analysis, he draws the conclusion that dynamic instability exists whenever the cutting force drops with increasing cutting speed. This causes self-excited vibrations.

Our practical conclusion from the above discussion is that chatter can be prevented and economical cutting conditions obtained by setting limits to depth of cut, feed rate, and cutting speed. The limits—minimum and maximum values—will be set by considering vibration, machine stability, surface finish, and tool material.

12.3 Operation Establishment

Specifying which operations are required to machine a part is straightforward if each segment of the part is machined in one pass, that is, its dimension is reduced from that of the raw material to that of the finished part.

This can seldom be done due to surface finish tolerances, power, and forces. If more than one cut per segment is needed, the questions of how to divide the depth of cut between the passes and when and how to combine the several segments to be machined in one operation arise. Figure 11.3 demonstrates that there are many alternatives for a simple part. This section is devoted to this problem and develops a method that will select the best alternative.

Depth of Cut for Finish Cut

Process planning requires a rule by which a decision is made on whether a rough cut is needed or the segment can be machined by one finish cut. The trend in this problem today can be seen in J. R. Crockall's (11) model. He states that this decision is affected by the surface finish required. He sets an arbitrary limit of surface finish of 100 CLA and rules that below this limit the surface should be machined by rough cut followed by a finish cut.

We prefer to use a different approach that will consider surface finish as well as part geometry, raw material, and the conclusion reached in Section 12.2.

The basic surface finish equation is eq. 12.19:

$$R_T = \frac{1}{8} \cdot \frac{S^2}{R}$$

The conclusion of the analysis made in the previous section is that feed rate, tool nose radius, and depth of cut must have a certain ratio. Let us use the following notation:

$$M = \frac{A}{S} \quad \text{and} \quad N = \frac{A}{R}$$

where A = depth of cut (mm)
S = feed rate (mm per revolution)
R = tool nose radius (mm).

We replace the variables in eq. 12.19 with our new symbols:

$$R_T = \frac{1}{8} \cdot \frac{S^2}{R} = \frac{1}{8} \cdot \frac{(A/M)^2}{(A/N)} = \frac{AN}{8M^2} \tag{12.22}$$

Solving for the depth of cut, we obtain

$$A = 8 \cdot \frac{M^2}{N} \cdot R_T \times 10^{-6} \qquad (12.23)$$

where R_T is given in RMS (μin.).

When the recommended values of M and N are inserted into eq. 12.23, the following equation is derived:

$$A = \frac{R_T}{3,000} \text{ to } \frac{R_T}{1,000} \qquad (12.24)$$

The limits are due to the flexibility in defining the values for M and N. Actually, the lower limit is intended for high-quality surface finish and the upper limit for normal surface finish. A straight-line equation bridging the two limits results in low depth of cut values in the middle of the range. A hyperbolic curve (see Figure 12.4) gives better practical results and can be expressed by the following equation:

$$A = \frac{(R_T/60)^5}{11.7 + 20(R_T/60)^4} \qquad (12.25)$$

This equation is used to determine the depth of cut value for each segment of the part. Initially the required depth of cut is computed by subtracting the required finished dimension from the present dimension. If the required value is below the maximum value, a finish cut operation is specified. If the required value is greater than the maximum, one or more rough cuts and finish cuts are needed.

The question is how to divide the required depth of cut, that is, which values should be assigned to the rough cut and which to the finish cut. The above decision is affected by the following:

1. No strength constraint on the part exists.
2. There is a strength constraint.
3. It is possible to extend the rough cut length to other segments.
4. It is impossible to extend the rough cut length to other segments.

Let us examine each of these cases:

1. When there are no strength constraints, control is transferred from the part to the tool, and a maximum feed rate can be used. Since the depth of cut does not effect machining time, the method by which the total depth of cut is divided between rough and finish cut is not really important. The only consideration is, therefore, the depth of cut to feed rate ratio. Another consideration might be increasing the safety factor of the finish cut. A low depth of cut value in the finish cut reduces the probability of

12.3 Operation Establishment

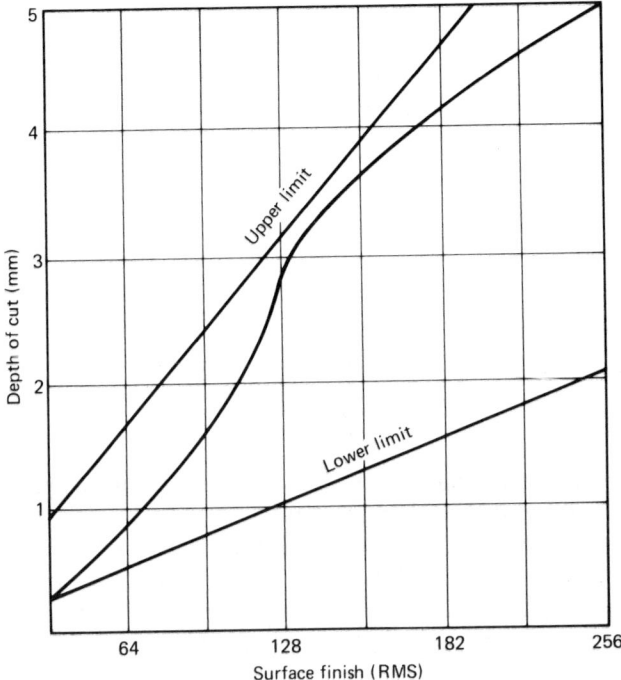

Figure 12.4 Depth of cut as a function of surface finish.

chatter and allows an increase in cutting speed. Both factors are beneficial in maintaining good surface finish.

It was decided arbitrarily that the depth of cut of the finish cut would be half of its maximum value.

2. When there is a strength constraint in the rough cut and the finish cut operation, the remedy is not dependent on the depth of cut distribution. Other means should be applied. If the constraint exists only in the rough cut, it is possible to increase the depth of the finish cut, thereby reducing that of the rough cut and the cutting forces. The distribution is controlled by the strength of the part, and thus the constraint is overcome. If by working at the maximum depth of cut value of the finish cut the strength constraint is not removed, other means should be applied (e.g., more than one rough cut or increasing the allowable forces by a more rigid gripping of the part).

3. Machining time is composed of machine time and handling time (adjust tool, engage feed, etc.) and can be expressed as

$$T_P = T_M + T_H = \frac{L}{NS} + T_H$$

where T_p = total operation time (minutes)
 T_M = machine time (minutes)
 T_H = handling time (minutes)
 L = operation cutting length (mm)
 N = spindle speed (RPM)
 S = feed rate (mm per revolution).

The total part time is

$$T_C = \sum_{i=1}^{C} T_M + CT_H + T_K \qquad (12.26)$$

where T_C = total part machining time (minutes)
 C = number of operations per part
 T_K = extra interoperation time (minutes).

The finish cut is defined according to the part drawing, where each segment imposes its own requirements.

Rough cuts are an intermediate operation that leaves planners some degree of freedom in their decisions. This degree of freedom can be used to combine rough cuts of several segments into one operation. Figure 12.5 demonstrates this case. The upper depth of cut limit of segment 1 is below the lower limit of segment 2. Therefore, they cannot be combined into one rough operation. (If they are combined, an extra rough cut must be performed anyway for segment 1.) However, segments 2, 3, 4, and 5 can be combined. They can be machined by one combined rough cut followed by finish cuts.

If the operations were combined, the total part machining time, as expressed in eq. 12.26, is

$$T_C = \sum_{i=1}^{C} T_{M_1} + (C - K)T_H + T_K \qquad (12.27)$$

It is advantageous to combine rough cuts only if the total part time as computed by eq. 12.27 is lower than that of eq. 12.26. There are two cases to consider:

a. When there are no strength constraints. In such a case, the maximum feed rate is used, $T_M = T_{M_1}$, and $(C - K)T_H < CT_H$. This means that rough cuts should be combined whenever possible (see Figure 12.5).

b. There are strength constraints. In such a case, $T_{M_1} \neq T_M$. The inequality is due to the fact that the feed rate is computed as a function of the average depth of cut, which is then reduced in some segments and increased in others.

12.3 Operation Establishment

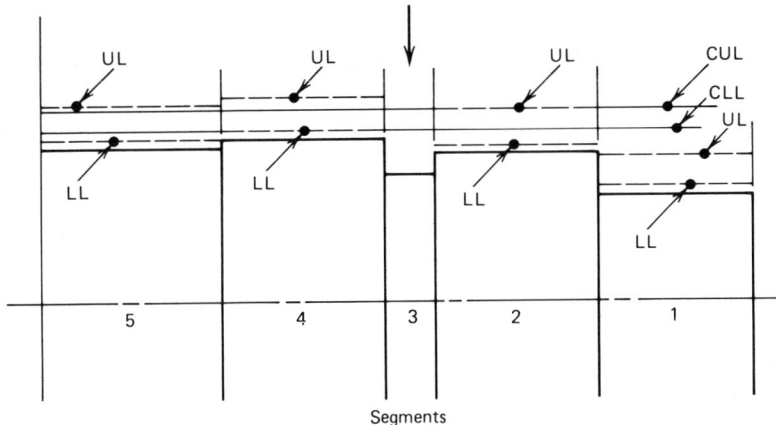

Figure 12.5 Depth of cut limits and combination. Notation: LL, lower depth of cut limit; UL, upper depth of cut limit; CLL, combined LL; CUL, combined UL.

If the value of $T_{M_1} - T_M$ is greater than the savings in handling time $(C - K)T_H$, segments should not be combined.

Some general rules can be derived from the following example. Assume a case of two segments that can be combined, and assume that the feed rate is reduced by 50% on segment 1, while remaining unchanged on segment 2. Furthermore, let us assume that L_2 is negligible compared to L_1. A. A. Hadden (23) gives standard handling time T_H as

Engage or release feed	0.0005 hour
Adjust each tool per each cut	0.0021 hour
	0.0026 hour = 0.156 minute

If the machining time of segment 1 by itself is T_1, it is advantageous to combine the two segments when

$$0.5 T_1 \leq 0.156$$

or

$$T_1 = \frac{0.156}{0.5} = 0.3 \text{ minute}$$

This means that if the machining time of a single operation is less than or equal to 0.3 minute, it is always advantageous to combine operations. It is not convenient to base this decision on time, since time is not one of the basic data. Therefore, a translation into length of cut will be made.

In the previous section (see eq. 12.14) it has been proved that in order to economically overcome a strength constraint, the feed rate can be reduced to a minimal value of 0.21–0.3 mm per revolution. Below this value, it is advantageous to split the cut (reduce the depth of cut). Machining time can be expressed as

$$T = \frac{L}{NS} = \frac{\pi DL}{VS}$$

This expression gives

$$DL = \frac{VST}{\pi}$$

It is known that when the length to diameter ratio is 2:1 there is no strength constraint. Using this value to replace D we arrive at

$$DL = \frac{L}{2} \times L = \frac{L^2}{2}$$

Substituting this in the above equation,

$$DL = \frac{L^2}{2} = \frac{VST}{\pi}$$

For the case of $V = 100$ m per minute, $S = 0.3$ mm per revolution, and $T = 0.9$ minute (arbitrarily chosen time limit),

$$L^2 = \left(\frac{2 \times 100 \times 1{,}000 \times 0.3 \times 0.9}{3{,}14}\right)^{1/2} = (1.7 \times 10^4)^{1/2} = 132 \text{ mm}$$

This means that it is advantageous to combine any segments if their length is less than or equal to 132 mm.

Using the above values, a general rule can be derived:

$$L^3 \leq 1.7 \times 10^4 = C \times \sum_{i=1}^{C} L_i \qquad (12.28)$$

where C is the number of segments that can be combined.

This equation is convenient, since the values of feed rate, cutting speed, and machine time are not required (and are unknown at the moment) in order to make a decision on segment combination and depth of cut. The lengths of the segments are known values.

If the L^3 in eq. 12.28 is greater than the right side of the equation, it does not mean that the segments should not be combined. It suggests that detailed computations should be made.

4. When the difference in diameters of adjoining segments is twice the

maximum depth of cut of the lower diameter, it is not advantageous to combine rough cuts. The rules that govern the depth of cut decision are those specified in cases 1 and 2.

When more than one rough cut must be used for a segment, limits are set to each pass, and the rules mentioned in cases 1–4 apply to them.

12.4 Chucking Type and Location

The effect of chucking location can be studied by eq. 12.8:

$$S = \left(\frac{3EID}{CA^x L^3}\right)^{1/y} = \left(\frac{3 \times 3.14 \times E \times D \times D_A^4}{64 A^x L^3}\right)^{1/y}$$

where S = feed rate (mm per revolution)
E = modulus of elasticity (N per mm²)
D = allowable deflection (tolerance)(mm)
D_A = part diameter (mm)
A = depth of cut (mm)
L = free length of the part (mm)
x, y = exponents.

It seems that the free length has a tremendous effect on feed rate. Prior to investigating this effect, let us get a feeling of the magnitude involved.

If the maximum and minimum feed rate values are $S_{max} = 0.6$ and $S_{min} = 0.04$, respectively, the ratio is $S_{max}/S_{min} = 15:1$. If all parameters except feed rate and free length are kept constant, eq. 12.8 can be written as

$$S = \frac{K}{L^{3/y}} \qquad (12.29)$$

For the extreme cases we have

$$\frac{S_{max}}{S_{min}} = \frac{K}{L_1^{3/y}} \cdot \frac{L_2^{3/y}}{K} = \left(\frac{L_2}{L_1}\right)^{3/y} = 15$$

or

$$\frac{L_2}{L_1} = 15^{y/3}$$

For $y = 0.65$, $\frac{L_2}{L_1} = 1.77$

For $y = 1$, $\frac{L_2}{L_1} = 2.26$

This means that a change of 70–100% of the free length can result in a change from minimum to maximum feed rate. The effect of chucking location on machining time can be seen by using eq. 12.10 when a constant depth of cut is assumed:

$$T = \frac{K_2 A^q}{S^{1-p}} = \frac{K}{S^{1-p}}$$

Substituting the length from eq. 2.29 for feed rate results in

$$T = K_4 L^{(3/y)(1-p)} \tag{12.30}$$

For a 70% change in the free length, $p = 0.26$, and $y = 0.65$, the time ratio will be

$$\frac{T_2}{T_1} = \left(\frac{L_2}{L_1}\right)^{(3/y)(1-p)} = 1.7^{3.4} = 6$$

The above demonstration shows that the sensitivity of free length is much greater than that of depth of cut. The feed rate range for depth of cut was shown before to be around 0.21–0.3. The equivalent in free length is $(0.6/0.21)^{y/3} = 1.25$. This means that a 25% change in free length has the same effect on machine time as a 100% change in depth of cut.

Two possible methods can be applied to control the chucking effect:

1. Change the chucking location.
2. Change the chucking type.

These two cases are studied in the following sections.

Change of Chucking Location

Before discussing changing the chucking location, we must explain how and why the initial location was selected.

Chucking the part in the machine is a time-consuming operation. This is why the first priority is to keep the number of chuckings to a minimum, that is, once for each machining side. (If later it is decided to use more than one machine per side, the same rule will be applied.) The machining side is dictated by the required tool direction movement and access. Figure 12.6 shows the required tool direction movement for each segment and points to the chucking location. The chucking location is checked for validity with respect to width, shape of the segment, and space for the chuck. When the preferred chucking locations of both sides of the part coincide (see Figure 12.6), an algorithm is used to decide which side will be chucked on the preferred location and which will be moved to the

12.4 Chucking Type and Location

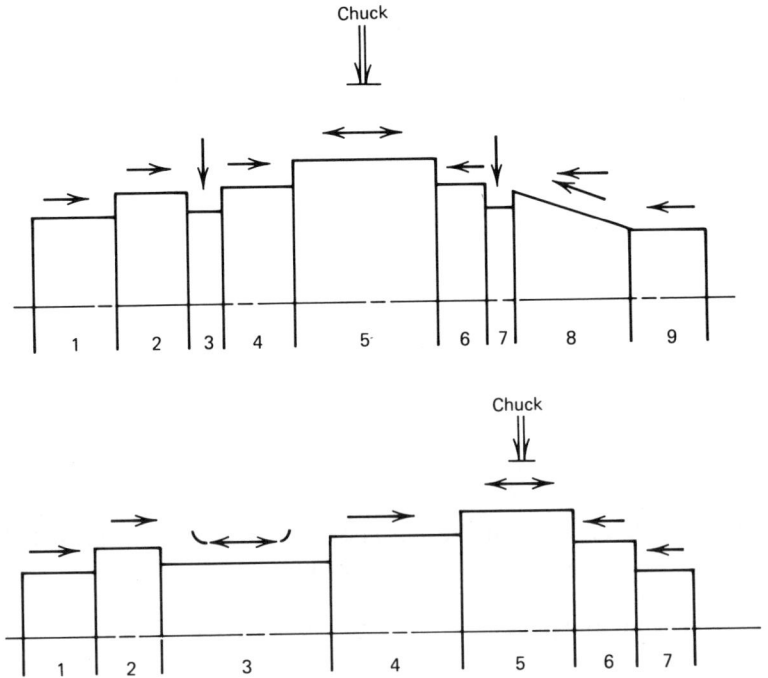

Figure 12.6 Tool direction movement and chucking location.

second best location. This algorithm is based on the free part length on each side, shape and diameter of the part, chucks available, and raw material. The machine spindle bore imposes some constraints on this decision. However, at this stage the machine to be used for the job is unknown, and, therefore, this consideration can be ignored. In stage two—machine selection—the spindle bore will be known and considered. Time and cost will be adjusted for each specific machine. Thus the spindle bore will affect the decision of machine selection. The chucking location selected is thus the most economical. The question is not whether to change the chucking location, but rather whether it is economical to add an extra chucking and thus to change the location. The alternative is to reduce feed rate. This alternative can be put in a mathematical form as

$$T_D - T \geq T_{DF}$$

where T_D = machine time with the original chucking location
T = machine time with changed chucking location
T_{DF} = added chucking time.

Dividing all terms by T_D results in

$$1 - \frac{T}{T_D} \geq \frac{T_{DF}}{T_D} \rightarrow \frac{T}{T_D} \leq 1 - \frac{T_{DF}}{T_D} \qquad (12.31)$$

The meaning of this equation is that if the added chucking time (T_{DF}) is greater than the original machine time, there is no possible change that will result in a more economical machine time. The initial chucking location should be maintained.

To examine other time ratios, let us use eq. 12.30:

$$\frac{T}{T_D} = \frac{K_4 L^{(3/y)(1-p)}}{K_4 L_D^{(3/y)(1-p)}} = \left(\frac{L}{L_D}\right)^{(3/y)(1-p)}$$

Inserting this value into eq. 12.31, we obtain

$$\frac{L}{L_D} \leq \left(1 - \frac{T_{DF}}{T_D}\right)^{1/(3/y)(1-p)} \qquad (12.32)$$

This equation holds only in the permissible range of feed rates. It has previously been shown that when the computed feed rate is below its minimum value, the length ratio is about 0.5.

Using the above exponent values, we compute eq. 12.32:

$$\left(1 - \frac{T_{DF}}{T_D}\right)^{1/(3/y)(1-p)} = 0.5$$

$$\frac{T_{DF}}{T_D} = 1 - 0.5^{(3/y)(1-p)} = 0.92$$

$$T_D = \frac{T_{DF}}{0.92} \qquad (12.33)$$

The feed rate must be maintained within a defined range. By using eq. 12.29 and the given range, the feed rate can be expressed as a function of chucking location

$$\frac{L}{L_D} = \left(\frac{S}{S_{max}}\right)^{y/3} \leq \left(1 - \frac{T_{DF}}{T_D}\right)^{1/[(3/y)(1-p)]}$$

Solving for feed rate S, we obtain

$$S \leq S_{max}\left(1 - \frac{T_{DF}}{T_D}\right)^{\{1/[(3/y)(1-p)]\} \times [1/(y/3)]} = S_{max}\left(1 - \frac{T_{DF}}{T_D}\right)^{1/(1-p)} \qquad (12.34)$$

From the above, it can be concluded that it is advantageous to change chucking location if the following conditions obtain:

- The original machine time is greater than the added chucking time.

12.4 Chucking Type and Location

- The computed feed rate in the case of the original chucking location is smaller than the one computed by eq. 12.34.
- The free length in the case of the changed chucking location must be greater than the machined segment length.

Change of Chucking Type

Initially, the program considers chucking one side of the part while the other is hanging free. This is the least time-consuming chucking type. However, this type can result (due to low force resistance) in unsatisfactory cutting conditions. In such a case, the program will consider supporting the part by a tailstock center or by a steady rest. A rough estimate indicates that a tailstock center type of chucking can increase allowable feed rate 36 times over that attained with a free-hanging type. If this increase is still not sufficient (the computed feed is below the minimum value), a steady rest can be added. Theoretically, no force limitation exists in such a case, and the maximum feed rate can be employed. However, these two types of chucking require a longer chucking time and a countersink hole; the countersink operation time should be added to the chucking time and its advantages computed.

Changing the chucking type should affect all operations performed from the same side, on the same machine, and by the same chucking type and location. Whether the change of chucking type is economical depends on part geometry, external and internal shape, and material. Therefore, it is impossible to construct standard rules. The conventional method specifies that if the length to diameter ratio of the part is greater than 5:1, a tailstock center chucking type should be used. Although this rule might be true on a statistical basis, since we intend to use computer power and speed in our program, we can allow for computation of each part and case separately instead of using a statistical rule of thumb. Our approach is as follows:

1. Assume a simple chucking type (one side in chuck, the other free) and compute the cutting conditions as a function of tolerance (part deflection). If the computed feed is below the minimum feed and the cause is part deflection, set a flag and recompute the operation with a split in depth of cut. Repeat this process until the feed rate is within limits. Store machining time.
2. If a flag is set, recompute all part operations assuming tailstock center type chucking.
3. If again the computed feed rate is below the minimum, set a flag to examine steady rest type of chucking.

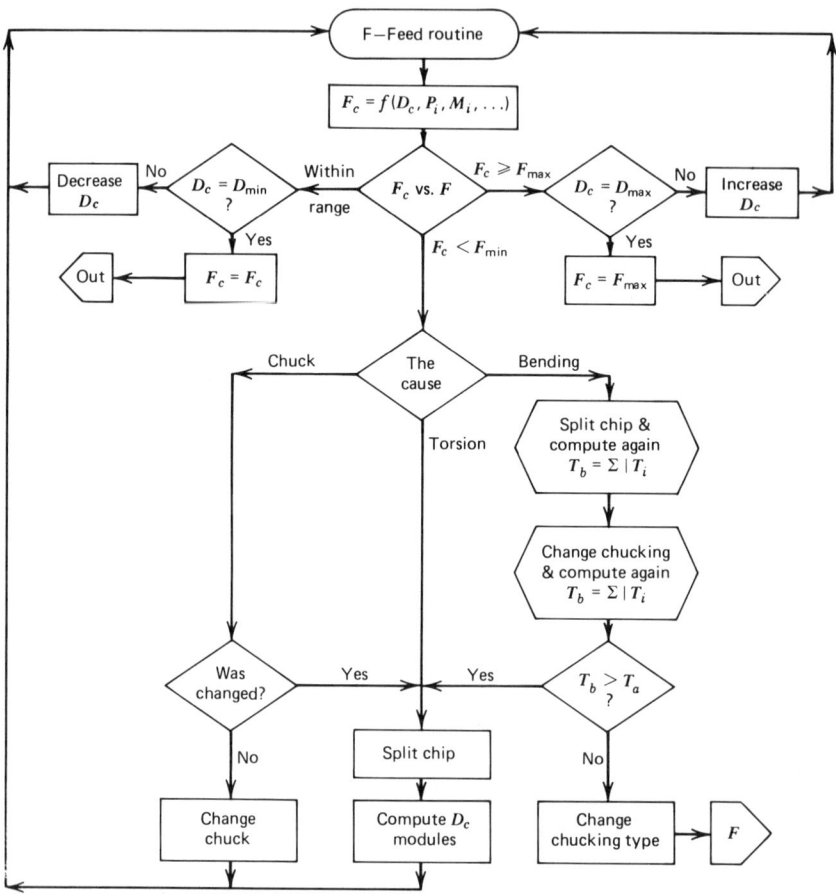

Figure 12.7 Flow chart of feed rate module. Notation: F, feed rate; F_c, computed feed rate; F_{max}, upper limit for feed rate; F_{min}, lower limit for feed rate; D, depth of cut; D_c, computed depth of cut; D_{max}, upper limit for depth of cut; D_{min}, lower limit for depth of cut; T_a, total time for all operations with simple chucking type; T_b, total time for all operations with changed chucking type; T_i, time of a single operation.

4. If a flag for steady rest is set, store total machining time with tailstock center and recompute the part.
5. Compare total machining times, and select the chucking type that results in a minimum.
6. Recompute the part with the appropriate chucking type.

Figure 12.7 shows the logic of this module. In this figure two other cases are considered:

1. When the cutting forces may break the part due to torsion forces. In such a case, the forces must be reduced and the depth of cut must be split into several cutting operations.
2. When the cutting forces are greater than the gripping forces in the chuck. In such a case, a bigger chuck can be used. If no other chuck is available, if the proportion between the chuck and the part is not acceptable, or if the gripping forces might distort the part, the depth of cut will be split in order to reduce forces.

12.5 Summary

The aim of the above study was to construct a generative process planning algorithm that would not result in an endless loop of computations and would not be restricted by artificial constraints. The process computed should be practical from an engineering standpoint and theoretical from a shop standpoint (since it does not consider available machinery). Based on the above discussion, the algorithm is as follows:

1. Compute the chucking location—free chucking type with minimum overhang.
2. Assign the depth of cut limits based on the surface finish.
3. Define the type of cut required for each segment. Attempt to combine as many segments as possible to be machined in one rough cut operation. Use the depth of cut limits to achieve this goal.
4. Ensure tool access to each specified operation. Change the sequence of operations if required.
5. Establish the feed rate limits based on the surface finish and depth of cut.
6. Compute the allowable forces as a function of the chucking type and location, surface finish, and tolerances. Use the force of least magnitude as the controlling one (bend, torsion, or chuck).
7. Compute the feed rate based on the allowable forces.
8. Try to reach the maximum feed rate by:
    a. Division of depth of cut between rough and finish cuts.
    b. Change of rough depth of cut and of combination of segments.
    c. Change of chucking location.
    d. Change of chucking type.
9. Check the effect of step 8 on previous operations and change them if necessary.

10. Compute the economical cutting speed. Check against cutting speed limits.

11. Compute the forces, power, and time for each operation.

The above sequence of computations ensures an economical process by specifying:

- Minimum number of chuckings.
- Minimum handling time.
- Minimum number of operations.
- Economic cutting conditions.

References

1. Taylor, F. W., "On the Art of Cutting Metals," ASME Transactions (1907).
2. Brewer, R. C., "On the Economics of the Basic Turning Operation," *Transactions of the ASME* (October 1958), pp. 1479–1489.
3. Lambert, B. K., "An Analysis of the Reliability of Tool Life Prediction," Ph.D. dissertation, Department of Industrial Engineering, Texas Technological College (1967).
4. Bhattacharyya, A., I. Ham, and A. Ghosh, "Analysis of Tool Wear. Part 2—Application of Flank Wear Models," *Transactions of the ASME* (February 1970), pp. 109–114.
5. Opitz, H., "Tool Wear and Tool Life," *International Research in Production Engineering* (1963), p. 107.
6. Kronenberg, M., *Machining Science and Application,* Pergamon Press, Elmsford, NY (1966).
7. Waxen, R. "Teknish Tidskrift," *Mahamite, H. 4* (1931), p. 431.
8. Lentz, E., and Y. Koren, "Optimization and Identification of Cutting Process. *TME 132* (October 1971).
9. Depiereux, W.-R., "Die Ermittlung Optimaler Schnittbedingungen, Insbesondere im Hinblick auf die Wirtschaftliche Nutzung Numerisch Gesteuerter Werkzeugmaxdhinen," Doktor-Ingenieurs, Genehmigte Dissertation, Rheinisch-Westfaelischen Technischen Hochschule, Aachen (December 12, 1969).
10. Peters, J., "The Strategy in Introducing Constraints in the Computer Models for Optimalizing Machining Conditions," Lecture No. 71C3 Technion, Haifa (1971).
11. Crockall, J. R., "The Performance Envelope Concept in the Economics of Machining," CIRP General Assembly, Nottingham, England (September 6–13, 1968).
12. Shaw, M. C., "Study of Machined Surfaces," DECD Proceedings of the Seminar on Metal Cutting, Presentation 11, Paris (September 1–2, 1966), pp. 235–268.
13. Rasch, F. O., and A. Rolstadas, "Selection of Optimum Feed and Speed in Finish Turning," *Annals of the CIRP,* Vol. 19 (1971), pp. 787–792.
14. Shaw, Milton C., *Metal Cutting Principles,* MIT Press, Cambridge, MA.
15. Connolly, R., and C. Rubenstein, "The Mechanics of Continuous Chip Formation in Orthogonal Cutting," *Journal of Machine Tool Design Research,* Vol. 8 (1968), pp. 159–187.

16. Bhattacharyya, A., and Ham, I. *Design of Cutting Tools*, American Society of Tool and Manufacturing Engineers, Dearborn, MI. (1969).
17. Graham, W., and C. Rubenstein, "An Investigation into the Degree and Depth of Work Hardening Produced at the Surface of a Workpiece by Turning," 7th International MTDR Conference, University of Alabama in Birmingham, Birmingham, AL (1966).
18. Rooks, B. W., "The Effect of Vibrations on Tool Wear," Phd. dissertation, Department of Mechanical Engineering, University of Alabama in Birmingham, Birmingham, AL (May 1964).
19. Tobias, S. A., *Machine-Tool Vibration*, Blackie and Sons Ltd., Glasgow (1965).
20. Hahn, R. S., "Metal Cutting Chatter and its Elimination," *Transactions of the ASME* (1953), p. 1073.
21. Doi, S., "Chatter of Lathe Tool," *Journal of the Society of Mechanical Engineers (Japan)* (1937), p. 94.
22. Tlusty, G. and M. Polacek, "The Stability of Machine Tools Against Self-Excited Vibration in Machining," *International Research in Production Engineering* (1963), p. 465.
23. Hadden, A. A., and V. K. Genger, *Handbook of Standard Time Data*, Ronald Press, New York.

Chapter Thirteen
Generative Process Planning— Mathematics

Process planning is a decision-making process. The process planner has to specify which machines to use, which operations to perform on each machine and their sequence, and which tooling and cutting conditions to use. It is a series of decisions, where each decision imposes constraints on the following ones. These constraints are artificial, since they are dependent on the sequence of decisions. Our aim is to generate the most economical process plan; therefore, no artificial constraints should affect the decisions. To accomplish this, it was decided to separate the program into two phases. The first module, as discussed in Chapter 12, handles the engineering stage and is limited only by engineering technology.

A theoretical process plan was derived; it is theoretical from the shop standpoint (machine is not available), but practical from the engineering standpoint, since all real strength and technical constraints have been taken into consideration.

The problem in phase two is how to adjust the process to the facilities available in a particular shop. It is basically a mathematical problem, where all alternatives can be generated and computed in order to arrive at the best decision. The only technical knowledge required is how to make the specified operations comply with a particular machine constraint.

This adjustment can easily be constructed. Thus the problem of process planning is transformed from an engineering problem into a mathematical problem. Chapter 12 presented a demonstration of the research and the engineering work required for metal cutting–lathe work. The technique discussed in this chapter is a general solution for all types of manufacturing and can actually be used independently of the engineering phase (if a retrieval-type process plan is preferred).

13.1 Definition of the Mathematical Problem

Given a list of operations to be performed and a list of available facilities, a decision is required as to which machine (or machines) to use, which operation(s) to perform on each machine, what their sequence should be, and what cutting conditions to employ. The optimization criterion is either maximum production or minimum cost. These two are the generally accepted criteria in the metal-cutting field.

The operations in the list are the absolute optimum, that is, theoretical with respect to the existing facilities, however, they are not arranged according to any reasonable machining sequence. If the theoretical number of machines (chuckings) is also the practical one, the sequence of operations will not have a significant bearing on part cost and machining time. However, if the practical process calls for more machines, the effect of the sequence of operations is meaningful.

Some of the operations specified in the list must precede others. Figure 13.1 shows the operations as they will appear in the list. It is clear that for the same segment rough operations must precede finishing operations. Thus an operation sequence of 1, 2, 3, 6; 1, 2, 4, 7; and 1, 5, 8 must be maintained, even though any number of operations can be inserted (e.g., 1, 13, *2*, 16, 17, *3*, 4, 10, *6*). Another consideration is tool access. Clear tool access will save machining time. This requirement restricts the sequence of the operations shown in Figure 13.1 to 1, 2, 3, 4, 5; 1, 2, 3, 5, 4; and 1, 2, 5, 3, 4, while finishing operations 6, 7, and 8 can come in any

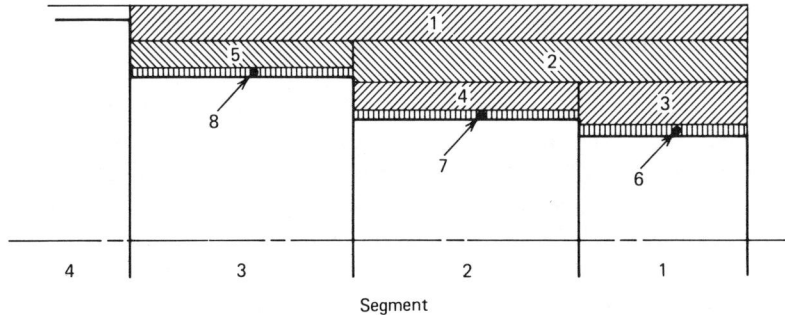

Figure 13.1 Operation sequence as specified in module one.

sequence as long as the previous conditions prevail (e.g., 1, 2, 5, 3, 6, 4, 8, 7; 1, 2, 5, 3, 4, 7, 6, 8; . . .)

If the number of machines in the practical process is greater than that specified in the theoretical process, extra time and cost should be added to cover extra set up, chucking, and transfer of parts between machines; additional complications in capacity planning and job recording; additional inspection; and so on.

The savings gained by adding another machine to the process must be greater than the additional expenses. The savings, however, must not be in one particular operation.

The extra expenses, later to be referred to as transfer time, are of two types: additional time for each part in the batch (e.g., chucking) and additional time for the batch (e.g., set up). Thus the transfer time is a function of the quantity to be produced. It is therefore possible that different process plans will result for different quantities. The bigger the batch quantity, the lower the transfer time and thus the higher the profitability of selecting the best machine for each specific operation. Naturally, in each case the sequence of operations might be different. Machine time and cutting conditions must be practical. The specified spindle speed must be available on the selected machine and so must the power and feed.

The cutting speeds in the operation list are computed on the basis of a 60-minute tool life. (We did so, since the parameters needed to compute economical tool life, e.g., machine hourly rate, were unknown.) In this module, where machine details are known, the economical cutting speed for each individual machine can be computed. The economical cutting speed will vary from machine to machine, according to its hourly rate and the optimization criterion. Power is a linear function of cutting speed; thus a particular operation might call for a different cutting speed and power when performed on a different machine. The power required must be available on the machine; if not, the cutting conditions should be modified to suit the particular machine. All of these modifications will result in a different machine time and cost for each operation when performed on different machines.

Thus the mathematical problem can be defined as follows: Given M machines and N operations that must be performed, machine time and cost of each operation on each machine is known (that differ from one another); the sequence of operations is unknown but must conform to certain constraints. The problem is to determine the sequence of operations required to meet minimum cost or maximum production criteria, which machines to use, and which operations to perform on each selected machine.

As previously shown, the number of combinations in a problem of this type is $N! \times M^N$, which is almost infinite. The first step is to find a way to solve this problem in a short time and economically.

13.2 Mathematical Methods Review

The mathematical field that handles problems of the nature stated in Section 13.1 is operations research. The allocation problem as handled by R. L. Ackoff (1) resembles the problem at hand. A typical allocation problem is as follows: Given a number of jobs to be performed and a finite number of resources, assign resources to the jobs in a way that will result in minimum cost or maximum returns.

The problem is presented in matrix form, where one side of the matrix represents jobs, while the other side represents resources (see Figure 13.2). The content of the matrix is the cost of each job with respect to each resource. The best method to solve the matrix problem is dynamic programming. This method is based on R. Bellman's theory, which states that the solution is a series of decisions, starting from the last row (i) of the matrix and working all the way up to the first row. In order to use this technique, it is necessary that the payoffs from each decision be additive and that no matter how a state arose, the consequences for the future are the same.

The following example demonstrates this logic. At an intermittent point of the series of decisions it was decided that job 3 is to be performed on resource 4 (see Figure 13.2). At this stage we can ignore how and why we reached this decision. The problem at hand is: To where should we proceed from this point, that is, on which resource should we perform job 2. This problem can be expressed mathematically as finding which of the following expressions has the minimum value:

$$T_{2,1} = T_{3,4} + C_{2,1}$$
$$T_{2,2} = T_{3,4} + C_{2,2}$$
$$T_{2,3} = T_{3,4} + C_{2,3}$$
$$\cdot \quad \cdot \quad \cdot$$
$$\cdot \quad \cdot \quad \cdot$$
$$\cdot \quad \cdot \quad \cdot$$
$$T_{2,j} = T_{3,4} + C_{2,j}$$

This is a finite problem and can be solved easily and fast. The number of combinations to be solved by this method is $N \times M$.

Generative Process Planning—Mathematics

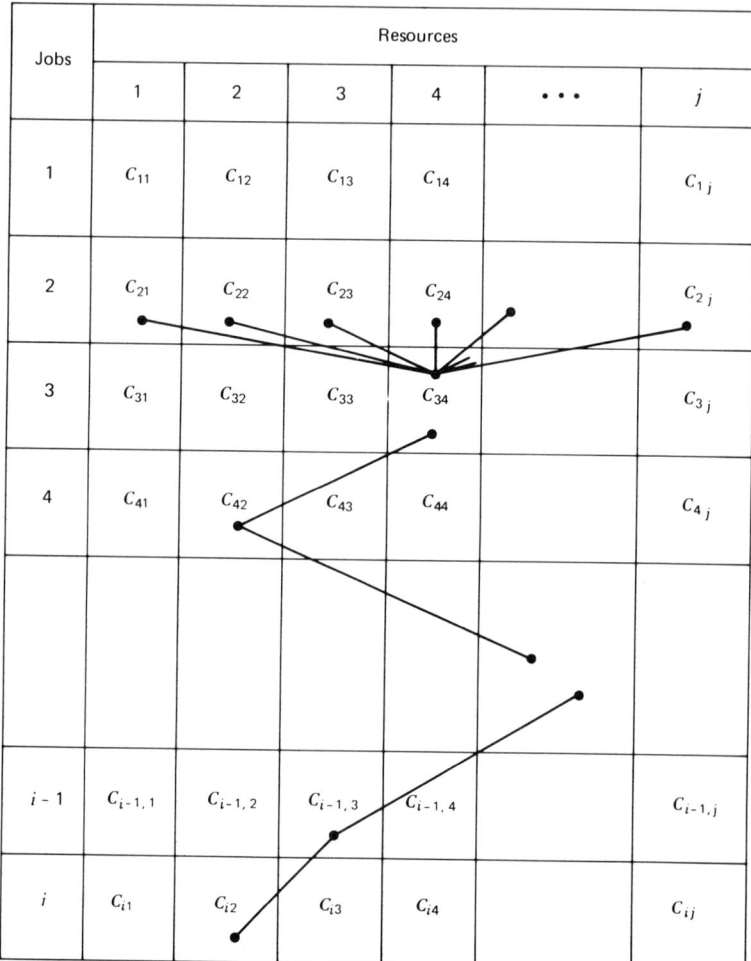

Figure 13.2 Dynamic programming procedure.

Bellman's theory is based on two assumptions:

1. The total value is an accumulation of the individual job values. This assumption is true in our case, with the exception of transfer time.
2. The total value at any point is independent of the path by which it was arrived at. This assumption does not hold in our case. Since the jobs (operations) are interdependent and their sequence may or must be changed in order to reach an optimum, the path is one of the decisions to be reached by the solution.

Another approach might be using the sequencing and coordination problem (1). This method attempts to solve PERT capacity planning and queuing problems. However, the existing method is capable of solving only two-resource problems. This area is not yet fully developed mathematically.

Another approach is to apply the routine problems in network (1). The problem that this method attempts to solve is to find the path between two or more points when the optimization is a desirable variable value. One may insert such constraints on the path as: it is allowed to cross any point only once or the path must pass through all the points. The classical application of this method is with respect to the traveling salesman problem. Attempts to find an analytical solution to this problem were not very successful. The solution is arrived at by a matrix and a special algorithm, which leaves much to be desired.

It seems clear that in order to solve the problem at hand, a machine–operation matrix must be used. A discrete solution must be composed. It might, however, be based on some of the ideas discussed.

13.3 Constructing the Machine–Operation Matrix

The matrix will be used to solve the problem of machine and sequence of operations selection. The machines (j) will be listed horizontally and the required operations (i) vertically. The time or cost of operation T_{ij} will be entered accordingly. The operations as specified in the first module did not (and could not) consider machine constraints. At this point, each machine is handled separately. When all of its characteristics are known, adjustment to the theoretical operation can take place. The entry T_{ij} in the matrix is a practical time or cost for each operation.

The adjustments of the operations are discussed in the following sections.

Adjusting Economical Cutting Speed

The cutting speed in the first module was based on a 60-minute tool life. This was done because the economical cutting speed is machine-dependent. The cutting speed was computed by eq. 12.2:

$$V = \frac{C_v(B/5)^f}{(1{,}000\ R)^z (T_L/60)^y}$$

where the value of $(T_L/60)^y$ was assumed to be one.

The economical tool life for the minimum cost criteria is given by eq. 11.6:

$$T_L = \frac{C_3}{C_2}\left(\frac{1}{n} - 1\right)$$

However, for the maximum production criteria it is

$$T_L = \left(\frac{1}{n} - 1\right)t_c \qquad (13.1)$$

where t_c is the time to replace the cutting tip.

These values can now be computed and the cutting speed adjusted by

$$V_{ij} = V \frac{1}{(T_L/60)^y} \qquad (13.2)$$

where V_{ij} = specific operation and machine cutting speed
V = theoretical cutting speed (equal for all machines)
T_L = computed value for a specific machine (by eq. 11.6 or 13.1)
y = exponent.

The adjusted cutting speed is then converted to spindle speed for further adjustments.

First Power Adjustment

Power is a linear function of cutting speed:

$$Q_{ij} = KV_{ij}$$

Thus the adjustment factor used for cutting speed should be applied to adjust power. Thus

$$Q_{ij} = Q_i \frac{1}{(T_L/60)^y}$$

where Q_{ij} = power required for operation i on machine j
Q_i = theoretical power (equal for all machines).

Final Cutting Speed and Power Adjustments

The adjustments of cutting speed and power previously made are theoretical or semipractical. They were computed to match a particular machine, but it was not checked if the required speed and power were available.

13.3 Constructing the Machine–Operation Matrix

The required spindle speed is adjusted to the available speeds on the machine. It is permissible to increase speed up to 10%. If such a speed does not exist, use the nearest lower available speed. If the required speed is above the range of the machine, the maximum machine speed is selected. If this speed is below the lower cutting speed limit, the machine is not capable of performing this operation (see following discussion on machine accuracy). The machine remains as an alternative, because it might be best suited to another operation.

The power is adjusted according to the practical selected speed. If the resulting power is equal to or less than the machine power, the selected cutting conditions are in effect. However, if the required power is above that available, a further adjustment must be made—the power must be reduced. There are several courses of action. Rules must be made to keep the process at optimum for that machine. The machining power requirement can be computed by the general equation

$$Q = K_4 V P \tag{13.3}$$

where Q = power (hp)
K_4 = coefficient that includes the efficiency factor
V = cutting speed (m per minute)
P = cutting force (N).

The cutting speed can be expressed as a function of spindle speed:

$$V = \pi D N = K_1 N \tag{13.4}$$

where D = machined segment diameter (mm)
N = spindle speed (RPM)

The cutting force, from metal-cutting theory, is

$$P = K_3 A^x S^y = K_2 S^y \tag{13.5}$$

where A = depth of cut (mm)
S = feed rate (mm per revolution)
x, y = exponents
K_2, K_3 = constants.

Introducing eqs. 13.4 and 13.5 into eq. 13.3, we obtain

$$Q = K_4 \times (K_1 \cdot N) \times (K_2 \cdot S^y) = K N S^y \tag{13.6}$$

This equation can be written in another form:

$$Q = K N S^y = K \frac{NS}{S} S^y = K \frac{NS}{S^{1-y}} \rightarrow NS = \frac{Q S^{1-y}}{K} \tag{13.7}$$

The machining time can thus be expressed as a function of power:

$$T = \frac{L}{NS} = \frac{KL}{QS^{1-y}} \tag{13.8}$$

where T = machining time (minutes)
L = machining length (mm).

Equation 13.8 tells us that as the power increases, machining time decreases. However, the tool and the strength of part material limit the maximum power that can be exerted in any single operation. This maximum power is the one specified by the adjustment previously made. A practical conclusion from eq. 13.8 is that there is no advantage in employing a machine that has power in excess of requirements.

When machine power is less than that required by the theoretical optimum, the cutting speed or feed rate can be reduced in order to arrive at a practical optimum.

If Q_N designates the available machine power and is used as a limiting value, the machining time as expressed in eq. 13.8 becomes

$$T_1 = \frac{KL}{Q_N S^{1-y}} \tag{13.9}$$

In this case it is assumed that the reduction of machine power is achieved by reducing the cutting speed. Another approach is to maintain the cutting speed and reduce the feed rate to a value S_N. At this feed rate the resulting power is Q_N. (The third approach—reducing the depth of cut—is not considered here, since it means adding an operation. This case was previously discussed. It will be used only if the reduction in cutting speed and feed rate decreases their values below their lower limits.

In such a case, the machining time is

$$T_2 = \frac{KL}{Q_N S_N^{1-y}} \tag{13.10}$$

If we take the ratio of the machining times that result from the above two methods, we obtain

$$\frac{T_2}{T_1} = \frac{KL}{Q_N S_N^{1-y}} \times \frac{Q_N S^{1-y}}{KL} = \left(\frac{S}{S_N}\right)^{1-y} \tag{13.11}$$

Since, by definition, S is always greater than S_N, T_2 will always be greater than T_1. This means that it is more profitable to reduce the cutting speed than to reduce the feed rate. Reducing the cutting speed will also result in an increased tool life. A practical conclusion is that power adjustment should be made according to the following priorities:

13.3 Constructing the Machine–Operation Matrix

1. Reducing the cutting speed down to its lower limit value.
2. Reducing the feed rate down to its lower limit value.
3. Reducing the depth of cut, that is, split the cut into more than one pass and readjust cutting speed and feed rate.

Spindle Bore Constraint

If the spindle bore diameter of the machine is less than the part diameter, the part cannot be inserted. This means that the free length in chucking is not controlled by the chucking location, but rather by the machine. In such a case, the allowable bending forces should be adjusted to the new conditions, and the cutting conditions have to be recomputed. This will affect the machining time and thus the selection of machine and sequence of operations. The spindle bore constraint is a go–no go constraint; thus the first module computes the cutting conditions for the two cases. The adjustment of cutting speed and power previously discussed will use the appropriate cutting condition set.

Maximum Depth of Cut

Each individual machine has its maximum depth of cut value. Chatter might appear above this value. This is a constraint we must accept. If the operation depth of cut is greater than the maximum value, it must be reduced. The reduction can be made by changing the cut distribution or by splitting the operation into several steps. In either case, the machining time is affected and must be adjusted accordingly.

Maximum Torque Constraints

Some machines are defined by power and maximum allowable torque on the spindle. On such machines the operation torque should be examined. If it is higher than allowed, it must be reduced. A first attempt should be made by reducing the feed rate. If this does not work, reduce the depth of cut. The machining time is affected by this constraint and should be adjusted accordingly.

Machine Accuracy Constraint

Some machines might be worn out and thus their accuracy capabilities below standard. The machine detail record includes such information. If the part tolerance is lower than machine capabilities, the machine cannot perform that operation. However, the machine might be suitable for other operations (rough cut). Therefore, this machine remains in the matrix. As a common practice, for all incapabilities a value of $T_{ij} = 999$ is inserted. This will eliminate the selection of this machine for the performance of an operation it is incapable of, while at the same time considering it for other operations. The inserted value is much higher than the transfer time (TRN); thus

$$T_{i-1,j} + T_{i,j} + T_{i+1,j} > T_{i-1,j} + TRN_{i,j+n} + T_{i,j+n} + TRN_{i+n,j} + T_{i+1,j}$$

and the lower value path is selected.

Handling Time

The specified operation list contains an operation code. A handling time table is constructed in such a way that the operation code and machine code number are the search arguments.

This handling time can be added to the machining time in the matrix if so desired. Thus the machine selection solution will take into account total machine time (which is especially important if NC turret lathe and universal machines are available).

Time and Cost Conversion

The optimization criterion can be either maximum production or minimum cost. Maximum production actually means minimum machining time. Thus both criteria call for a path that results in a minimum value. In order to use one solution method, the content of the matrix is altered from time to cost. Time is an engineering value and is determined by the cutting conditions. Cost is the multiplication of time by the hourly rate. The machine detail record contains this information.

13.4 Preliminary Machine Selection

Solving the problem of machine selection is basically a mathematical combinatory problem. The straightforward solution is to generate all the alternative combinations and then—according to the optimization criterion—select the optimum alternative. It is, as previously shown, a huge job. In order to lessen the number of alternatives and thus reduce solution time, a preselection of machines will be made. This preselection should be made so as not to affect the optimum solution.

13.4 Preliminary Machine Selection

First Step in Machine Selection

The first step in machine selection is based on the type of machine and its physical dimension. For round symmetrical parts, for example, only the lathe machine group is considered. A lathe machine is partially defined by the swing over bed, swing over cross slide, total cross slide travel, and so on. These are the physical dimensions of the machine. A part with a radius greater than the swing over bed cannot be gripped on the machine. A part whose length is greater than the machine bed cannot be turned on that machine.

In the first step of machine selection, all the machines that cannot accommodate the part, or cannot perform even one of the required operations, are excluded from further consideration. No machine is excluded because of a single incapability, such as small spindle bore, insufficient accuracy, or power. If machines are excluded for such reasons, the optimum can be jeopardized or no machines will be left after the first step in selection. The optimum we are looking for is a part optimum and not a single-operation optimum. It involves a "trade-in," that is, the best compromise between operations. The mathematical module and not the preselection of machines should choose the compromise. Hence, in the first step of machine selection only those machines that absolutely cannot perform any one of the operations are excluded from further consideration.

Second Step in Machine Selection

Equation 13.8 expresses machining time as a function of power:

$$T = \frac{KL}{QS^{1-y}}$$

This equation, eqs. 13.10 and 13.11, and the conclusions from the power adjustment lead us to the following selection rules:

- A machine whose power is lower than the minimum required does not stand a chance of being selected for the job. Therefore, it can be excluded from further consideration, without affecting the optimum process plan, unless there are no other machines.
- A machine with more power than required has no advantage over lower-power machines. Therefore, such a machine can be excluded from further machine selection consideration, unless its hourly rate is lower than the lower-power machine or it has a spindle speed higher than that of a lower-power machine and this spindle speed is required by one of the operations.

Operation (i)	Machine (j)					
	1	2	3	...	j	
	Left side					
1	$T_{1,1}$	$T_{1,2}$	$T_{1,3}$		$T_{i,j}$ →	
2	$T_{2,1}$	$T_{2,2}$	$T_{2,3}$		$T_{2,j}$ →	
3	$T_{3,1}$	$T_{3,2}$	$T_{3,3}$		$T_{3,j}$ →	
⋮						
n	$T_{n,1}$	$T_{n,2}$	$T_{n,3}$		$T_{n,j}$ →	
		Side total	three machines: k_n, l_n, and m_n			
	Right side					
n + 1	$T_{n+1,1}$	$T_{n+1,2}$	$T_{n+1,3}$		$T_{n+1,j}$ →	
⋮						
i − 1	$T_{i-1,1}$	$T_{i-1,2}$	$T_{i-1,3}$		$T_{i-1,j}$ →	
i	$T_{i,1}$	$T_{i,2}$	$T_{i,3}$		$T_{i,j}$ →	
		Side total	three machines: k_i, l_i, and m_i			
$T_j = \sum_{i=1}^{i} T_{ij}$	T_1	T_2	T_3		T_j	
	Three machines: k, l, and m					

Three machines for each operation: k_j, l_j, and m_j

Figure 13.3 Third step in machine selection—unlimited.

Third Step in Machine Selection

This third step is employed only if the number of remaining machines is too high. An estimation of a machine's chance of being chosen serves as the basis of this step. The chances are estimated by comparing the machining time or cost for the given machine to those of other machines (see Figure 13.3). Two approaches can be used:

1. *Unlimited selection.* In this approach those three machines that give the lowest values for each operation and for total operations value are selected. The steps are as follows:
   a. Consider the first operation. List the machining times or costs in ascending order ($T_{1,j}$). Select the three machines conforming to the three first times listed.
   b. Store the machine numbers in a list.
   c. Repeat steps a and b for each operation in the matrix.

13.4 Preliminary Machine Selection

d. Sum up the total machining time on each machine:

$$T_j = \sum_{i=1}^{i} T_{i,j}$$

e. Arrange the above totals in ascending order.
f. Select the three machines conforming to the three first times listed.
g. Store the machine numbers.
h. Sum up the total machining time of all operations to be performed by chucking the part on its left side (for each machine):

$$T_j = \sum_{i=1}^{n} T_{i,j}$$

i. Arrange the above totals in ascending order.
j. Select the three machines conforming to the three first times listed.
k–m. Repeat steps h–j for operations to be performed by chucking the part on its right side:

$$T_j = \sum_{i=n+1}^{i} T_{i,j}$$

n. Examine the list of selected machines and eliminate duplication. The machines remaining on the list are considered for the final step in machine selection.

2. *Limited selection.* By this approach a predetermined number of machines is to be selected. Experience has proved that 8–15 machines selected by the following rules give reasonable results. The selection procedure is as follows:

   a. Sum up total machining time on each machine:

$$T_j = \sum_{i=1}^{i} T_{i,j}$$

   b. Arrange the above totals in ascending order.
   c. Select the three machines conforming to the three first times listed.
   d. Sum up the total machining time of all operations to be performed by chucking the part on its left side (for each machine):

$$T_j = \sum_{i=1}^{n} T_{i,j}$$

   e. Arrange the above totals in ascending order.
   f. Select the three machines conforming to the three first times listed. Add these machines, if they were not previously selected, to the

selected machine list and keep track of the total number of machines on the list.

g. Repeat step d for operations to be performed by chucking the part on its right side:

$$T_j = \sum_{i=n+1}^{i} T_{i,j}$$

h–i. Repeat steps e and f.

j. Arrange the operations in descending order of theoretical machining time. The following steps will be repeated, starting from the top of the series (from the operation that consumes the largest amount of time down to the last operation, i.e., the shortest).

k. Arrange the machining times or costs of the operation under consideration in ascending order.

l. Select the three machines conforming to the three first times listed.

m. Add these machines, if they were not previously selected, to the selected machine list. Keep track of the number of machines on the list.

n. If the number of machines on the list equals the predetermined number, stop this procedure. If not, repeat steps k, l, and m down to the last operation.

Figure 13.4 shows the above machine selection procedure in block diagram form.

13.5 Operation Arrangement in Matrix

The list of operations resulting from module one is by no means arranged in any reasonable machining sequence. Yet the solution of the matrix is heavily dependent on the initial operation arrangement. Some of the operations are interdependent. A countersink operation must precede a drill operation. If the part is chucked with tailstock center, the countersink operation must precede the external turning operation. With a free-end chucking, the external operations are independent of the internal ones, and the external turning operation can precede the countersink. This information is vital in solving the matrix, since the decision whether to add transfer time or to alter the sequence of operations depends on it.

The interdependence of operations is treated by assigning a priority number to each operation. A priority number is the operation number that must precede the particular operation. A priority number of zero means that the operation may be the first one. Assigning the priority number is

13.6 Matrix Solution

done by scanning all the operations on the list and checking for tool access. An example will demonstrate this procedure. Figure 13.5 shows a part and machining operations to be performed on it according to module one. Since part chucking requires a tailstock center, the countersink operations 14 and 18 must be the first operations on their respective sides; they are therefore assigned a priority number of zero. Thus operation 1 can be performed only after operation 14 and operation 2 after operation 1. Operation 5 can be machined after operation 2; however, operation 4 can be machined only after operation 3. The internal operations can be machined independently of the external operations. Thus after the countersink operation 14, either operation 1 or operation 15 can be performed.

The above operation sequence is somewhat ambiguous and may cause logical difficulties in solving the matrix. Rearranging the operation sequence in the matrix might be of some assistance.

Since process planning handles machining of each side of the part separately (each chucking), let us first arrange operations by machining side. External and internal operations are usually independent; therefore, they can be arranged in separate groups. Finish cuts are usually independent from one another. Therefore, it was decided to arrange the operations in the matrix by side and group as follows:

Side one	Rough external
	Finish external
	Rough internal
	Finish internal
Side two	Rough external
	Finish external
	Rough internal
	Finish internal

In each group the operations were arranged by ascending priority number. Figure 13.6 shows the operations of Figure 13.5 in the sequence in which they are introduced into the matrix.

13.6 Matrix Solution

The matrix solution should specify machine selection and the sequence of operations. The solution is a result of generating all alternatives and selecting the optimum one. The approach described earlier introduces a term "absolute optimum" that was missing in production studies. This is a theoretical optimum that probably cannot be reached, but can be used as a yardstick. This theoretical optimum can be used to determine whether a

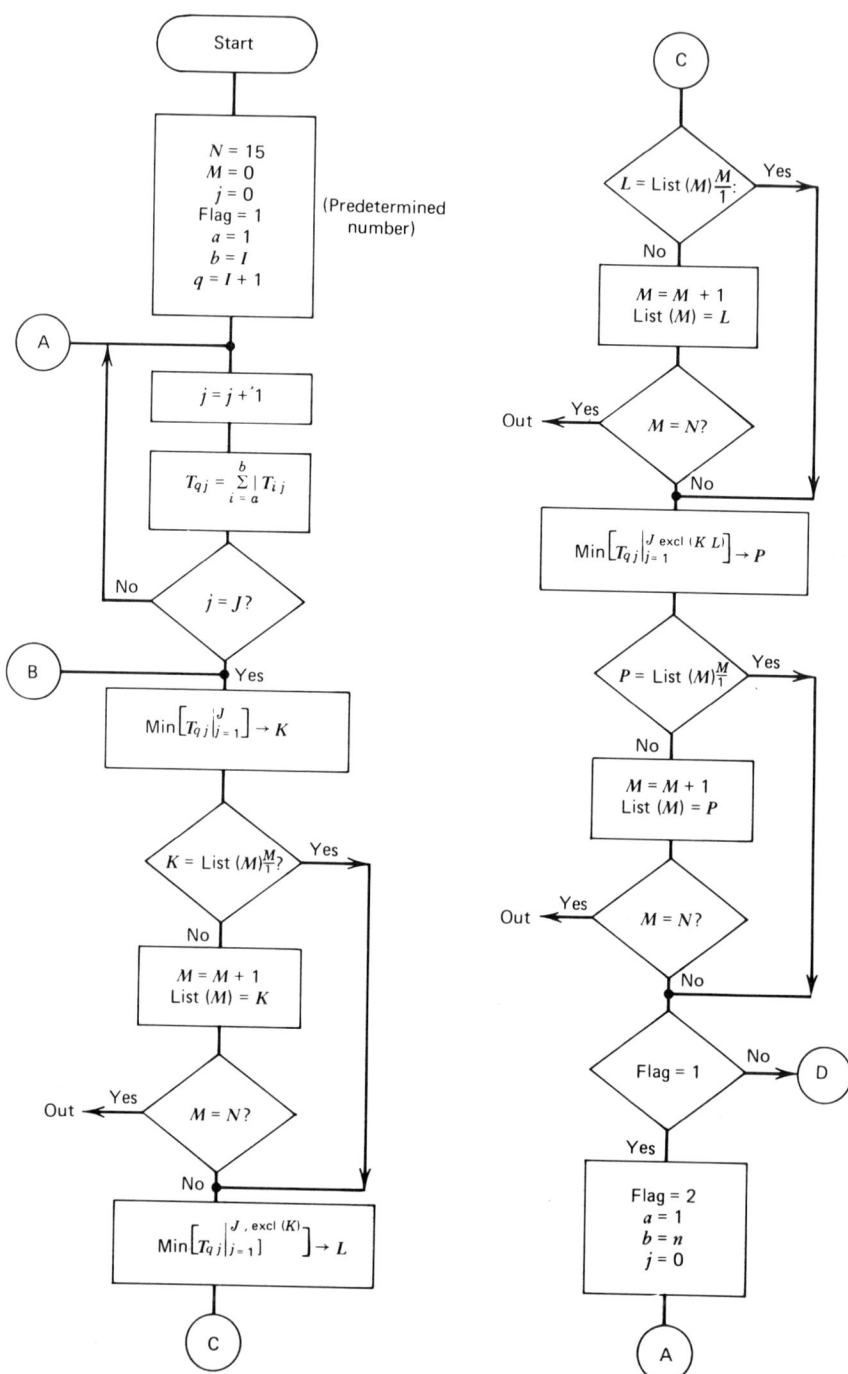

Figure 13.4 Third step in machine selection—limited.

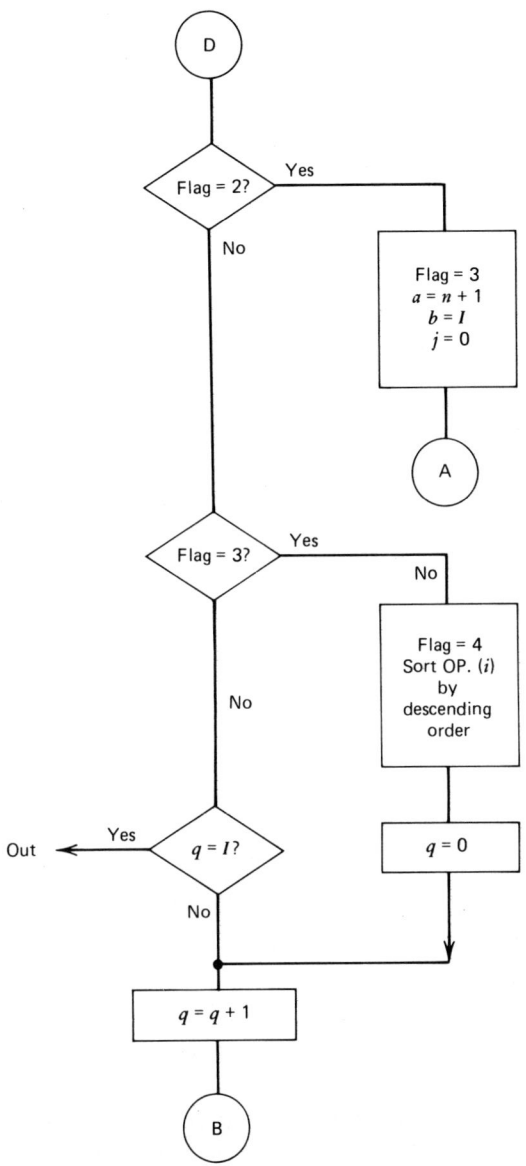

Figure 13.4 *(Continued)*

356 Generative Process Planning—Mathematics

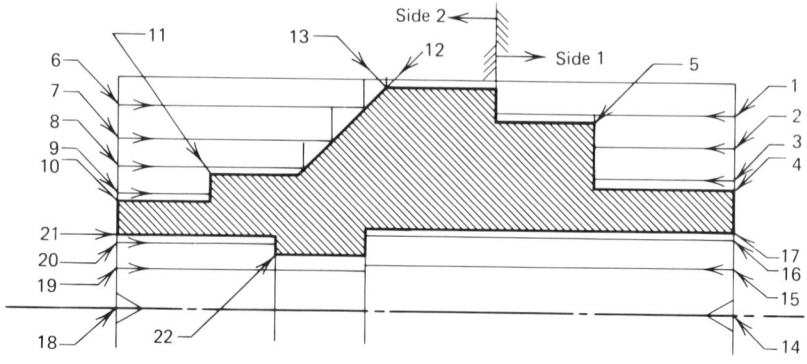

Operation number	Side	Priority number	Operation number	Side	Priority number
1	1	14	12	2	11
2	1	1	13	2	6
3	1	2	14	1	0
4	1	3	15	1	14
5	1	2	16	1	15
6	2	18	17	1	16
7	2	6	18	2	0
8	2	7	19	2	18
9	2	8	20	2	19
10	2	9	21	2	20
11	2	9	22	2	20

Figure 13.5 Priority of operations.

single machine exists that gives an optimum process and thus whether there is any sense in generating and evaluating all other alternatives. If such a machine does not exist, the general matrix solution is employed.

Single-Machine Solution

An imaginary machine is added to the matrix. It is assumed that this is an ideal machine, that is, it has the properties most desirable in any machine present in the matrix. Thus this imaginary machine has a value T_{ij} equal to that of the machine with the lowest value. The minimum value of each operation is thus inserted into this machine column (as shown in Figure 13.7). The vertical sum

$$T_M = \sum_{i=1}^{I} (T_i)_{\min}$$

13.6 Matrix Solution

Operation			Priority		
Number	Initial number	Side	Number	Initial number	Group
1	14	1	0	0	
2	1	1	1	14	I
3	2	1	2	1	
4	3	1	3	2	
5	5	1	3	2	II
6	4	1	4	3	
7	15	1	1	14	III
8	16	1	7	15	
9	17	1	8	16	IV
10	18	2	0	0	
11	6	2	10	18	I
12	7	2	11	6	
13	8	2	12	7	
14	9	2	13	8	
15	13	2	11	6	II
16	10	2	14	9	
17	11	2	14	9	
18	12	2	16	11	
19	19	2	10	18	III
20	20	2	19	19	
21	21	2	20	20	IV
22	22	2	20	20	

Figure 13.6 Rearrangement of operations shown in Figure 13.5.

is thus the theoretical optimum. This is the minimum value (time or cost) we can expect for this part.

The total part value (see Figure 13.7)

$$T_j = \sum_{i=1}^{l} T_i$$

is computed for each machine. Within these total values there exists a minimum—$(T_j)_{min}$. This minimum points to a single-machine solution.

The general solution attempts to arrive at a lower value than $(T_j)_{min}$. However, for each additional machine a transfer time must be added. Therefore, the condition that must prevail in order for a single-machine

Operation (i)	Machine (j)					Imaginary machine
	1	2	3	....	J	$J+1$
1	$T_{1,1}$	$T_{1,2}$	$T_{1,3}$		$T_{1,J}$	$T_{1,\min}$
2	$T_{2,1}$	$T_{2,2}$	$T_{2,3}$		$T_{2,J}$	$T_{2,\min}$
3	$T_{3,1}$	$T_{3,2}$	$T_{3,3}$		$T_{3,J}$	$T_{3,\min}$
I	$T_{I,1}$	$T_{I,2}$	$T_{I,3}$		$T_{I,J}$	$T_{I,\min}$
Total value of a single machine	$(\Sigma\vert T_i\vert)_{,1}$	$(\Sigma\vert T_i\vert)_{,2}$	$(\Sigma\vert T_i\vert)_{,3}$		$(\Sigma\vert T_i\vert)_{,J}$	$(\Sigma\vert T_i\vert)_{,\min}$
	T_1	T_2	T_3		T_J	T_M

Figure 13.7 Addition of an imaginary machine ($J+1$) to the matrix (T_M is the theoretical optimum).

solution to be the optimum is

$$T_M + \mathrm{TRN} \geq (T_j)_{\min}$$

or in another form

$$(T_j)_{\min} - T_M \leq \mathrm{TRN}$$

where TRN is the transfer time value.

General Matrix Solution

The general solution is based (mathematically) on a dynamic programming technique. The basic feature of dynamic programming is that the optimum is reached stepwise, proceeding from one stage to the next. An optimum solution set is determined, given any conditions in the first stage. This optimum solution set, from the first stage, is then integrated with the second stage to obtain a new optimum solution, given any conditions. Then, in a sense ignoring the first and second stages as such, this new optimum solution is integrated in the third stage to obtain still further optimum solutions and so on until the last stage. It is the optimum solution that is carried forward rather than all previous stages.

In the problem at hand the stages are referred to as operations, and decisions are made by choosing the optimum path between any two

13.6 Matrix Solution

operations (stages). However, since the sequence of operations listed in the matrix (and thus the defining stages) is not fixed, this sequence can be changed. One of the problems to be solved is which sequence of operations will result in an optimum solution. Therefore, the general dynamic programming solution procedure has to be modified in order to handle the problem at hand.

The proposed solution is divided into two phases. The first phase is from the bottom up, that is, from the last operation up to the first. It will proceed stage by stage (operation by operation), determining the optimum path (machine selection) for each stage independently of the previous stages. However, at each stage a review of all previous optimum decisions is made in order to examine the effect of the sequence of stages (operations). The sequence that results in a total path optimum is selected.

The second phase is from the top down, that is, from the first operation down to the last. It reviews the optimum achieved by examining the effect of the sequence of stages from any stage up to the first stage (operation). The sequence that results in a total path optimum will be used.

The following describes how this technique is applied. For convenience, two auxiliary matrices are constructed. Thus the total solution uses three machine (j)–operation (i) matrices:

1. Element value T_{ij}
2. Total downward value C_{ij}
3. Path pointers P_{ij}

The computation starts with operation $I - 1$ and machine 1. The problem is as follows: To which machine should we proceed from this point in order to arrive at a minimum value. Since this is the last operation, the transfer time (R_{ij}) should be added when the machine is changed. Thus the alternatives are

$$S_1 = T_{I-1,1} + T_{I,1}$$
$$S_2 = T_{I-1,1} + R_{1,2} + T_{I,2}$$
$$S_3 = T_{I-1,1} + R_{1,3} + T_{I,3}$$
$$S_j = T_{I-1,1} + R_{1,j} + T_{I,j}$$

The chosen path will be where S_j is the minimum value. This minimum value is placed in the total matrix as $C_{I-1,1}$. The path matrix lists the machine number of operation I that results in the above minimum value. Thus $P_{I-1,1} = K$. These values are shown in Figure 13.8.

This process is repeated for operation $I - 1$ and machine 2 (resulting in values for $C_{I-1,2}$ and $P_{I-1,2}$) and so on until machine J and all values of

360 Generative Process Planning—Mathematics

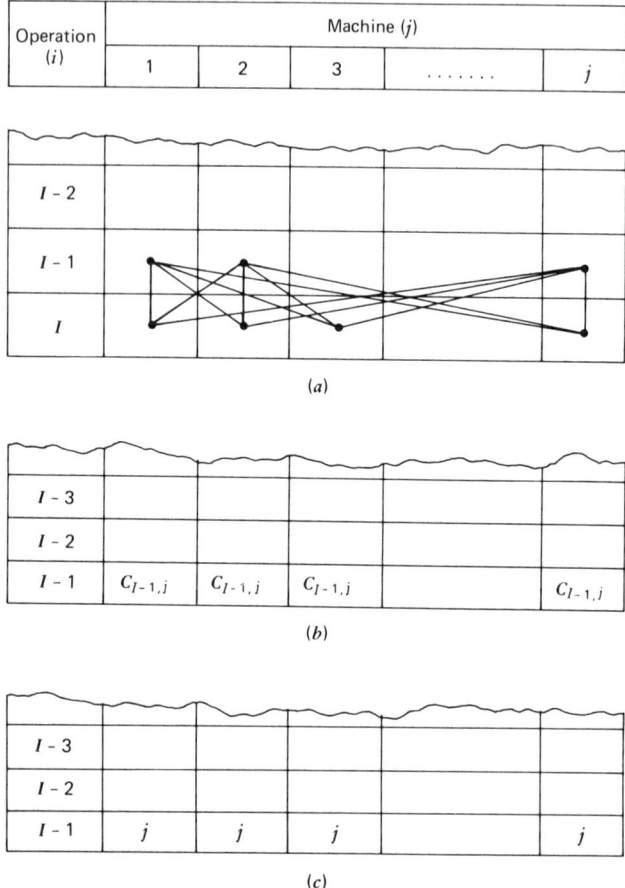

Figure 13.8 General solution matrices. (a) element value matrix $T_{i,j}$; (b) total downward value matrix $C_{i,j}$ (j = machine that results in minimum value); (c) path pointers matrix $P_{i,j}$.

$C_{I-1,j}$ and $P_{I-1,j}$ are computed. Covering all junction points of operation $I - 1$, the solution process proceeds upward to handle all junction points of operations $I - 2$, $I - 3$, and so on until the first operation.

The general junction alternatives to be evaluated can be expressed as

$$S_j = T_{i,j} + C_{i+1,k} \Big|_{k=1}^{J} + R_{j,k}$$

The junction to be evaluated is operation i on machine j. Its time or cost is T_{ij}. From this point it is possible to proceed downward to perform opera-

13.6 Matrix Solution

tion $i + 1$ with one of the available machines. The optimum solution for each machine in operation $i + 1$ is the total $C_{i+1,k}$ and is independent of the path by which it was reached.

The term R_{jk} is the transfer time covering the expenses caused by shifting the work from machine j to machine k. The addition of the transfer time is path-dependent. It is possible that by changing the sequence of operations no transfer time should be added. This case occurs either when the path from the current operation down to the last operation passes through the current machine and the operation that uses that machine can be shifted upward (to be performed right after the current operation) or when the current operation can be shifted downward to be performed right before the other operation. In such cases, transfer time has already been added to the total and no extra transfer time is required.

The information on whether an operation can be shifted upward or downward is made available by the priority code. Figure 13.9 demonstrates such a case (upward phase). The optimum path from operation 10, machine 6 is (i,j) 10, 6; 11, 7; 12, 4; 13, 4, as shown by the heavy line in the figure. Transfer time has been added twice so far. The junction of operation 9, machine 4 is evaluated. One of the alternatives is to proceed to operation 10 on machine 6. This calls for adding $T_{9,4} + C_{10,6} + R_{4,6}$. Scanning down the path reveals that machine 4 is selected for operations 12 and 13. If the sequence of operations can be altered to read 9, 12, 13, 10, 11 (Part 2 of Figure 13.9), only two machine transfer times occur in this path, and no extra transfer time should be added. Thus the total value should be $T_{9,4} + C_{10,6}$. If this is the best alternative for the junction of operation 9, machine 4, the operation sequence will be altered.

This computation and path checking are performed for any alternative. Thus when evaluating the alternative to proceed from operation 9, machine 4 to operation 10 on machine 2, transfer time must be added (see Figure 13.9), since machine 4 does not participate in the above path.

The priority number indicates whether the sequence of operations can or cannot be changed. The values in the first column of the total matrix $(C_{1,j})$ represent the total cost or time required to produce the part when starting with any one of the available machines. The machine chosen for the first operation is the one with the minimum value of $C_{1,j}$. The path matrix will then lead us through the machine selected for the other operations and to the sequence of operations.

The above solution considered changing the sequence of operations by looking downward and saving transfer time. It could not predict the machine selection of the upper part of the matrix. To improve the solution, a second phase of computation is used. In this phase the operations are examined from the first operation down to the last one on the com-

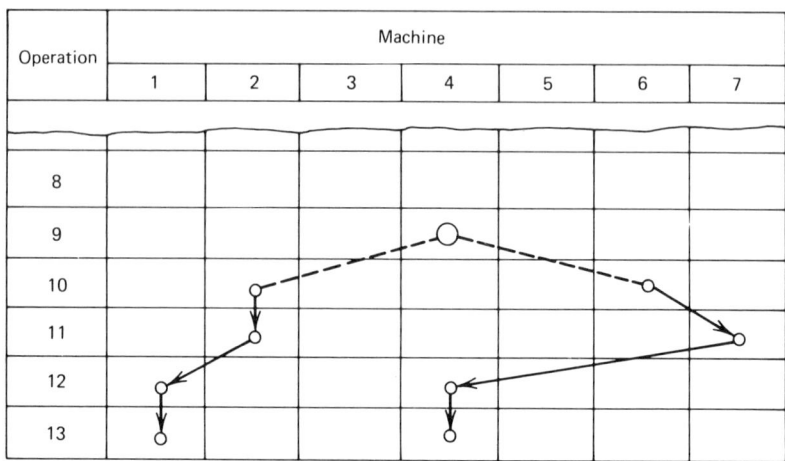

Part 2—Alter sequence of operations

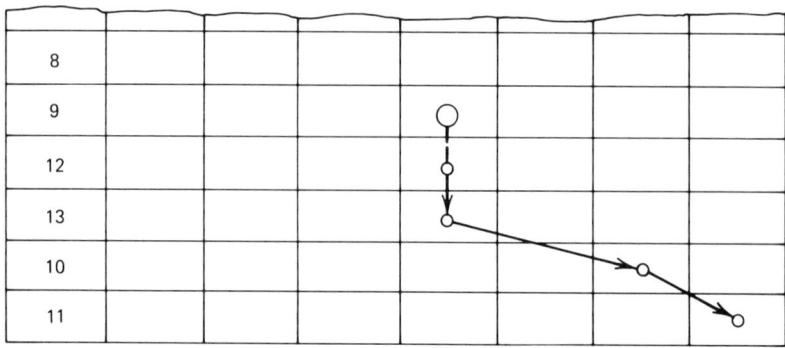

Figure 13.9 Selection of the sequence of operations—upward phase.

puted path to check whether a change in sequence of operations will reduce total machining time and not by eliminating transfer time. Figure 13.10 demonstrates this case. Scanning the total value of operation 1 indicates that machine 4 results in a minimum value. Thus machine 4 is selected for operation 1. The path matrix leads to machine selection for the other operations. This path is shown by a heavy line in Figure 13.10.

Operation 4 has a lower value when performed on machine 4 than when performed on machine 2: $T_{4,4} < T_{4,2}$. However, it was not selected because

$$T_{3,2} + T_{4,2} < T_{3,2} + T_{4,4} + R_{2,4}$$

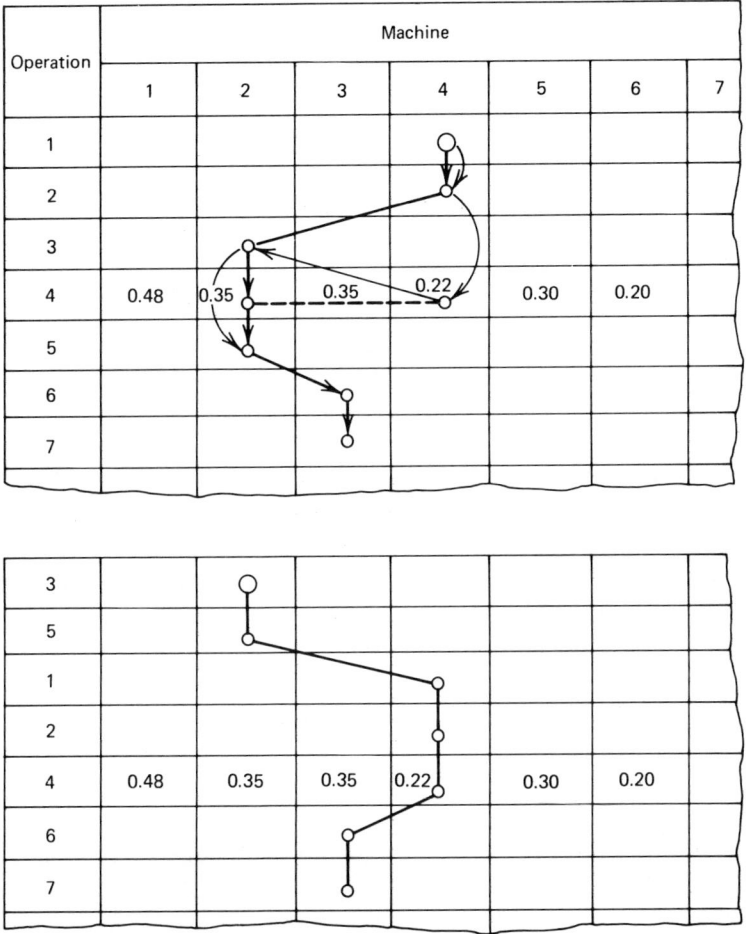

Figure 13.10 Selection of the sequence of operations—downward phase.

The transfer time $R_{2,4}$ must be added, since machine 4 is not available on the lower side of the path. Looking from the top down, we know that machine 4 is available, and if according to the priority code operation 4 can be moved forward, no transfer time should be added. Examining junction 3, 2 results in

$$T_{3,2} + T_{4,2} > T_{3,2} + T_{4,4} + 0$$

Thus the sequence of operations should be modified to read $i = 1, 2, 4, 3, 5, 6, 7, \ldots$, as shown in Figure 13.10.

If operation 4 cannot be processed before operation 3, this change of sequence is not allowed. However, it might be possible to machine operations 3 and 5 prior to operations 1 and 2.

This means that operation 4 is not going to be pulled up, but rather that operations 1 and 2 are going to be pulled down. This will result in a machine selection and sequence of operations as shown in the bottom part of Figure 13.10.

13.7 Summary

The proposed method results in optimum machine selection and sequence of operations. The selection and decision process is purely mathematical and is not affected by intuition or rule of thumb as used today. (An example of a rule of thumb is to use old inaccurate machines for rough cut and accurate machines for finish cut.) The decisions and thus process planning take the following into consideration:

- *Quantity.* The quantity affects the transfer time. The larger the quantity, the lower the value of the transfer time. Thus it is more economical to split the process among several machines, each one suited to some of the operations.
- *Machine capabilities.* Machining time and cost are adjusted for each available machine.
- *Machine incapabilities.* This is introduced by inserting high values for machining time and cost. Thus it is always more advantageous to add transfer time and bypass this machine for operations that require accuracy.
- *Machining type.* Machining time is the same in any machine type, depending only on the power, speed, and so on. The difference lies in handling time. The NC machines perform such operations as adjust tool or change speed much faster than a universal manual machine. Automat requires a long set up, but possesses the ability to do fast return and change tools in a short period of time. The handling time table can have many columns, one for each type of machine. Thus the general matrix solution can handle different machine types.

It was found that this method of solution is easily adapted to computing optimum process planning for various quantities. It calls merely for the computation of a new value of transfer time and resolution of the matrix problem.

Alternatives are easily computed. It merely calls for the removal of a machine from the matrix and resolution of the matrix problem.

This leads to capacity planning, where loaded machines can be removed from the matrix, thereby using alternative process plan for some of the parts and resulting in an even machine loading. This topic will be further discussed in Part III.

References

1. Ackoff, R. L., and M. W. Sasieni, *Fundamentals of Operations Research,* John Wiley & Sons, Inc., New York (1968).

Chapter Fourteen
Generative Process Planning—Summary

Generative process planning is far more powerful than a retrieval-type process planning system. It is a fully automatic system, in which quantity, part tolerances, surface finish, optimization criteria, and machine constraints are integrated. It works with a specially developed part description module based on the traditional engineering drawing technique. It uses shop-dependent data as input and produces a complete process plan automatically. As such, it leads to a new manufacturing concept where engineering design (CAD) and production planning are combined into one system.

A retrieval-type system depends absolutely on classification and coding, which are quite expensive to construct and install, and does not lead to any meaningful automatic system. It is an interactive system in which manual decisions are required. Decisions of a computational nature are made on an intuitive basis, whereas the computer can be used to compute and evaluate all alternatives and thus reach a true optimum decision. For computational-type decisions, the machine (computer) is far superior to humans. For technical and engineering-type decisions, humans are superior to the computer. The GPPP is aware of and utilizes this fact, but in an approach different from that of the retrieval interactive system. In a way, the GPPP is based on human experience and judgment (at least in those areas where science does not present an answer), but on a microlevel. It does not present the problem on a part basis as a retrieval system does, but rather on a detail level. For example, given a 12-mm hole diameter, length 20 mm, and 64 RMS surface finish, how do you machine it? Why? What are the limits? (and so on). The problem and answer are interactive, but in the stage of forming the GPPP. The system will recognize the individual element in any part and, as if interactive, use the reply already existing in the program. This approach follows the GT concept of "a single solution can be found to a set of problems, thus saving time and effort." There is no sense in repeating interactive questions when the answer previously given can be used again and again.

The main features of the GPPP are the separation of computational from noncomputational elements and following a nonloop sequence of decisions. Each decision is a module in the decision tree and can be replaced without disruption of the whole program.

The concepts for the construction of a GPPP are independent of machining type. The mathematical section can be universally used, while the engineering section has to follow the preliminary study of the particular manufacturing field. Turning is used merely for demonstration of the type of preliminary study required.

14.1 Generative Process Planning— Example Demonstration

A computer program based on the above concept and logic was prepared. The inputs to the system are:

- Part definition (geometry and material).
- Raw material geometry.
- List and descriptive parameters of available machines (can be stored in a file).

The output is the optimum practical process plan.

For turning, the system requires about 110K bytes core memory (can be virtual), printer, card reader, and a few tracks of disk space (up to 100, depending on the number of parts to be processed in one batch) or two tape drives. On the IBM 370/148 it takes about 15–20 seconds elapsed time to generate one process.

System Flow

The system consists of seven sections (programs):

1. *Data validity check.* This section ensures that the input data lie within acceptable fields. (Written in Cobol.)
2. *Data table preparation.* This section arranges the input data into a specified form for use in the operation specification section. (Written in Cobol.)

3. *Part drawing (optional).* This section allows a visual check of component shape. (Written in Fortran.)
4. *Operation specification.* This section specifies the theoretically economical operations that are to be performed (the majority of the constraints are eliminated). The limitations of each operation are also coded. (Written in Fortran.)
5. *Coding and sorting.* This section arranges the data into specified arrays for the final manipulation. (Written in Fortran.)
6. *Machine selection and output printing.* This is the major section in the package; it evaluates the optimum manufacturing sequence. (Written in Fortran.)
7. *Summary printing.* (Written in Fortran.)

These seven sections (programs) are linked together so that a complete package is formed, as shown in Figure 14.1. The job control language (JCL) deck of cards to run the job in a batch mode is shown in Figure 14.2.

Sample

The user should describe the part to be process planned (Figure 14.3) on a system form, as shown in Figure 14.4. The form is keypunched, and the cards are inserted into the proper location in the JCL deck. The output of the computer run is a summary report, as shown in Figure 14.5, and, if desired (indicated on a parameter card PC-2), a detail operations list, as shown in Figure 14.6. This list includes machine name, chucking number and type, operations, and their cutting conditions for each machine. The intermittent reports generated by the system are:

- Error report (Figure 14.7).
- Listing of input cards (Figure 14.8).
- Part drawing (Figure 14.9).
- Theoretical operations list (Figure 14.10).
- Machine list and error report (Figure 14.11).
- Handling time table (Figure 14.12).
- Matrix printout (Figure 14.13).

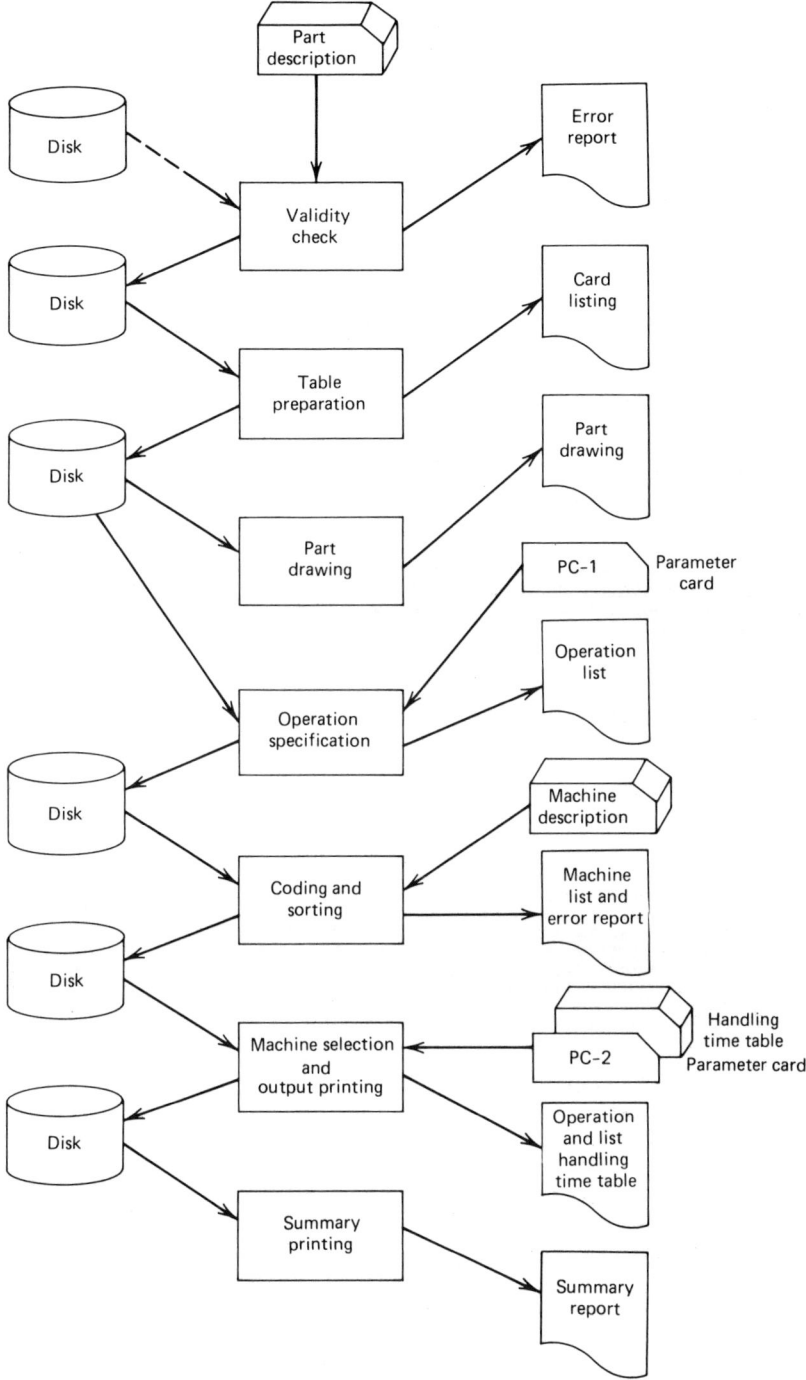

Figure 14.1 System flow.

```
// JOB 6900BEN1 CLLI    81 6
// EXEC SEP
// OPTION NODUMP
// EXEC EHUCS                                              00000200
// ASSGN SYS006,X'151'
// ASSGN SYS005,X'151'
// ASSGN SYS004,X'0CC'
// DLBL UOUT,'HCMAR',70/001
// EXTENT SYS014,SD0000,1,0,19,57
// EXEC CDDK
// UCD TR,FF,A=(80,80),B=(80,1680),E=(3330)
// END
030678MACHST01 ST50     010450001160B10170001000C0000              01G
030678MACHST1C0100200125                                         F 02G
030678MACHSTGA00400C1100 03000050 063      015     3               10G
030678MACHSTGA0025002100 03500050 063 030          2               11G
030678MACHSTGA00100C2    C4000025 063                              12G
030678MACHSTGA0053002    03000            030030   4               13G
030678MACHSTGB00420C1    03700    063       020   3                14G
030678MACHSTTB0025001    02000050 063                              15G
030678MACHSTTB00100.02   C3000                                     16G
030678MACHSTTB0135002    C0000                                   F 17G
030678MACHST    01700 0                  017000                    ZZ99G
010278-1291 02    4340 010160001260B10308600300P0000             F 01G
010278-1291 GA0065202100 00500100 125                              11G
010278-1291 GB0237401100 01400100 125                            F 12G
010278-1291    030860                    000000                    ZZ99G
KONUS-01037801        CK15010650001200B100900000050P0000           01G
KONUS-01037GA0050001100 06000050 125                               10G
KONUS-01037GB0040001100 04000              060000   5              11G
KONUS-010378  CC9000                      000000                   ZZ99G
 RMS-040278 01     ST50 011100001200B10120000020P0000              01G
 RMS-040278 100100100063                                         F 02G
 RMS-040278 GA0040001    10000            025   2                  10G
 RMS-040278 GA0040002    C8000            020   2                  11G
 RMS-040278 GD0040001    06000                                     12G
 RMS-040278 TA00450C1    C4000                                     13G
 RMS-040278 TA0015002    05000                                     14G
 RMS-040278 TB0060001    C3000                                   F 15G
 RMS-040278   012000                      012000                   ZZ99G
/*
*       SCRT
// ASSGN SYS001,X'151'
// ASSGN SYS002,X'151'
// ASSGN SYS003,X'151'
// DLBL SORTIN1,'HOMAR',70/001
// EXTENT SYS002,SD0000,1,0,19,57
// DLBL SORTWK1,'WORK',72/001
// EXTENT SYS003,SD0000,,,779,513
// DLBL SORTOUT,'HOMOR',70/001
// EXTENT SYS001,SD0000,1,0,190,57
// EXEC SORT
 SORT FIELDS=(1,12,A,78,2,A),FORMAT=BI,WORK=1
 RECORD TYPE=F,LENGTH=(80,80,80)
```

Figure 14.2 JCL deck for batch mode.

```
 INPFIL BLKSIZE=1680
 OUTFIL BLKSIZE=1680
 END
// RESET ALL
/*
*        BDIKOT   LOGIOT
// DLBL SYS021,'HJMOR'
// EXTENT SYS001,SDCCOC,1,0,190,57
// DLBL SYS032,'HOMER',70/0C1
// EXTENT SYS014,SD0000,1,0,76,57
// EXEC GDD001,SIZE=(AUTO,50K)
/*
*        JOB BNIT TAVLA
// DLBL SYS022,'HUMER'
// EXTENT SYS014,SDCCOC,1,0,76,57
// DLBL SYS023,'TAVLA',73/0C1
// EXTENT SYS014,SD0000,1,0,133,247
// EXEC GDD005
/*
*        JOB SRTOT HALAK
// ASSGN SYS005,X'151'
// DLBL IJSYS05,'TAVLA'
// EXTENT SYS005,SD0000,1,0,133,247
// EXEC GDD020,SIZE=L0
/*
*        JOB ADPASA
// DLBL TAPEIN,'TAVLA'
// EXTENT SYS014,SDCCOC,1,0,133,247
// EXEC GDD007
/*
*        JOB HISOV PEOLOT
// ASSGN SYS005,X'151'
// ASSGN SYS006,X'151'
// DLBL IJSYS05,'TAVLA'
// EXTENT SYS005,SDCOOC,1,0,133,247
// DLBL IJSYS06,'MOCN',73/0C1
// EXTENT SYS006,SD0000,1,0,380,399
// EXEC GDD011,SIZE=L0
7877085
/*
*        JOB SIMCN IMION   GDD030
// ASSGN SYS005,X'151'
// ASSGN SYS006,X'151'
// DLBL IJSYS05,'MOCN'
// EXTENT SYS005,SD0000,1,0,380,399
// DLBL IJSYS06,'SOFI',73/001
// EXTENT SYS006,SD0000,1,0,779,513
// EXEC GDD030,SIZE=L0

725ATLAS        1245       10''   LATHE 258 H.P.                                01    00G
725250120 60  250  545 160  25525002066001 0      52 700 54105  5 56 605080           101G
725  52  62   75 125 174 237 370 450 540 700                                          911G
727PROGRESS     2515       10''   LATHE 258 H.P.                                01    00G
727250120 60  250  545 160  255250100230015         221600 45    02050  52080         101G
727  22  30   40  60 100 130 190 260 380 500 750 900110014001630                      111G
```

Figure 14.2 (Continued)

```
727    2   3   4   5   6   7   8   9 10 11 12 13 14 15 16 17 18 19 20 21 22 23 24 25    12G
727   26  27  28  29  30  31 32 33 34 35 36 37 38 39 40 41 42 44 46 48 50              913G
709 COLCHESTER  1800              6.5 25 CENTER LATHE 105H.P                         01   00G
709335210145  635  710    95 400100100150014           221400 45       02050  52080     101G
709   22  30  40  60 100 130 190 260 380 500 750 90011001400                            111G
709    2   3   4   5   6   7   8   9 10 11 12 13 14 15 16 17 18 19 20 21 22 23 24 25    12G
709   26  27  28  29  30  31 32 33 34 35 36 37 38 39 40 41 42 44 46 48 50              913G
803CHGAL CM            BF 3         12" LATHE 30HP                                   01   00G
803300200100075012000200150030001590001        71400 40     006201    1100             101G
803    7  70 140 175 280 350 500 700 850100012001400                                    111G
803    6  11  16  21 26  31 36 41 46 51 56 61 66 71 76 81 86 91 96101106111116121       12G
80312613113614114615115616116617117618118619119620 1                                   913G
705ATLAS            1245         10'' LATHE  5 H.P                                   01   00G
705250120  60 250 445  60 255 50 20200012         521700 54105  5 66 601080            101G
705   52  62  75 125 174 237 370 450 540 90012501700                                    911G
801VDF            BF01            10" LATHE  25HP                                    01   00G
8012501500750600100002000600250010850012         482310 40     006201    1100          101G
801   48 120 240 360 600 770 9001200154018002000 2310                                   111G
801    6  11  16  21 26  31 36 41 46 51 56 61 66 71 76 81 86 91 96101106111116121       12G
80112613113614114615115616116617117618118619119620 1                                   913G
708 COLCHESTER 1800              6.5 25 CENTER LATHE 75H.P                           01   00G
708335210145  635  810    95 400350020500015        221600 45       02050  52080       101G
708   22  30  40  60 100 130 190 260 380 500 750 900110014001600                        111G
708    2   3   4   5   6   7   8   9 10 11 12 13 14 15 16 17 18 19 20 21 22 23 24 25    12G
708   26  27  28  29  30  31 32 33 34 35 36 37 38 39 40 41 42 44 46 48 50              913G
802SOMUA           BF02           12" LATHE  30HP                                    01   00G
80230020010007501200020008103000159000 12       71400 40     006201    1100            101G
802    7  70 140 175 280 350 500 700 850100012001400                                    111G
802    6  11  16  21 26  31 36 41 46 51 56 61 66 71 76 81 86 91 96101106111116121       12G
80212613113614114615115616116617117618118619119620 1                                   913G
804 CASANEUVE                    LATHE 7.5 HP                                        01   00G
8044003501001000090001250600075010800012     00461750072      004065    1050           101G
80404600650110016002300280031003900650094013001750                                      111G
804  04005006007008009010011012013014015016017018019020021022023024025026027            12G
804  28029030031032033034035036037038039040041042043044045046047048049050051            13G
804  520530540550560570580590600610620630640650660670680690700710720730740750           914G
/*
*          JOB MACHINE SELECTION
// ASSGN SYS007,X'151'
// ASSGN SYS008,X'151'
// DLBL IJSYS06,'SOFI'
// EXTENT SYS006,SD000C,1,0,779,513
// DLBL IJSYS05,'GMRI',73/001
// EXTENT SYS005,SD0000,1,0,1292,513
// DLBL IJSYS07,'TVLAT MCONOT',72/149
// EXTENT SYS007,SD0000,1,,1805,57
// DLBL IJSYS08,'RCCS',76/001
// EXTENT SYS008,SD0000,1,0,1862,57
// EXEC GDD040,SIZE=LO
   75   77
    1ROUGH CUT          ROUGH BORING                                                  HNTG
    2FINISH CUT         FINISH BORING                                                 HNTG
    3                                                                                 HNTC
    4WIDE SLOT ROUGH WIDE SLOT ROUGH                                                  HNTG
    5WIDE SLOT FINISHWIDE SLOT FINISH                                                 HNTG
```

Figure 14.2 *(Continued)*

```
 6REV. ROUGH CUT   REV. ROUGH BORE           HNT G
 7REV. FINISH CUT  REV. FINISH BORE          HNT G
 8                                           HNT G
 9                                           HNT G
10                                           HNT G
11                                           HNT G
12                                           HNT G
13                                           HNT G
14CROSS ROUGH CUT  CROSS ROUGH BOR           HNT G
15CROSS FINISH CUT.CROSS FIN. BORS           HNT G
16REV. ROUGH CUT   REV. ROUGH BORE           HNT G
17REV. FINISH CUT  REV. FINISH BORE          HNT G
18                                           HNT G
19                                           HNT G
20                                           HNT G
21            ROUGH DRILL                    HNT G
22            FINAL DRILL                    HNT G
23            DRILL                          HNT G
24                                           HNT G
25                                           HNT G
26                                           HNT G
27                                           HNT G
28                                           HNT G
29                                           HNT G
30                                           HNT G
31            FIRST REAM                     HNT G
32            FINAL REAM                     HNT G
33                                           HNT G
34                                           HNT G
35                                           HNT G
36                                           HNT G
37                                           HNT G
38                                           HNT G
39                                           HNT G
40                                           HNT G
41                                           HNT G
42                                           HNT G
43                                           HNT G
44FACE CUT                                   HNT G
45SLOT                                       HNT G
46                                           HNT G
47                                           HNT G
48                                           HNT G
49                                           HNT G
50                                           HNT G
51RADIUS                                     HNT G
52RADIUS                                     HNT G
53RADIUS                                     HNT G
54RADIUS                                     HNT G
55RADIUS                                     HNT G
56RADIUS                                     HNT G
57RADIUS                                     HNT G
58RADIUS                                     HNT G
59                                           HNT G
60                                           HNT G
```

Figure 14.2 *(Continued)*

```
 61ROUGH STEP CONE                                              HNT G
 62                                                             HNT G
 63                                                             HNT G
 64                                                             HNT G
 65CONE BY FORMES T.                                            HNT G
 66                                                             HNT G
 67                                                             HNT G
 68                                                             HNT G
 69                                                             HNT G
 70                                                             HNT G
 71CONE ROUGH CUT                                               HNT G
 72TAPER FINAL CUT                                              HNT G
 73                                                             HNT G
 74                                                             HNT G
 75                                                             HNT G
 76                                                             HNT G
 77                                                             HNT G
 78                                                             HNT G
 79                                                             HNT G
 80                                                             HNT G
 81                                                             HNT G
 82                                                             HNT G
 83                                                             HNT G
 84                                                             HNT G
 85             COUNTERSINK                                     HNT G
 86                                                             HNT G
 87                                                             HNT G
 88                                                             HNT G
 89                                                             HNT G
 90SET UP TIME     MIN. DIV BY 10  2500400099000200             HNT G
 91HANDLING WORK   LIROT DIV BY 10 1200050010009900             HNT G
 92TRANSFER EXPENS.LIROT DIV BY 10 0500050005000500             HNT G
 93                                                             HNT G
 94                                                             HNT G
 95                                                             HNT G
 96                                                             HNT G
 97                                                             HNT G
 98                                                             HNT G
 99                                                             HNT G
100                                                             HNT G
101CHUCH              100         0077007700770077              HNT G
102CHUCH              250         0110011001100110              HNT G
103CHUCH              450         0130013001300130013 0         HNT G
104CHUCH             1200         018001800180018 0             HNT G
105CHUCH             2250         022002200220022 0             HNT G
106CHUCH             5000         0300030003000300              HNT G
107CHUCH             9999         0370037003700370              HNT G
108                                                             HNT G
109                                                             HNT G
110GAGE-MEASURE                   024002400010001 0             HNT G
111GAGE-MEASURE                   012001200010001 0             HNT G
112                                                             HNT G
113                                                             HNT G
114                                                             HNT G
115                                                             HNT G
```

Figure 14.2 *(Continued)*

```
116                                             HNTG
117                                             HNTG
118                                             HNTG
119                                             HNTG
120START-STOP MACH.     C04800480010 0010       HNTG
121ENGAGE FEED          0030003000100010        HNTG
122ADJUST TOOL          0126003000100010        HNTG
123REV.CROSS FEED       0084002000100010        HNTG
124CHNGE TOOL BIT       0390039003900390        HNTG
125INDEX 4P HOLDER      0036003600100010        HNTG
126CHANGE TOOL CHUK     0204015000100010        HNTG
127TAILSTOCK PO.        0270001000100010        HNTG
128CROSE STEADY RST     0030003000100010        HNTG
129SET COM.TO ANGLE     0816010000100010        HNTG
130SQUAR CORNER         0102010201200102        HNTG
131CHNGE FEED CR SP     0030003000100010        HNTG
132RETURN PER INCH      0012001200120012        HNTG
133                                             HNTG
134                                             HNTG
135                                             HNTG
136                                             HNTG
137                                             HNTG
138                                             HNTG
139                                             HNTG
140                                             HNTG
141                                             HNTG
142                                             HNTG
143                                             HNTG
144                                             HNTG
145                                             HNTG
146                                             HNTG
147                                             HNTG
148                                             HNTG
149                                             HNTG
150                                             HNTG
/*
*       JOB HADPASAT RIKUZ
// ASSGN SYS008,X'151'
// DLBL IJSYS08,'RCOS',76/001
// EXTENT SYSOC8,SDCQ00,1,0,1862,57
// EXEC GDD050,SIZE=LO
/*
/&
```

Figure 14.2 (Continued)

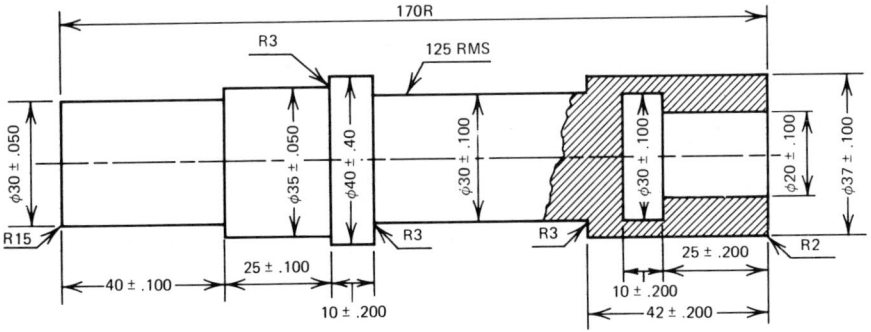

Figure 14.3 Part drawing: Part no. 030678 MACHST; Material, St 50; surface finish, 63(125) RMS; hardness, 160 RB.

Part number: `0 3 0 6 7 8 M A C H S T` (1-12)

HAL-T1
Page 1   G-Halevi
Date 16.1.1979
Programmer

Part Description Form

Entry type	Side	Length dimension X Length xxxx.x	Code	Tolerance .xxx	Side	Diameter dimension Y Length xxxx.x	Code	Tolerance .xxx	Side	Dimension Z Length xxxx.x	Code	Tolerance .xxx	Surface finish	Shape code	Dimension type	Special code dimensions (50-77)	Sequence number 78-79	End code 80
13	14	15-19	20	21-23	24	25-29	30	31-33	34	35-39	40	41-43	44-46	47-48	49			
G	R	170.0	R	200								100	063	01	1	5 T 50	01	G
M	L	40.0	1	100		40.0	1	100						11	2	15	02	
M	L	25.0	2	100		30.0	2	050						11	2	30	03	
M	L	10.0	2			25.0	2	050									04	
M	R	42.0	1			40.0	2	040						11	2	20	05	
M	R		2			37.0								11	1	30	06	
V	R	25.0	1			30.0							125				07	
V	R	10.0	2			20.0											08	F
						30.0												

Figure 14.4 Part description on a system form.

376

SUMMARY REPORT

PART NUMBER	QUANTITY	CRIT	ALT	SIDE 1	MACHINE	TIME	COST	SIDE 2	MACHINE	TIME	COST
030678MACHST	50.	P	0		708	2.41	2.01		801	5.46	7.73
				8.87	MIN.			11.55	LI.		
030678MACHST	50.	P	1		708	2.41	2.01		803	5.69	8.53
				9.10	MIN.			12.39	LI.		
030678MACHST	1000.	P	0		708	2.41	2.01		801	5.46	7.73
				7.92	MIN.			9.84	LI.		
030678MACHST	1000.	P	1		708	2.41	2.01		803	5.69	8.53
				8.15	MIN.			10.64	LI.		
030678MACHST	50.	C	0		705	3.04	1.01		709	6.02	1.50
									705	1.27	0.42
				11.84	MIN.			4.42	LI.		
030678MACHST	50.	C	1		705	3.04	1.01		727	3.39	1.30
									705	3.14	1.05
				11.07	MIN.			4.91	LI.		
030678MACHST	1000.	C	0		709	1.15	0.29		709	6.02	1.50
					705	1.68	0.56		705	1.27	0.42
				10.22	MIN.			2.87	LI.		
030678MACHST	1000.	C	1		727	0.98	0.37		727	3.39	1.30
					705	1.68	0.56		705	3.14	1.05
				9.28	MIN.			3.38	LI.		

Figure 14.5 Summary report.

ALT. 0   PART NO 030678MACHSI   OPERATIONS LIST   FOR 1000. PIECES. CRITERION C   16/ 1/79 PAGE 1
                                                                                                                DD/MM/YY

SIDE 2   ON MACHINE NO 709   COLCHESTER 1800   6.5 25 CENTER LATHE 105H.P   1   CHUCK NO 3   TYPE 0

OPER	DESCRIPTION	SEGMTS	TO DIA.	LENGTH	DEPTH	FEED	RPM	M.TIME	H.TIME	T.TIME	C.SPEED
	CHUCK								0.22	0.22	
10	ROUGH CUT	3– 7	40.80	106.50	2.10	0.50	900.	0.24	0.19	0.42	121.24
20	WIDE SLOT ROUGH	4– 4	34.44	54.50	3.18	0.13	500.	1.14	0.31	1.44	59.06
30	FINISH CUT	5– 7	37.00	43.50	1.90	0.24	1400.	0.13	0.30	0.43	171.00
40	RADIUS	7– 7	0.0	2.00	0.0	0.24	1400.	0.01	0.10	0.11	0.0
50	WIDE SLOT FINISH	4– 4	30.00	54.50	2.22	0.07	1400.	0.86	0.32	1.17	141.64
60	RADIUS	4– 4	0.0	3.00	0.0	0.07	1400.	0.03	0.10	0.13	0.0
70	RADIUS	4– 4	0.0	3.00	0.0	0.07	1400.	0.03	0.10	0.13	0.0
80	COUNTERSINK	1– 7	0.0	10.75	6.75	0.08	1400.	0.10	0.55	0.65	29.67
90	ROUGH DRILL	6– 7	18.00	36.50	9.00	0.31	260.	0.45	0.47	0.93	22.04
100	CROSS FIN. BORS	6– 6	30.00	11.50	6.00	0.17	1100.	0.06	0.32	0.38	124.34

TOTAL TIME ON MACHINE NO 709   3.04   2.98   6.02   1.50   0.01   0.01

SIDE 2   ON MACHINE NO 705   ATLAS   1245   10** LATHE   5 H.P   1   CHUCK NO 3   TYPE 0

OPER	DESCRIPTION	SEGMTS	TO DIA.	LENGTH	DEPTH	FEED	RPM	M.TIME	H.TIME	T.TIME	C.SPEED
	CHUCK								0.22	0.22	
10	FINISH CUT	3– 3	40.00	11.50	0.40	0.11	1700.	0.06	0.30	0.36	215.66
20	FINISH BORING	7– 7	20.00	26.50	1.00	0.14	1700.	0.11	0.30	0.41	112.10
30	FACE CUT	0– 0	0.0	8.50	2.00	0.16	900.	0.06	0.22	0.28	

TOTAL TIME ON MACHINE NO 705   0.23   1.04   1.27   0.42   0.01   0.01

Figure 14.6 Detail operations list.

ALT. 0 PART NO 03067BMACHST OPERATIONS LIST FOR 1000. PIECES. CRITERION C 16/ 1/79 PAGE 2
DD/MM/YY

SIDE 1 ON MACHINE NO 709 COLCHESTER 1800 6.5 25 CENTER LATHE 10 5H.P 1 CHUCK NO 3 TYPE 0

OPER	DESCRIPTION	SEGMTS	TO DIA.	LENGTH	DEPTH	FEED	RPM	M.TIME	H.TIME	T.TIME		C.SPEED
	CHUCK								0.22	0.22		
10	ROUGH CUT	1- 2	40.40	66.50	2.30	0.50	750.	0.18	0.13	0.30		100.56
20	ROUGH CUT	1- 2	35.80	66.50	2.30	0.50	750.	0.18	0.19	0.36		89.73
30	ROUGH CUT	1- 1	31.30	41.50	2.25	0.37	1400.	0.08	0.19	0.27		147.49

TOTAL TIME ON MACHINE NO 709 0.43 0.72 1.15 0.29 0.01 0.01 0.00

SIDE 1 ON MACHINE NO 705 ATLAS 1245 10** LATHE 5 H.P 1 CHUCK NO 3 TYPE 0

OPER	DESCRIPTION	SEGMTS	TO DIA.	LENGTH	DEPTH	FEED	RPM	M.TIME	H.TIME	T.TIME		C.SPEED
	CHUCK								0.22	0.22		
10	FINISH CUT	1- 1	30.00	41.50	0.65	0.13	1700.	0.19	0.30	0.49		163.61
20	RADIUS	1- 1	0.0	1.50	0.0	0.13	1700.	0.01	0.10	0.11		0.0
30	FINISH CUT	2- 2	35.00	26.50	0.40	0.13	1700.	0.12	0.30	0.42		188.97
40	RADIUS	2- 2	0.0	3.00	0.0	0.13	1700.	0.01	0.10	0.12		0.0
50	FACE CUT	0- 0	0.0	15.00	2.00	0.16	900.	0.10	0.22	0.33		

TOTAL TIME ON MACHINE NO 705 0.43 1.25 1.68 0.56 0.01 0.01 0.00

PART NO 03067BMACHST TOTAL PART TIME 10.22 MIN. TOTAL PART COST 2.87 LI.

Figure 14.6 *(Continued)*

```
DATE PRINTED 16/01/79                    E R R O R   R E P O R T                              PAGE NO.  1

PART NO.  MAT.  LENG  TOLERANCE COD     NO.  SA  SURF  DIMS   EXT  COD MATR COD    OVER ALL LNG   CARD
          TYPE  DIM.  DIAM.             MAT. PE  FNS   DIAM   SURF SRF HRDN DIMS                  COD
                      EXT INI LNGT S              FNS  EXTRN  FINIS FIN            EXTER  INTER

RMS-040278      004000  100 100 1            063  06000                            GD+12 G        2

KONU$-010378 01                         CK15 01        06500  01        200  B  00.9000  01 G

D10278-1291                                                                                    CARD 10 MISSIN
030678MACHST                                                                                   ACCEPTED
```

E R R O R   S Y M B O L S

1 NUMERIC (* NEAR APPROPRIAE NO.)
2 WRONG CODE (+ BY APP. NUMBER)
3 MORE THEN ONE INT. DIM. MISSING
4 MORE THEN ONE EXT. DIM. MISSING
5 END CODE MISSING
6 LENGTH ERROR DIM OR OVERAL
7 SEQUENCE ERROR OR A CARD IS MISSING
8 LENGTH CODE NOT 2 AFTER A MISSING DIMENSION

Figure 14.7 Error report.

```
                              LISTING  OF  CARDS
C30678MACHST01  ST50      01 0450001160B1 0170001000C0000                    01G
030678MACHST100100200125                                                  F  02G
C30678MACHSTGA004000110010300005O 063      015     3                        GA10G
030678MACHSTGA0025002100103500050 063 030          2                        GA11G
030678MACHSTGA0010002200204000025 063                                       GA12G
030678MACHSTGA0053002200203000100 125 030030       4                        GA13G
030678MACHSTGB0042001200203700100 125      020     3                        GB14G
030678MACHSTTB0025001200202000050 063                                       TB15G
030678MACHSTTB0010002200203000100 125                                       TB16G
030678MACHSTTB0135002200200000100 125                                      FTB17G
030678MACHST    017000                017000                                ZZ99G
```

Figure 14.8 Listing of input cards.

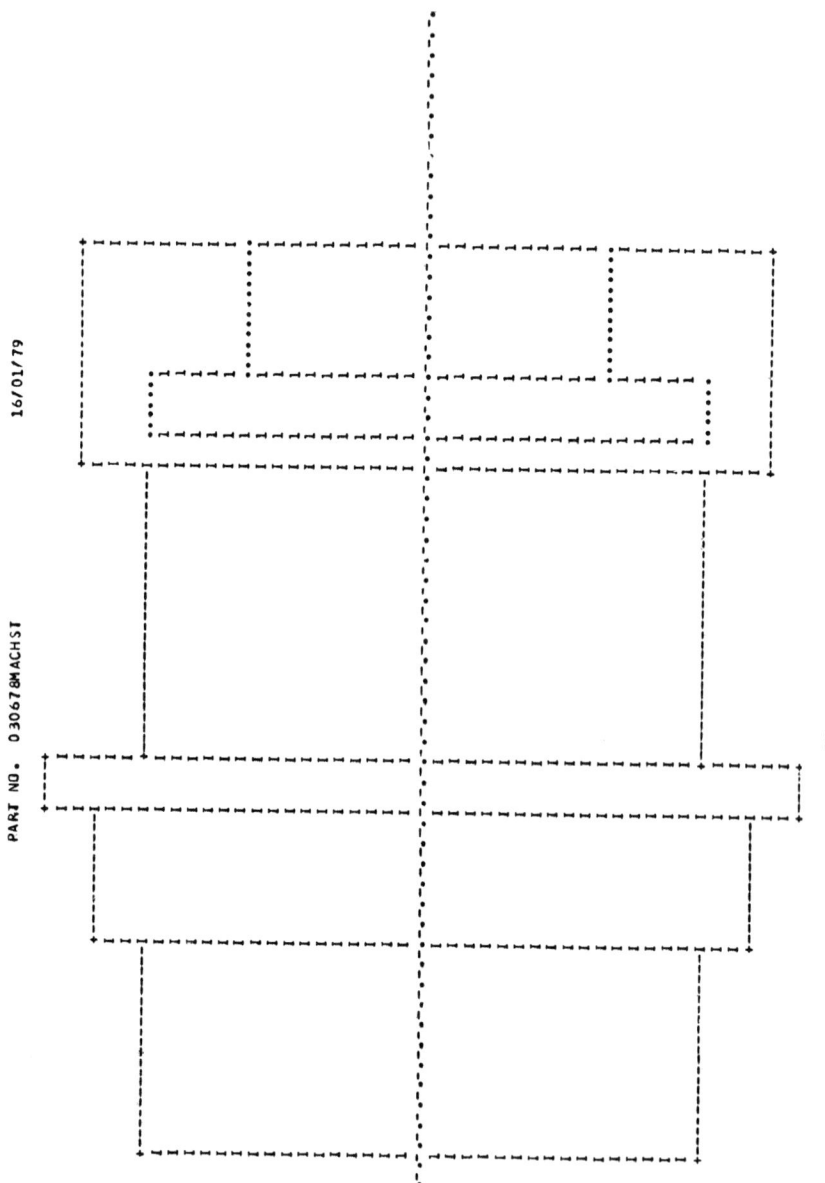

Figure 14.9 Part drawing.

382

PART NO 03067BMACHST OPERATIONS LIST DATE 16/ 1/79 DD/MM/YY

CHUCK SIDE 1 ON SEGMENT 3 CHUCK NO 3 CHUCK SIDE 2 ON SEGMENT 2 CHUCK NO 3 START FROM SIDE 2
SIDE 1 CHUCK TYPE FOR ROUGH CUT 0 FOR FINISH CUT 0 FOR SIDE 2 ROUGH 0 FINISH CUT 0

SEGM I/O	OPER NO.	SIDE	SEG	TOOL R MM	TOOL MM TYPE	FEED MM	VELOC. M/MIN	DEPTH MM	FORCE KG	HP	TIME MIN	LENGTH	RPM	DIA. AFTER				
1	3	2	10	2	85	0.0	850 0.0789	37.	6.75	0.	0.0	0.08	10.75	1765.	0.0	9	9	3	7
6	3	2	20	2	21	18.00	800 0.3080	21.	9.00	532.	2.03	0.32	36.50	376.	18.00	9	9	3	7
6	2	2	30	2	15	0.0	150 0.1600	207.	6.00	360.	21.76	0.33	11.50	2748.	30.00	9	9	5	6
7	2	2	40	2	2	0.60	100 0.1300	291.	1.00	56.	4.76	0.04	26.50	4867.	20.00	9	9	5	7
3	1	2	50	2	1	0.06	50 0.6000	138.	2.10	306.	12.51	0.17	106.50	1023.	40.80	0	3	1	7
4	1	2	60	2	4	0.06	150 0.5818	132.	3.18	448.	17.48	0.38	54.50	1115.	34.44	0	3	1	4
5	1	2	70	2	2	0.02	50 0.2500	202.	1.90	158.	9.47	0.11	43.50	1652.	37.00	0	1	2	7
7	1	2	80	2	54	2.00	0 0.0	30.	0.0	0.	0.0	0.05	2.00	258.	0.0	9	9	2	7
4	1	2	90	2	5	0.02	150 0.2500	197.	2.22	184.	10.77	0.41	54.50	1952.	30.00	0	1	2	4
4	1	2	100	2	53	3.00	0 0.0	30.	0.0	0.	0.0	0.06	3.00	318.	0.0	9	9	2	4
4	1	2	110	2	54	3.00	0 0.0	30.	0.0	0.	0.0	0.06	3.00	318.	0.0	9	9	2	4
3	1	2	120	2	2	0.01	50 0.1212	340.	0.40	22.	2.24	0.04	11.50	2681.	40.00	0	3	2	3
1	1	1	130	1	1	0.05	50 0.5454	140.	2.55	347.	14.34	0.12	66.50	1047.	39.91	0	3	1	2
1	1	1	140	1	1	0.05	50 0.5468	144.	2.05	283.	12.03	0.10	66.50	1209.	35.80	0	3	1	2
1	1	1	150	1	1	0.06	50 0.6000	136.	2.25	327.	13.24	0.05	41.50	1295.	31.30	0	3	1	1
1	1	1	160	1	2	0.01	50 0.1260	313.	0.65	36.	3.36	0.10	41.50	3249.	30.00	0	3	2	1
1	1	1	170	1	52	1.50	0 0.0	30.	0.0	0.	0.0	0.03	1.50	318.	0.0	9	9	2	1
2	1	1	180	1	2	0.01	50 0.1212	340.	0.40	22.	2.24	0.07	26.50	3060.	35.00	0	3	2	2
2	1	1	190	1	51	3.00	0 0.0	30.	0.0	0.	0.0	0.07	3.00	273.	0.0	9	9	2	2

Figure 14.10 Theoretical operations list.

MACHINE DETAILS　　　　NO.　3　　　16/ 1/79

*09 COLCHESTER　　1800　　　　6.5 25 CENTER LATHE 135H.P

NO.	DIAMTER O.	LENGTH			CROS	HOL	HP	TOL	H.COST		SPEED				FEED			
	BED	CROS	TALL	CENTR	CROS	MOVE	HEAD	X10 X100	F4.2	NO	F3.2 MIN	MAX	NO X100Y MIN		MAX		SPE. C	
709	335	210	145	635	710	95	400	100 100	1500	14	0	22	1400	45	0	2	50	5	2	9	3G

709	22	30	40	60	100	130	190	260	380	500	750	900	1100	1430							
	0	0	0	0	0	0	0	0	0	0	0	0	0	0	0	0	0	0	0	0	0
	0	0	0	0	0	0	0	0	0	0	0	0	0	0	0	0	0	0	0	0	0

709	2	3	4	5	6	7	8	9	10	11	12	13	14	15	16	17	18	19	20	21	22	23	24	25
	26	27	28	29	30	31	32	33	34	35	36	37	38	39	40	41	42	44	46	48	50	0	0	0
	52	54	57	60	63	66	0	0	0	0	0	0	0	0	0	0	0	0	0	0	0	0	0	0

Figure 14.11 Machine list and error report.

MSPR	T A O R		TIME			HAND	E	TABLE	16/ 1/79	
1	ROUGH CUT	ROUGH BORING	0.19	0.09	0.03	0.03	0.0	0.0	0.0	0.0
2	FINISH CUT	FINISH BORING	0.30	0.20	0.15	0.13	0.0	0.0	0.0	0.0
4	WIDE SLOT ROUGH	WIDE SLOT ROUGH	0.31	0.15	0.05	0.05	0.0	0.0	0.0	0.0
5	WIDE SLOT FINISH	WIDE SLOT FINISH	0.32	0.16	0.05	0.05	0.0	0.0	0.0	0.0
6	REV. ROUGH CUT	REV. ROUGH BORE	0.22	0.13	0.04	0.04	0.0	0.0	0.0	0.0
7	REV. FINISH CUT	REV. FINISH BORE	0.34	0.24	0.16	0.14	0.0	0.0	0.0	0.0
14	CROSS ROUGH CUT	CROSS ROUGH BOR	0.31	0.15	0.05	0.05	0.0	0.0	0.0	0.0
15	CROSS FINISH CUT	CROSS FIN. BORS	0.32	0.16	0.05	0.05	0.0	0.0	0.0	0.0
16	REV. ROUGH CUT	REV. ROUGH BORE	0.22	0.13	0.04	0.04	0.0	0.0	0.0	0.0
17	REV. FINISH CUT	REV. FINISH BORE	0.34	0.24	0.16	0.14	0.0	0.0	0.0	0.0
21		ROUGH DRILL	0.47	0.16	0.02	0.02	0.0	0.0	0.0	0.0
22		FINAL DRILL	0.49	0.17	0.02	0.02	0.0	0.0	0.0	0.0
23		DRILL	0.47	0.16	0.02	0.02	0.0	0.0	0.0	0.0
31		FIRST REAM	0.47	0.16	0.02	0.02	0.0	0.0	0.0	0.0
32		FINAL REAM	0.49	0.17	0.02	0.02	0.0	0.0	0.0	0.0
44	FACE CUT		0.22	0.13	0.04	0.04	0.0	0.0	0.0	0.0
45	SLOT		0.31	0.15	0.05	0.05	0.0	0.0	0.0	0.0
51	RADIUS		0.10	0.10	0.12	0.10	0.0	0.0	0.0	0.0
52	RADIUS		0.10	0.10	0.12	0.10	0.0	0.0	0.0	0.0
53	RADIUS		0.10	0.10	0.12	0.10	0.0	0.0	0.0	0.0
54	RADIUS		0.10	0.10	0.12	0.10	0.0	0.0	0.0	0.0
55	RADIUS		0.10	0.10	0.12	0.10	0.0	0.0	0.0	0.0
56	RADIUS		0.10	0.10	0.12	0.10	0.0	0.0	0.0	0.0
57	RADIUS		0.10	0.10	0.12	0.10	0.0	0.0	0.0	0.0
58	RADIUS		0.10	0.10	0.12	0.10	0.0	0.0	0.0	0.0

Figure 14.12 Handling time table.

MSPR		T A O R	TIME			HANDLING TIME		
							TABLE	16/ 1/79
61	ROUGH STEP CONE		1.01	0.20	0.04	0.0	0.0	0.0
65	CONE BY FORMES T		0.32	0.16	0.05	0.0	0.0	0.0
71	CONE ROUGH CUT.		1.00	0.19	0.04	0.0	0.0	0.0
72	TAPER FINAL CUT		1.01	0.20	0.04	0.0	0.0	0.0
85		COUNTERSINK	0.55	0.24	0.04	0.0	0.0	0.0
90	SET UP TIME	MIN. DIV BY 10	2.50	4.00	9.90	0.0	0.0	0.0
91	HANDLING WORK	LIROT DIV BY 10	1.20	0.50	1.20	0.0	0.0	0.0
92	TRANSFER EXPENS.	LIROT DIV BY 10	0.50	0.50	0.50	0.0	0.0	0.0
101	CHUCH	100	0.08	0.08	0.08	0.0	0.0	0.0
102	CHUCH	250	0.11	0.11	0.11	0.0	0.0	0.0
103	CHUCH	450	0.13	0.13	0.13	0.13	0.0	0.0
104	CHUCH	1200	0.18	0.18	0.18	0.0	0.0	0.0
105	CHUCH	2250	0.22	0.22	0.22	0.0	0.0	0.0
106	CHUCH	5000	0.30	0.30	0.30	0.0	0.0	0.0
107	CHUCH	9999	0.37	0.37	0.37	0.0	0.0	0.0
110	GAGE-MEASURE		0.24	0.24	0.01	0.0	0.0	0.0
111	GAGE-MEASURE		0.12	0.12	0.01	0.0	0.0	0.0
120	START-STOP MACH.		0.05	0.05	0.01	0.0	0.0	0.0
121	ENGAGE FEED		0.03	0.03	0.01	0.0	0.0	0.0
122	ADJUST TOOL		0.13	0.03	0.01	0.0	0.0	0.0
123	REV.CROSS FEED		0.08	0.02	0.01	0.0	0.0	0.0
124	CHNGE TOOL BIT		0.39	0.39	0.39	0.0	0.0	0.0
125	INDEX 4P HOLDER		0.04	0.04	0.01	0.0	0.0	0.0
126	CHANGE TOOL CHUK		0.20	0.15	0.01	0.0	0.0	0.0
127	TAILSTOCK PO.		0.27	0.01	0.01	0.0	0.0	0.0

Figure 14.12 *(Continued)*

HANDLING TIME TABLE   16/ 1/79

MSPR	T A B R		TIME						
128	CROSE STEADY RST	0.03	0.03	0.01	0.01	0.0	0.0	0.0	0.0
129	SET COM.TO ANGLE	0.82	0.10	0.01	0.01	0.0	0.0	0.0	0.0
130	SQUAR CORNER	0.10	0.10	0.12	0.10	0.0	0.0	0.0	0.0
131	CHNGE FEED OR SP	0.03	0.03	0.01	0.01	0.0	0.0	0.0	0.0
132	RETURN PER INCH	0.01	0.01	0.01	0.01	0.0	0.0	0.0	0.0

Figure 14.12 *(Continued)*

	725	727	709	803	705	801	708	802	804	0
0.12	15.94	10.25	8.93	17.51	10.63	16.86	13.31	17.51	19.44	0.0
0.12	5.37	1.56	1.22	5.95	2.15	5.51	3.39	5.95	8.29	0.0
0.12	3.00	2.00	3.00	3.00	3.00	3.00	3.00	3.00	3.00	0.0
0.10	11.70	7.58	6.93	12.29	7.50	11.85	9.73	12.29	14.63	0.0
0.10	5.37	1.56	1.22	5.95	2.15	5.51	3.39	5.95	8.29	0.0
0.10	5.00	2.00	3.00	5.00	5.00	5.00	5.00	5.00	5.00	0.0
0.05	8.80	6.02	5.71	8.56	5.35	8.21	7.17	8.56	8.92	0.0
0.05	3.76	0.98	0.67	3.52	1.29	3.17	2.13	3.52	3.88	0.0
0.05	5.00	5.00	5.00	5.00	5.00	5.00	5.00	5.00	5.00	0.0
0.10	11.73	************		11.96	4.05	9.98	7.58	11.96	9.85	0.0
0.10	8.32	999.00	999.00	8.54	1.63	6.56	4.16	8.54	6.43	0.0
0.10	5.00	5.00	3.00	5.00	5.00	5.00	5.00	5.00	5.00	0.0
0.03	4.34	3.48	3.31	5.59	2.43	5.42	4.05	5.59	5.63	0.0
0.03	1.35	0.49	0.32	2.60	0.42	2.43	1.06	2.60	2.64	0.0
0.03	5.00	5.00	5.00	5.00	5.00	5.00	5.00	5.00	5.00	0.0
0.07	7.72	999.67	999.44	8.30	2.00	6.95	4.78	8.30	6.77	0.0
0.07	6.50	999.00	999.00	7.08	1.40	5.73	3.56	7.08	5.55	0.0
0.07	3.00	2.00	3.00	3.00	5.00	3.00	3.00	3.00	3.00	0.0
0.07	1.68	0.67	0.44	3.67	0.60	3.41	1.26	3.67	3.92	0.0
0.07	1.68	0.67	0.44	3.67	0.60	3.41	1.26	3.67	3.92	0.0
0.07	1.00	2.00	3.00	4.00	5.00	6.00	7.00	8.00	9.00	0.0
0.17	26.89	21.74	20.38	27.82	22.45	27.20	24.35	27.82	31.20	0.0
0.17	4.70	1.46	1.06	4.66	2.14	4.24	2.82	4.66	8.45	0.0
0.17	3.00	2.00	3.00	3.00	3.00	3.00	3.00	3.00	3.00	0.0
0.38	28.96	19.18	18.54	25.76	22.84	25.70	22.94	25.76	26.95	0.0
0.38	13.25	3.47	3.61	10.05	7.92	9.99	7.23	10.05	11.24	0.0
0.38	3.00	3.00	3.00	3.00	3.00	3.00	3.00	3.00	3.00	0.0
0.11	20.67	16.22	14.93	20.93	15.93	19.91	18.08	20.93	20.35	0.0
0.11	6.03	1.58	1.07	6.29	2.07	5.28	3.44	6.29	5.72	0.0
0.11	3.00	3.00	3.00	3.00	3.00	3.00	3.00	3.00	3.00	0.0
0.05	15.75	14.83	13.85	17.22	13.99	17.02	15.51	17.22	17.33	0.0
0.05	1.49	0.58	0.38	2.96	0.52	2.76	1.25	2.96	3.07	0.0
0.05	3.00	3.00	3.00	3.00	3.00	3.00	3.00	3.00	3.00	0.0
0.41	29.84	15.04	13.48	22.32	14.76	20.84	20.01	22.32	21.19	0.0
0.41	19.03	4.23	2.94	11.52	4.94	10.04	9.21	11.52	10.39	0.0
0.41	5.00	5.00	3.00	5.00	5.00	5.00	5.00	5.00	5.00	0.0
0.06	11.87	10.87	10.54	13.97	9.82	13.71	11.55	13.97	14.22	0.0
0.06	1.58	0.58	0.38	3.67	0.51	3.41	1.26	3.67	3.92	0.0
0.06	5.00	5.00	3.00	5.00	5.00	5.00	5.00	5.00	5.00	0.0
0.06	11.36	10.36	10.16	13.46	9.31	13.20	10.15	13.46	13.71	0.0
0.06	1.58	0.58	0.38	3.67	0.51	3.41	1.26	3.67	3.92	0.0
0.06	5.00	5.00	3.00	5.00	5.00	5.00	5.00	5.00	5.00	0.0
0.04	13.50	************		14.18	8.80	13.45	11.60	14.18	13.23	0.0
0.04	4.94	999.00	999.00	5.62	1.20	4.89	3.04	5.62	4.67	0.0
0.04	3.00	3.00	3.00	3.00	5.00	3.00	3.00	3.00	3.00	0.0
0.08	14.66	8.91	7.77	16.68	7.59	15.94	11.78	16.68	14.86	0.0
0.08	8.18	2.44	1.62	10.20	2.10	9.47	5.30	10.20	8.38	0.0
0.08	5.00	5.00	3.00	5.00	5.00	5.00	5.00	5.00	5.00	0.0
0.32	12.55	7.39	6.15	15.99	5.49	15.18	11.56	15.99	15.22	0.0
0.32	8.71	3.55	2.32	12.16	2.64	11.35	7.72	12.16	11.38	0.0
0.32	5.00	5.00	5.00	5.00	5.00	5.00	5.00	5.00	5.00	0.0
0.04	8.01	******	999.95	8.81	2.85	7.46	5.22	8.81	7.17	0
0.04	6.27	999.00	999.00	7.08	1.37	5.73	3.49	7.08	5.44	0
0.04	3.00	2.00	3.00	3.00	5.00	3.00	3.00	3.00	3.00	0
0.33	4.56	1.38	0.95	5.54	1.48	4.95	3.00	5.54	5.20	0
0.33	4.56	1.38	0.95	5.54	1.48	4.95	3.00	5.54	5.20	0
0.33	1.00	2.00	3.00	4.00	5.00	6.00	7.00	8.00	9.00	0
6.89	32.35	************		37.32	9.65	32.33	18.96	37.32	39.01	0.0
17.13	1.00	2.00	3.00	4.00	5.00	6.00	7.00	8.00	9.00	0.0

Figure 14.13 Matrix printout.

0	0	0	0	0	0	0	0	0	0
0.0	0.0	0.0	0.0	0.0					
0.0	0.0	0.0	0.0	0.0	0.0	0.0	0.0	5.00	2.00
0.0	0.0	0.0	0.0	0.0	0.0	0.0	0.0	0.0	0.0
0.0	0.0	0.0	0.0	0.0					
0.0	0.0	0.0	0.0	0.0	0.0	0.0	0.0	5.00	5.00
0.0	0.0	0.0	0.0	0.0	0.0	0.0	0.0	0.0	0.0
0.0	0.0	0.0	0.0	0.0					
0.0	0.0	0.0	0.0	0.0	0.0	0.0	0.0	5.00	5.00
0.0	0.0	0.0	0.0	0.0	0.0	0.0	0.0	0.0	0.0
0.0	0.0	0.0	0.0	0.0					
0.0	0.0	0.0	0.0	0.0	0.0	0.0	0.0	5.00	5.00
0.0	0.0	0.0	0.0	0.0	0.0	0.0	0.0	0.0	0.0
0.0	0.0	0.0	0.0	0.0					
0.0	0.0	0.0	0.0	0.0	0.0	0.0	0.0	3.00	2.00
0.0	0.0	0.0	0.0	0.0	0.0	0.0	0.0	0.0	0.0
0.0	0.0	0.0	0.0	0.0					
0.0	0.0	0.0	0.0	0.0	0.0	0.0	0.0	1.00	2.00
0.0	0.0	0.0	0.0	0.0	0.0	0.0	0.0	0.0	0.0
0.0	0.0	0.0	0.0	0.0					
0.0	0.0	0.0	0.0	0.0	0.0	0.0	0.0	3.00	2.00
0.0	0.0	0.0	0.0	0.0	0.0	0.0	0.0	0.0	0.0
0.0	0.0	0.0	0.0	0.0					
0.0	0.0	0.0	0.0	0.0	0.0	0.0	0.0	3.00	3.00
0.0	0.0	0.0	0.0	0.0	0.0	0.0	0.0	0.0	0.0
0.0	0.0	0.0	0.0	0.0					
0.0	0.0	0.0	0.0	0.0	0.0	0.0	0.0	3.00	3.00
0.0	0.0	0.0	0.0	0.0	0.0	0.0	0.0	0.0	0.0
0.0	0.0	0.0	0.0	0.0					
0.0	0.0	0.0	0.0	0.0	0.0	0.0	0.0	3.00	3.00
0.0	0.0	0.0	0.0	0.0	0.0	0.0	0.0	0.0	0.0
0.0	0.0	0.0	0.0	0.0					
0.0	0.0	0.0	0.0	0.0	0.0	0.0	0.0	5.00	5.00
0.0	0.0	0.0	0.0	0.0	0.0	0.0	0.0	0.0	0.0
0.0	0.0	0.0	0.0	0.0					
0.0	0.0	0.0	0.0	0.0	0.0	0.0	0.0	5.00	5.00
0.0	0.0	0.0	0.0	0.0	0.0	0.0	0.0	0.0	0.0
0.0	0.0	0.0	0.0	0.0					
0.0	0.0	0.0	0.0	0.0	0.0	0.0	0.0	5.00	5.00
0.0	0.0	0.0	0.0	0.0	0.0	0.0	0.0	0.0	0.0
0.0	0.0	0.0	0.0	0.0					
0.0	0.0	0.0	0.0	0.0	0.0	0.0	0.0	3.00	3.00
0.0	0.0	0.0	0.0	0.0	0.0	0.0	0.0	0.0	0.0
0.0	0.0	0.0	0.0	0.0					
0.0	0.0	0.0	0.0	0.0	0.0	0.0	0.0	5.00	5.00
0.0	0.0	0.0	0.0	0.0	0.0	0.0	0.0	0.0	0.0
0.0	0.0	0.0	0.0	0.0					
0.0	0.0	0.0	0.0	0.0	0.0	0.0	0.0	5.00	5.00
0.0	0.0	0.0	0.0	0.0	0.0	0.0	0.0	0.0	0.0
0.0	0.0	0.0	0.0	0.0					
0.0	0.0	0.0	0.0	0.0	0.0	0.0	0.0	3.00	2.00

— — selection of sequence upward phase
——— final modified sequence of downward phase
·—·— change of machine

0.0	0.0	0.0	0.0	0.0	0.0	0.0	0.0	0.0	0.0
0.0	0.0	0.0	0.0	0.0	0.0	0.0	0.0	0.0	0.0

Figure 14.13 (*Continued*)

PART III
HAL-TECHNOLOGY

Chapter Fifteen
Hal-Technology Concepts

Developments in the use of computer as an aid to manufacturing have proceeded in a modular disjointed fashion. Systems designed and developed to solve a specific problem as expediently as possible were necessarily limited in scope. Integration of these systems has been attempted in some cases, but only as an afterthought. The resulting proliferation of disjointed computer systems tends to magnify manufacturing problems.

The evidence shows that the economic benefits to be gained from integration far exceed those benefits directly attributable to individual development efforts. This is particularly true in discrete part-batch manufacturing based industries because of such factors as the need to maintain both a flexible fabrication base and highly efficient controlled operations. Such companies comprise a high percentage of U.S. industry, but their individual outputs are relatively small. Planned integration of systems would evolve programs that consider not only advances in individual areas of manufacturing, but also the relationships between some of these areas. An all-embracing technology would appear to be superior to an integrated system, since it does not contemplate the relationships between individual areas and activities, but rather dissolves them into one single system.

Hal-Technology (Hal means all-embracing in Hebrew) is a computer-oriented manufacturing philosophy. It utilizes the power and capabilities of present-day computers to meet the requirements of the manufacturing process.

Hal treats the manufacturing process as one interactive problem starting from engineering design to product shipment. It considers the manufacturing process as a nucleus and satellites rather than as a chain of activities. The engineering activities are the nucleus, and the other activities are the satellites. Thereby Hal introduces engineering, value engineering, and technology into all phases of the manufacturing process. It broadens the scope of alternate solutions and eliminates the artificial constraints used as an interface between the engineering and production phases in the chain of activities. Thus a better optimum solution can be reached.

Hal-Technology evolved from the generative process planning program (GPPP). The scheduling module of Hal is based on the machine selection technique of the GPPP and combines the process and capacity planning functions in one dynamic interactive module. The other manufacturing functions (such as design and material requirement planning) of Hal use the pattern recognition techniques of the GPPP to increase the number and quality of alternate solutions.

The production information and control system (PICS) as well as the integrated manufacturing system (IMS) used today regard the engineering phases of manufacturing, that is, product design (bill of material), process planning, and methods (routing), as constraints. These constraints are artificial and exist only because the product designer, the process planner, and the production engineer do not (and practically cannot) communicate continuously and work as a team. The engineering problems in manufacturing are being transformed into mathematical problems at too early a stage, so that in spite of the impressive mathematical techniques and a true optimum, from a mathematical standpoint, a real optimum solution is not always obtained. The inherent logic of the IMS results in inefficiency, long lead times, and high levels of work-in-process.

Group-Technology (GT) is aimed at assisting the engineering phases. However, it depends heavily on classification and coding systems, on past performance, and on human effort, skill, and experience. It is basically a manual technology. Even with the use of computers, it calls for interactive dialogs, but does not lead to a fully automatic system.

This is probably the reason why GT and IMS are not merged into one complementary system that introduces engineering into the production stages; GT is a good philosophy, but very difficult to implement. Hal is a computer-oriented system. In a way, it is an extension of GT. It does not use any classification and coding system at all, yet grouping of families of parts can be made if and as required. Logically speaking, there are no groups at all. Physically, however, all the group combinations we could possibly dream of are available, but there is no need to anticipate them. Thus many GT features can be applied with Hal-Technology, and the desired merging of GT and IMS becomes possible.

15.1 Hal Concepts

The objectives of Hal are to increase productivity and reduce manufacturing costs. These objectives are the same as those of many other systems. The difference lies in the approach and concepts employed. The Hal approach makes use of the following notions:

- There are infinite ways of producing a workpiece.
- Any component can be produced by any available facilities. It can be produced with a 5D DNC machine, a universal machine, or manually with the aid of a chisel and file.
- The cost and lead time required to produce a component are functions of the process used.
- There are infinite ways of meeting design objectives.
- In any component about 75% of the dimensions (geometric shape) are nonfunctional (fillers). These dimensions can vary considerably without affecting the component performance.
- The cost and lead time required to produce a component are functions of its design. A minute change in fillets or dimensions to suit a standard tooling or an existing set up on a machine can result in significant cost variations.
- With present-day techniques, there will always be overloaded and underloaded machines. If the life of a machine is 5–10 years, then 10–20% of the machines are replaced every year, usually with better, more modern machines that can perform jobs more economically. These new machines are therefore preferred for the jobs and are often overloaded while the old machines are underloaded. The process plan has to be altered continuously to comply with these changes.
- With present-day techniques, competition between workpieces for facilities will always occur. The method and logic of resolving this competition, that is, pull forward or backward, defeat the main purpose of production planning, which is to meet delivery due dates and minimize the amount of capital tied down in production.
- There exists a theoretical manufacturing optimum that is theoretical from a specific shop standpoint, but practical from a technology standpoint.
- There are many good ideas and technologies today, but not all of them are implemented to their full extent because of a lack of tools.

Hal-Technology is based on two powerful tools that result from the GPPP:

1. The ability to generate a practical optimum process plan in seconds.
2. The ability of a computer to "see" a component, manipulate it, display it, change it, store thousands of components drawings, and retrieve them by key (predetermined characteristics) or by attributes (random unpredictable inquiries).

Today's techniques lack these tools and must establish and freeze many parameters at too early a stage of the manufacturing process; this results in decreased efficiency.

The basic philosophy of Hal is that all parameters in the manufacturing process are flexible, that is, any of them is subject to change if such change contributes to increased productivity in manufacturing the product mix required for the immediate period. The parameters, including process plan and product design, become fixed and frozen only at the last minute before starting the actual processing. In such a flexible and dynamic environment, the only constant parameters are the products to be manufactured and the facilities available at the shop. Product objectives are external to the manufacturing cycle and must be preserved. Change recommendations are welcome, but should not be carried out automatically—they must be approved by management.

Hal can be operated efficiently in any environment. It foresees the manufacturing technology required for the automatic factory, and it is believed that Hal is a must for that era. Hal, operating under a main computer, can alter processes and control DNC machines. It enables the computer to "see" the geometric shape of parts at any stage of manufacture and to generate complementary processes. Hal can be used as a guide to industrial robots. The robot will transmit the picture it sees by a television camera or other sensors. This picture will be matched against Hal files, decisions made, and instructions set forth.

Hal can suggest and evaluate different manufacturing environments. Manufacturing costs, capabilities, and deviation from the theoretical optimum can be computed for any environment and submitted as information to management. Management, according to its forecasts and financial considerations, can reach an intelligent decision as to the desirable manufacturing environment. Once such a decision is made, Hal will accept it as a fixed and frozen datum and will optimize the manufacturing process accordingly.

The main concepts of Hal are:

- Engineering phases are incorporated into the production phases.
- All phases of the manufacturing cycle work toward a single objective. Each phase considers the problems and difficulties of the other phases.

- The objective is to increase productivity and decrease the manufacturing cost of the product mix required in any period rather than to optimize any single product, component, or operation.
- No artificial constraints are created and considered.
- The manufacturing cycle is kept dynamic and flexible until the moment production starts.
- The objectives and approach of GT are adapted.
- Creating part families is not an objective, but only a means to achieve savings in manufacturing and shortening of lead times.
- Efforts are made to use standard tooling and existing machine setups in product design and process planning.
- A computer is a working machine. Automatic decisions are reached by employing a computer in every possible case. Interactive human–machine use is reserved to cases in which data or an algorithm are not available.
- Work-in-process is regarded as shaped raw material, and its original destination can be changed.

15.2 Hal-Technology System Architecture

Under Hal, the manufacturing cycle is divided into three main modules, as shown in Figure 15.1.

Theoretical Optimum Design and Planning

This module includes product design, process planning, and methods, time and motion study. It is called theoretical, since it optimizes each item separately and is subject to change, yielding higher-level optimization (i.e., that of a product mix required in any time period). It is commonly agreed that standards are useful and that the engineers responsible for each of the above tasks should consider the problems of the others. It is suggested that design and planning should be carried out in groups, that is, by committees. However, such committees can seldom achieve their goals and therefore lead to a decrease in manufacturing efficiency.

Operating under Hal, the above tasks are either carried out automatically or interactively. The computer will act as a committee member taking on the role of representing the absent professions. Chapter 16 demonstrates how this is done.

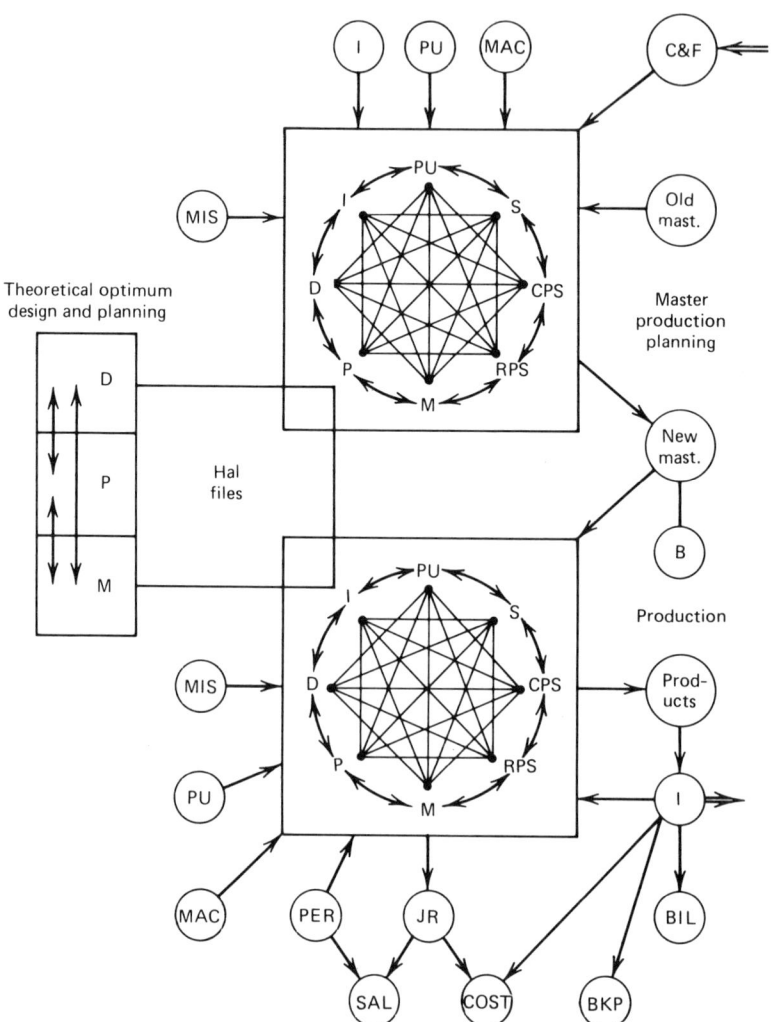

Figure 15.1 Hal manufacturing cycle. Notation: D, product design; P, process planning; M, methods, time, and motion study; RPS, requirement planning; CPS, capacity planning; S, shop; I, inventory; PU, purchasing; MAC, machine file; C&F, customer orders and forecasting; MIS, miscellaneous; B, budget; PER, personnel; JR, job recording; SAL, salaries; COST, costing; BKP, bookkeeping; BIL, billing.

Master Production Planning

Master production planning is a coordinating function between manufacturing, marketing, finance, and management. Its main objective is a realistic production program (see Part I, Section 5.3). In the IMS, the same scheduling is repeated three times for different purposes: in master production scheduling, in requirement planning, and in capacity planning. At each consecutive phase of the planning it considers more details and thus becomes more accurate. In a way, the master production schedule is not taken too seriously, since nobody really believes that it will actually be carried out. In practice, due to shop dynamics, changes must be made, and hence there is not much sense in going into too many details in planning for the future.

By its flexibility, Hal can overcome most problems caused by shop dynamics. Due to the speed of the computer, detailed planning is practical.

Hal combines master production scheduling, requirement planning, and capacity planning into one phase. The purpose of this combined phase is similar to that of the master production scheduling and requirement planning mentioned in Part I. Hal offers additional design features and options to work with (these will be discussed in Chapter 17). The purposes of the capacity planning phase are altered. Under Hal, no detailed machine loading is required. The loading is done on the work center or department level. The theoretical optimum design and planning are used as the initial capacity planning input data. The desired output is divided into three time elements. The first time element includes a few short-range time periods. Each period must have a load equal to the available capacity of the work center (not a machine). Deviations from the theoretical optimum are allowed if needed and are applied according to product mix optimization.

The output of this stage is a list of products and quantities that must be manufactured in a given time period in each department and work center. This list must be practical. The methods employed and deviations from optimum made in order to balance the load (i.e., to guarantee the possibility of producing the items on the list) must not be kept. The next phase will do the actual loading and will optimize the product mix on the list by starting with the theoretical optimum data.

The second time element is the medium range. In this range an attempt is made to balance the load. To do so, a limited deviation from the theoretical optimum data is allowed. Load balancing is not a must.

The third time element is the long range. In this range no deviation from the theoretical optimum data is allowed. Load balancing, if desired, will be done only in response to management considerations, such as forecasts, promised delivery dates, and lot sizes.

In a way, the output of the master production planning module is similar to that of the requirement planning and master production schedule phases of the IMS. The difference lies in the way in which it was computed. The use of the output in the administrative phases (purchasing, finance, etc.) remains the same as in the IMS.

Production

This module covers the actual manufacturing of items. The input is the product mix that must be produced in the given period. The output is the items. The purpose of this module is to make sure that all of the items planned for the period will actually be produced and in the most economical way possible. To achieve this, total flexibility concerning process plans, product design, inventory items, and so on is assumed, that is, deviation from the theoretical optimum input data is allowed. In this module, capacity planning is done simultaneously with the actual manufacturing. The facilities are fixed and all must be loaded. The list of products and items to be manufactured within the given period is fixed and assumed to be practical.

The flexibility lies in the sequence and method of production and in the allocation of available material or semifinished items. The first scheduling attempt is according to the theoretical optimum data and using the dispatching rule SIMSET (similar set up). If this fails to produce satisfactory results, the following measures are taken: alter process, contrive process to suit an existing setup of an unloaded machine, use and modify the in-process item, change filler dimensions in item design, and so on (see Chapter 17). The limiting factor in the use of the above measures is an economic consideration: The increase of the product mix cost due to the deviation from the theoretical optimum must remain below the expenses resulting from not meeting due dates. The output of this module is finished items and the administrative documents required. Billing, costing, salaries, and so on remain as in the IMS.

15.3 Part Description System

Hal-Technology is based on the ability of the computer to "see" geometric shapes, compare them, and manipulate them. All of these should be done internally in the computer system. Hal has the ability to display

15.3 Part Description System

shapes on a graphic display terminal and work interactively with a designer (CAD). This ability is of secondary importance, since it is used in the product design phase for data capture and validation in cases where automatic design is impractical and the design must be carried out in an interactive mode. All the other phases may not use the graphic display terminal at all.

Most CAD systems available at present are designed with the primary objective of display function. Comparisons and manipulations are carried out interactively by the worker who sits by the terminal. The computer can "see" only points, lines, arcs and circles, conics, splines, and interconnects. Hal requires the computer to "see" geometric shapes.

It was decided not to invent any new geometric system, but rather to teach the computer engineering drawing. The engineering drawing is the recognized engineering language. This language has been used for a long time, has been proven reliable, is unique (no ambiguity), and is universally recognized and accepted. So far, I have not heard of any tendency to change it. The drawing is the connecting link between the designer of the component and the worker who performs the work on the shop floor. In some phase or another, any classification and coding system or geometric language used will have to be translated into the engineering language, that is, into a drawing. We are working in an engineering environment; therefore, let us adopt its language and teach the computer to read engineering drawings.

Engineering drawing is a two-dimensional system. Bodies and volumes are described by projections or cross sections in as many views as required.

The initial step in drawing is the assignment of an origin to the coordinate system and a line of symmetry for bodies of rotation. The view is described by lines of reference to the coordinate system used: body lines, cross-section reference lines, and so on.

The Hal part description system is based on the rules of classical technical drawing. The first entry to the system is the "general data entry." It contains the item name, catalog number, and information pertaining to the item, such as raw material, standard tolerances and surface finish, standard fillet radius, and unit of measure (metric or inch). It establishes the origin of the coordinate system or a line of symmetry for bodies of rotation. Hal is a bounded geometry system, bound by fixed reference lines, two for each coordinate. These lines are for orientation and dimensioning. Additional reference lines at any distance can be used. They can be fixed or temporary, as shown in Figure 15.2. Any temporary reference line can be changed into a fixed reference line if required.

Any dimension given will be identified by the direction in which it

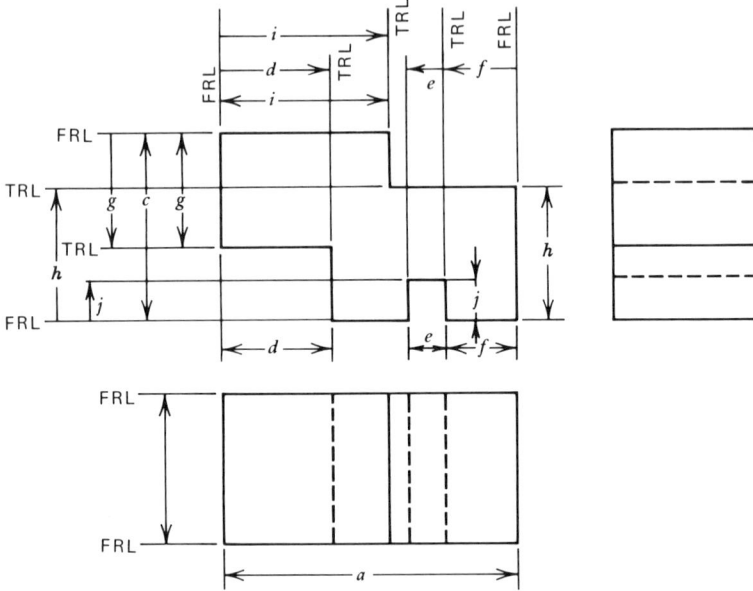

Figure 15.2 Definition of reference lines. Notation: FRL, fixed reference line; TRL, temporary reference line.

points and by whether it is given from a fixed or temporary reference line. The reference lines can be shifted and placed as required. Establishing additional coordinate systems or center lines is allowed. This feature is used to show a cross-section view other than the regular three views, or a cross-section of an assembly of mixed mode shapes (that is, concentric, flat, sheet metal or three-dimensional elements). An assembly can be real (made of several parts of the same or different materials) or virtual, that is, one part made of one piece (such as forged or cast items), but divided for ease of description. Each additional center line or coordinate system has its own fixed reference lines (two for each axis), and the correlation with the previous system or the original system is maintained. Figure 15.3 shows such additional systems. The center line can be straight, curved, or circular. A special mode enables one to define curved lines. A curved line might become a body line or a center line as defined by the user. Figure 15.4 shows curved and circular lines defined as center lines.

Body shapes are defined by lines or curves in a coordinate system. A code defines whether the description is of a body of rotation, a flat body, or a three-dimensional body.

For bodies of rotation it is assumed that the defined line or curve rotates

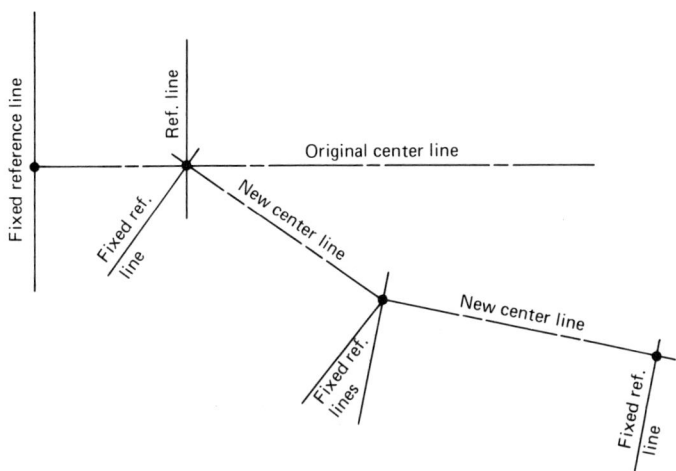

Figure 15.3 Original and newly established center lines.

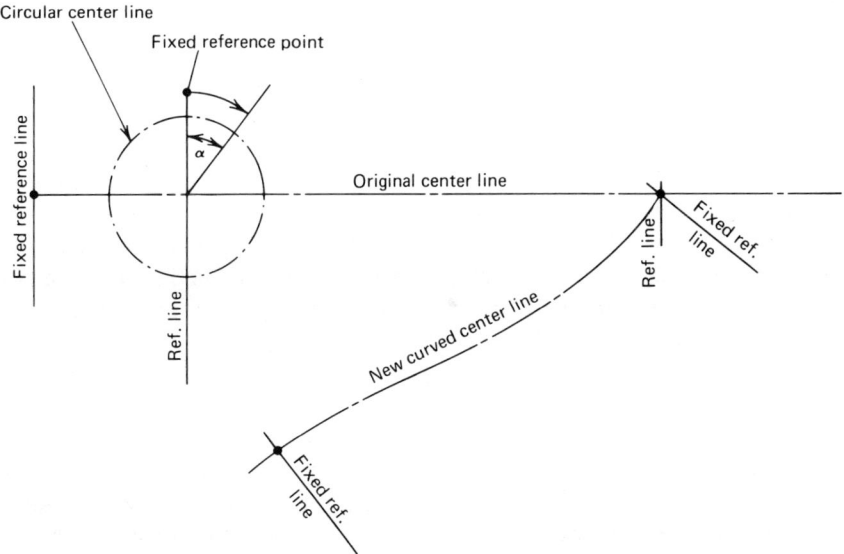

Figure 15.4 Original and new curved and circular center lines.

around the line of symmetry (polar coordinate system) and creates the area of an envelope. This envelope, when marked as material (M), becomes a solid body; when marked as void (V), it becomes a hole in a body; and when marked as sheet metal (S), the envelope line is widened by the thickness of the material and becomes a solid with a hole in it.

For three-dimensional bodies it is assumed that the line represents one side of an area on a plane. The area is constructed by connecting the edges of this line with the neighboring lines. If the other lines are not specified, it is assumed that they are straight lines perpendicular to the coordinate system presently in effect or lines of the adjoining segments. When this area is marked as sheet metal (S), no other views of the part are required. The thickness of the material is added perpendicularly to the defined area, thus creating a body. When this area is marked as a three-dimensional body (D), the other views will define the shape of the body. The defined volume can be solid (M) or a void (V) in the part.

Description of Parts

The unit of measure can be metric or inches to a given number of digits after the decimal point. Inch units can be defined as decimals or fractions. The accuracy can be defined by tolerance, concentricity, and angle of deviation. Surface finish can be defined in any of the three common scales. These options can be defined as standards for the complete part, separately for each segment, or intermixed where the segment definition supersedes the general definition.

Part description is divided into basic shape and special features. The basic shape is divided into segments. A segment is defined as a portion of the part in which the basic shape remains unchanged. The division into segments is done separately for each view, that is, top view, bottom view, and side views (see Figure 15.5). A segment is initially described by its basic data, that is, the overall dimensions of the contour envelope by a box. This defined outline is then transformed into or exchanged for the real shape. Such shapes can be tapers, triangles, radii, or any special shape. Figure 15.6 shows this transformation. Joining of the individual segments in the proper order creates the shape of the parts and the voids in it.

Special features can be defined as imposed on the basic segment or independent. In order to facilitate the description of such common features as slots, burrs, chamfers, and key ways an index (table) of special features has been compiled. Figure 15.7 shows some of these special features, the options, and the rules by which to define them and assign

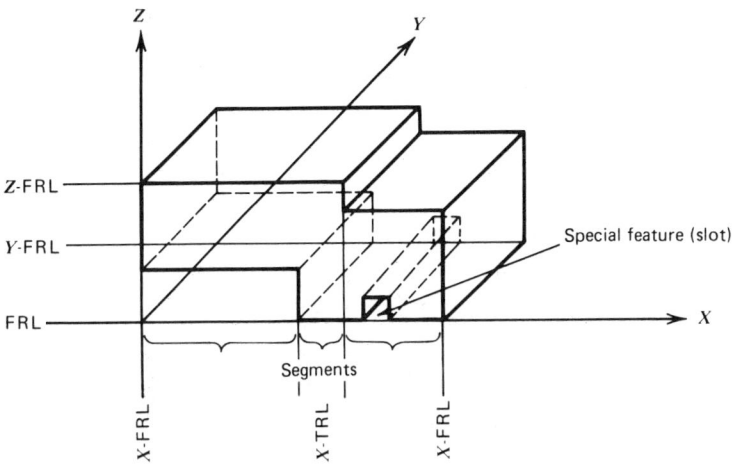

Figure 15.5 Three-dimensional part description. Notation: FRL, fixed reference line; TRL, temporary reference line.

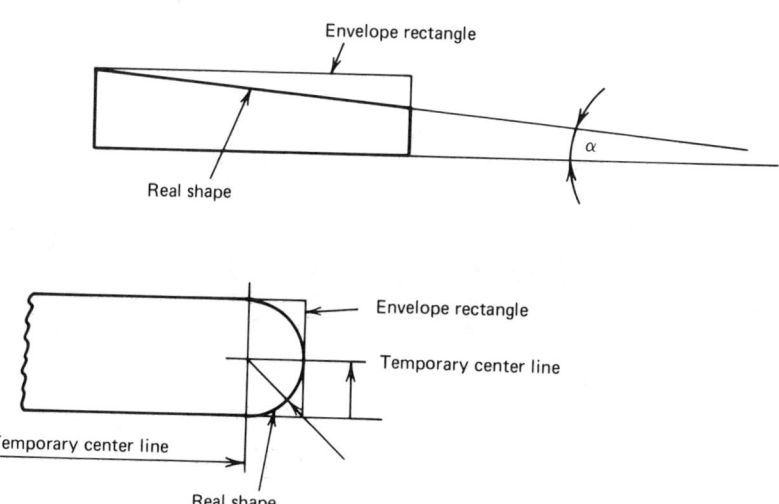

Figure 15.6 Transformation of outlines.

Special Features Table   Hal-T1   G. Halevi

Shape code 47-48	Dimen. type 49	Special code dimensions 50	51	52	53	54	55	56	57	58	59	60-77
01	1	x	x	x								
	2	x	x	x								
	3	x	x•	x	x							
02												
05	1	x	x	x	x• x							
05	2	x	x•	x	x/x							
05	3	x	x•x	x	x/x	x	x	x				
05	9									x•x	x x	

X = mandatory
Y = optional
Z = one entry mandatory

Accuracy of basic shape segment
Concentric to center line within ±
Square to center line with (in mμ) ±
Allowed deviation of angle ± Δα

Mirror image

Taper (cone)
— By two diameters D_R
— By angle α, C_α
— By angle, diameter defined on the right side α, C_α, D_R
— By identical diameter to adjoining segments

$C_\alpha = 1$ increasing angle $C_\alpha = 2$ decreasing angle

Figure 15.7 Special features table.

Special Features Table

Figure 15.7 *(Continued)*

General description of slots
(by three entries)

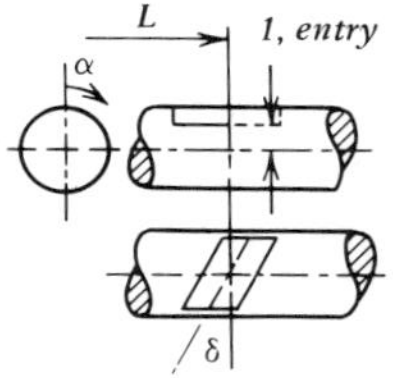

1, entry — geometrical location for:

Straight linear slot:
$L, C_L^{1)}, \alpha, Z, \delta, C_\delta, F_H^*, F_S^*$ S (×   see sketch 3. entry

* H = beginning of segment
  S = end   "   "

When $F_H = F_S$; F_S = blank

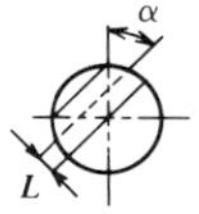

(× S: on the circumference parallel to the $\mathcal{C}_L = P$
        "   "   "   at right angle " $\mathcal{C}_L = R$
                                           facing $\mathcal{C}_L = F$
1) see table 1

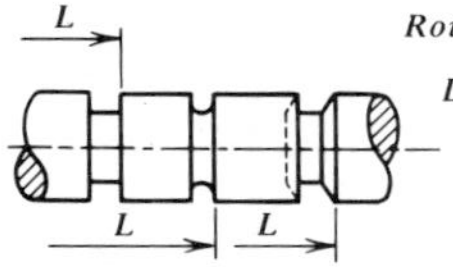

Round slot on circumference:
L, C_L^1

Round slot facing:
$D, C_D^{(15}$

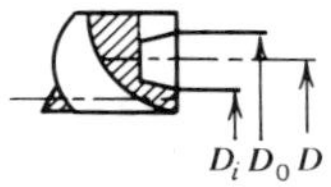

(15 see table 15)

Shape code 47–48	Dimen. type 49	Special code dimensions 50–77
30	0	entries in cols 52, 53, 54, 57, 59, 60, 61, 62, 63, 64, 66, 67, 69, 70, 71, 77
31	0	× × × ×•× / × (cols 50–54)
32	0	× × × ×•× / × (cols 50–54)

their dimensions. Threads and gears are regarded as special features. They can be either standard and thus defined by their nominal dimensions according to the engineering standards or nonstandard and have their dimensions given.

These special features are an integral part of the system and can be generally used. Space is reserved for any users' special features. Any shape or feature that is used repeatedly can be inserted into the system during customizing and becomes part of the system. In constructing the special feature, care is taken to follow the basic technique of the system. The concept of coordinates, center line, and reference line is applied. The data of each special feature generate a temporary center line, or temporary origin and coordinates, upon which a fixed reference line is defined. This is shown in Figure 15.8. The rules of defining a basic segment apply in defining the special feature. The above temporary center line and/or set of coordinates is in effect until a basic segment entry is encountered.

Additional coding space is available for defining temporary special features or curves. They are in effect only for the single part within whose entry data they were defined. Within that part they can be used as many times as required. Again, they are defined by using temporary center lines and reference lines, as called for by the basic technology. Such special features and curves are described by point-to-point sections as straight lines, arcs, radii, angles, and splines.

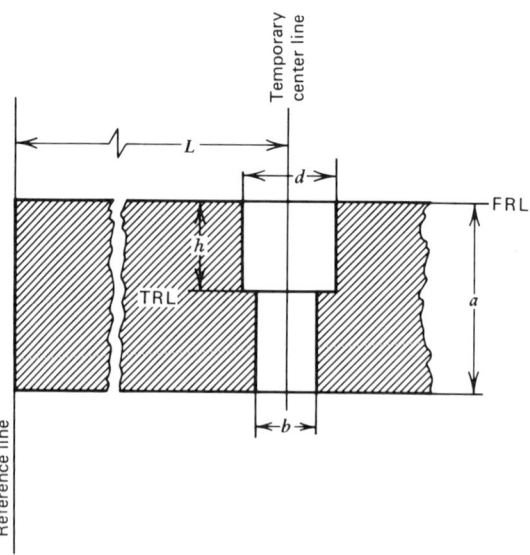

Figure 15.8 Coordinates for special features. Notation: FRL, fixed reference line; TRL, temporary reference line.

15.3 Part Description System

There are many possible shortcuts to reduce the work of definition. Whenever data have to be duplicated, for example, a mirror image feature can be used. Any group of basic segments can be looked on as a temporary special feature.

Any mixture of basic segments and standard and temporary special features can be defined as a subpart by looking on it as a temporary special feature to be used as many times as needed. A part can be looked on as an assembly of several parts.

As in engineering drawing, dimensions that can be computed may be omitted (e.g., the length of a basic segment or the junction point of a line tangent to a circle, etc.).

Nonmachining Treatment and Marking

Nonmachining treatment (heat treatment, cleaning, coating, painting, etc.) can be described for a complete part or for a single segment according to the following options:

- Description of the purpose of the treatment, such as preservation, decoration, lubrication, friction, and conductivity.
- Description of the final condition of the surface or the treated part, such as phosphated, painted, plated, coated, and hardened.
- Description of the treatment in detail, that is, the complete operation list required.

Any mark/trademark or text can be entered by its type and location.

The part description system can be used in an off-line mode or in an on-line interactive mode. For the off-line mode, the part description form shown in Figure 15.9 is used. The data entered on the form are keypunched, pass validation check, and are entered in the Hal data base files.

For the on-line interactive mode, use entries—similar to those described on the part description form—are entered on the terminal by the user. Each line is validated as it is entered, and the segment shape (or the special feature described) is displayed on the graphic terminal screen. With the last entry, the part is fully described and displayed, the designer checks it visually, and when approved the data are transferred to the Hal data bank. Figure 15.10 shows the entries that describe a part. The data bank is constructed to capture shapes rather than points and lines. Its purpose is to serve the following functions:

- Store at minimum disk space. Data compression techniques are used.
- Retrieval by key.

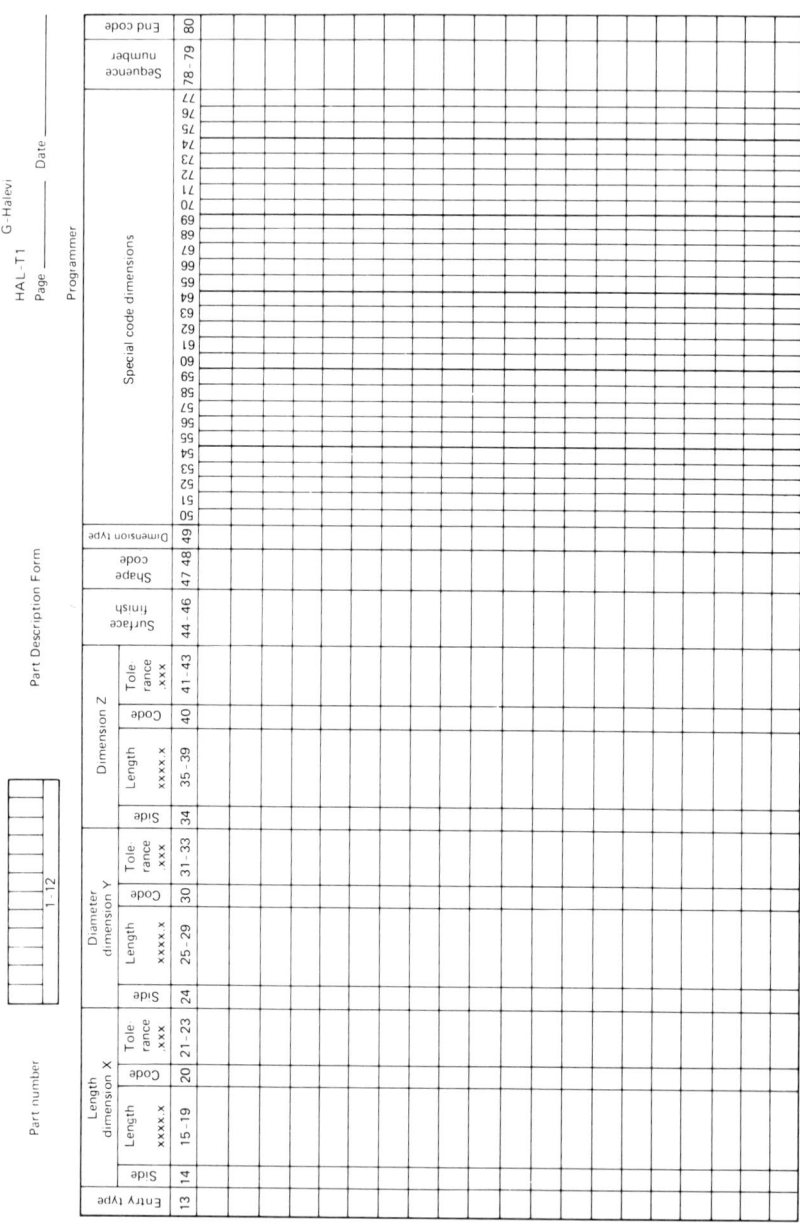

Figure 15.9 Part description form.

Part number: BOXSPANSOKET 1-12

Part Description Form

HAL T1
Page _____
Programmer G. Halevi
Date 20.4.78

Entry type	Side	Length dimension X Length xxxx.x	Code	Tolerance xxx	Side	Diameter dimension Y Length xxxx.x	Code	Tolerance xxx	Side	Dimension Z Length xxxx.x	Code	Tolerance xxx	Surface finish	Shape code	Dimension type	Special code dimensions	Sequence number	End code
13	14	15-19	20	21-23	24	25-29	30	31-33	34	35-39	40	41-43	44-46	47-48	49	50-77	78-79	80
G		00400	1	100		00340	1	100			1	100	125	02	1	CRN1 58RC	01	G
M	L	00400	1			00340								13	3		02	
V	L	00150	1											31	2	6 4 1500065 1	03	
V	R		1											31	2	6 6 2500120 1	04	
V														31	2	6 30 6 2500120 1	05	F

Box spanner socket 24 mm

Material: C, Ni
Plastic: Cr
Hardened: 58 RC

Figure 15.10 Part description entries.

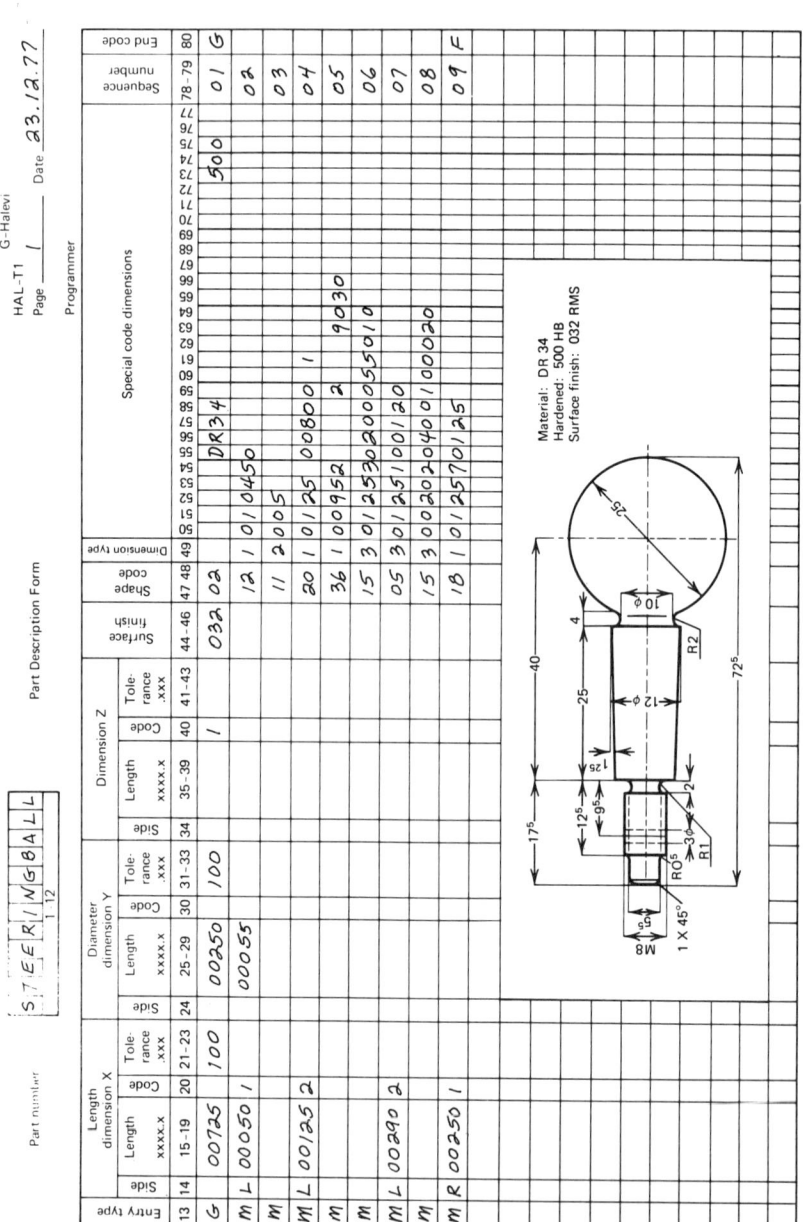

Figure 15.10 *(Continued)*

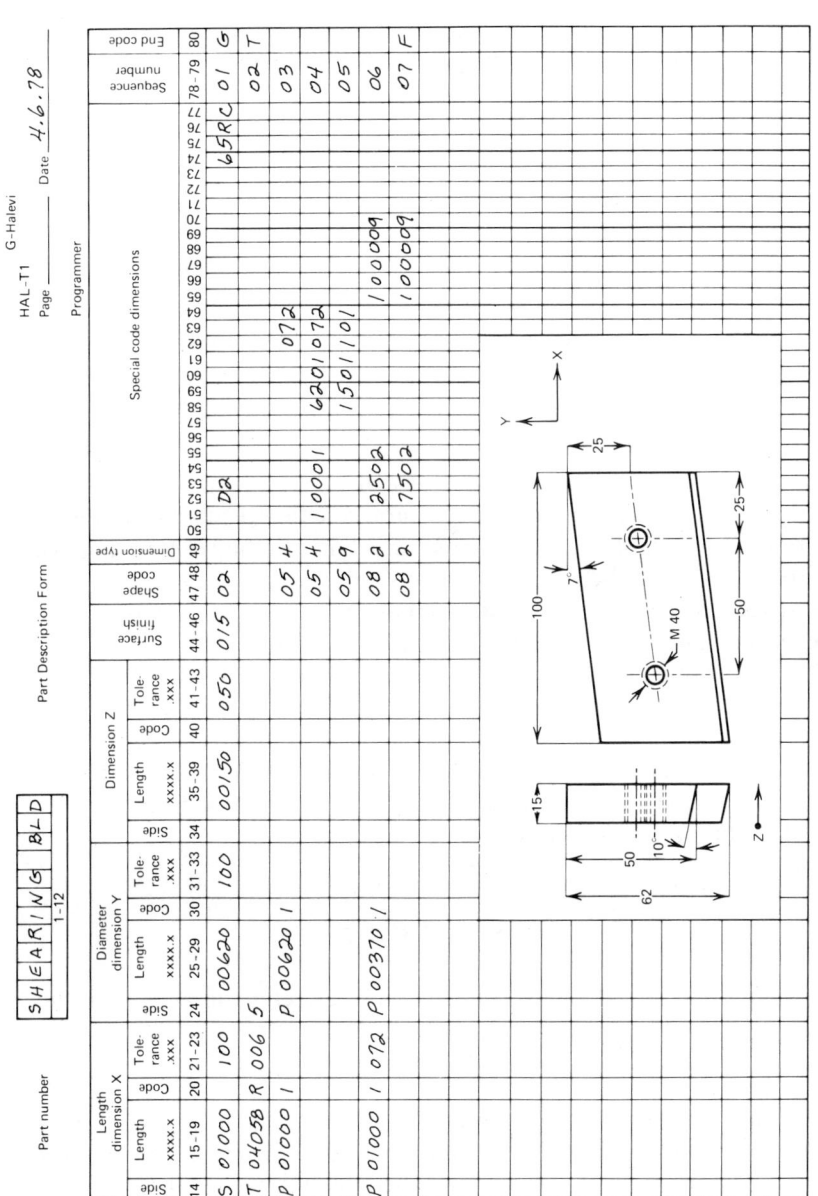

Figure 15.10 (Continued)

- Selective retrieval by attributes and parameters. Retrieval is to be fast to suit the interactive mode and with no limitation on permissible inquiries.
- Manipulation of data. Transformation of product design drawing to inspection, processing, assembly drawings, and so on.
- Change. A part drawing changed as per designer request, or a specific attribute change on all part drawings on the file. For example, scan the file and change all hole sizes that are within 9 to 12 mm to 10 mm.
- Supply data to process planning.
- Display. The ability to display any item on the terminal.

Notice that only the last purpose requires the graphic capabilities. It is the last and the least important function of the system. The data base construction method is of critical importance in Hal. Part data are not kept in the same form as entered by the designer. The details, however, are preserved. In describing a part, designers have the freedom to choose any coordinate system they prefer, to use any dimensioning option they wish, to look at and describe the part from any side they want, to use and define any special feature and to divide the part into segments in any way they like, to use any unit of measure they favor, and so on.

This freedom is desirable for the user, but hinders the function of comparing parts. User entry data have to be arranged in such a manner that the precise intentions expressed in the description will be maintained, while allowing the functions of comparison and retrieval by attributes to operate. The part description method proposed is not a must, it is just a tool facilitating the required data bank organization.

15.4 Hal Benefits

The characteristics of today's manufacturing technique are by no mean satisfactory, as is recognized by many. An article in the October 1978 issue of the *IEEE Spectrum* stated that:

> Less than 25% of the U.S. industrial output is the result of mass production. The rest is produced by batch manufacturing, a system plagued by long lead times, high in-process inventory, low machine utilization, and very little automation.

A CAM-I industrial survey found that, on the average, 40–60% of all batch jobs are delinquent when completed. Colin New, from the London Graduate School of Business Studies, puts it more bluntly:

15.4 Hal Benefits

The desirable characteristics of component production may be summarised as:

1. Short, reliable lead times.
2. High degree of flexibility.
3. Production of balanced sets of components.

Yet what do we find in practice?

1. Long, unreliable lead times.
2. Relatively inflexible control.
3. Production in 'economic' or established quantity lots totally unbalanced with regard to assembly requirements.

Hal-Technology improves the present situation. The benefits gained by using Hal include reduced batch flow times, lower inventory cost, smaller percentage of jobs completed behind schedule, and lower component cost. In general, it will increase the productivity and efficiency of the manufacturing plants. To be more specific and express the savings in dollars, let us examine a few specific areas.

Design

One of the Hal capabilities is to search for and retrieve existing parts. This specific feature is also available in GT by the use of classification and coding systems. In a paper delivered at a CAM-I's executive seminar, held at St. Louis Missouri, January 20–21, 1976 (P-76-PPP-01), p. 85), B. Guise of MDSI (Manufacturing Data Systems, Inc.) demonstrates the cost savings in the engineering field. One of the potential savings is the avoidance of the redesign of an existing part. To determine the significance of this, let us take a look at what happens when you create a new part. In product engineering, there is design time, detail drafting time, and mold and model shop or prototype time; occasionally, the test department is involved, there is documentation, and certainly drawing maintenance. From a service point of view, it has to be costed; again, documentation and inventory are affected.

When the new part reaches the manufacturing stage, many things happen. There is advance manufacturing engineering from a central location and possibly at a remote plant location. There is tool design, and tools have to be either made or bought. Time study is involved. Production control has to schedule the part, cost accounting is involved, and data processing, purchasing, quality control, and numerical control (N/C) programming are all affected—we could go on and on. The thing that I would like to convey here is that it is expensive to support new parts. These expenses can be avoided.

In one company, they have an active drawing base of 50,000 parts, and they create about 2,000 parts a year. They researched their costs to support these new parts. In engineering, they found that it cost about $250 to create a part, while in manufacturing, the costs were about $1,300 to support a new part—a total of $1,550. When you multiply this by the number of parts that they release a year, it costs $3,100,000 a year for them to support new parts. By the avoidance of redesigning existing parts, users have found reductions of 2–15% in their new part activity, with most of them reporting a 5% reduction. In the case just considered, a 5% reduction in new part activity is worth an annual savings of $155,000.

Experience has shown that if you provide a group of 20 manufacturing engineers with a blueprint and ask them to route that part, there is a fair chance that you might get 20 unique routings. You would certainly get more than one, and we can all agree that one of those routes would be the best—the least-cost method to manufacture that part. This least-cost method becomes the standard process of the family of parts, thereby improving the manufacturing methods of other members of the family.

In the case of the company considered above, the value added is $4,000,000 a year. By value added, I mean the cost of their parts that were internally manufactured minus the cost of material. In their case, an improvement of 3% of that value added amounts to an annual savings of $120,000 in direct labor.

Furthermore, when we find an improvement in a process, we can pass it on to every part in the family, thereby creating a dynamic state of optimization continuously occurring in our larger families. By utilizing this technique, a 5% improvement on the company's value added of $4,000,000 was achieved, resulting in a $200,000 annual savings in direct labor.

The savings experienced in this case history can be summarized as follows:

Engineering retrieval or avoiding new part design	$155,000
In the area of manufacturing by selecting the least cost method of a family	$120,000
By optimizing processes	$200,000
The cost of capital on the inventory reduction	$ 17,000
In the area of tooling: Design	$ 24,600
Expenses	$ 18,750
Total savings per year	$535,350

This is in a $30,000,000 year company.

15.4 Hal Benefits

Activities affected by introduction of a new part

Engineering activities:
Design and detail drafting
Prototype building
Testing and experimentation
Auditing and costing
Records and documentation

Manufacturing and finance activities:
Advance manufacturing engineering (central)
Manufacturing engineering (plant)
Tool design and tools and gauges
Time study and standards
Production control and scheduling
Cost accounting
Data processing
Purchasing and inventory
Quality control

Total cost range:	Low	Medium	High
	$1,300	$1,900	$2,500

Figure 15.11 Costs for introduction of a new part.

At the same CAM-I's seminar, (p. 192–218) Dr. I. Ham referred to the following savings: It has been reported that an average cost of introducing a new part into engineering and manufacturing systems is around $1,300 to $2,500 (average $1,900) per part, as shown in Figure 15.11. For example, a company reported that about 2,500 new parts were released annually (thus an average of 10 new parts every day), while about 30,000 active parts were in their design files. Therefore, it can easily be estimated that the annual cost of new part introduction becomes $4,750,000 per year (= 2,500 parts × $1,900 per part). So it is clear how much one can save by eliminating duplication of parts and thus reducing the number of new parts. It has also been reported that about 5 to 10% of annual new parts output could be avoided by the proper use of classification and coding systems. Thus a company can save about $237,500 to $475,000 in reduction of duplicated design alone.

Hal capabilities go much further than retrieval of items (see Chapter 16) and thus the savings potential is far greater.

Production

Production can be carried out in different environments, and Hal can be used in any one of them. In organizing work cells, Hal can serve as a tool to evaluate and construct the cells; in a working cellular organization, Hal can group families of parts to be produced in any single cell. Moreover, Hal can create groups by contriving a process plan to satisfy a particular cell.

The benefits to be gained from cellular organization are reduced lead times, set-up costs, and work-in-process. The average reduction in throughput time reported in practice is 70%, although in some cases it has been as high as 97%.

A mechanical engineering working party of the Institute of Production Engineers, NEDO, London (1975) reports the following:

1. A control valve manufacturer reduced average throughput time from 43 days to 17 days with much improved delivery performance.
2. A gauge manufacturer reduced lead times on items produced in one cell from 12 weeks down to 14 days.
3. An instrument manufacturer reduced throughput time by between 75% and 90% and as a result had stopped losing orders because of long lead times.
4. A major machine tool manufacturer reduced the lead time on machine beds from 9 months to 6 weeks and on gears from 13 weeks to 13 days.
5. A gear manufacturer reduced throughput times from 10 weeks to 3 weeks.

These are all actual results of the practical implementation of cell production methods. It is clear that in every case the use of group organization has enabled the company to offer short, reliable delivery times for component production.

In a functional organization, some of the benefits gained by the cellular organization can be achieved by similar set-up sequencing. The item unit cost is composed of set-up cost and machining cost. In small batch lots the set-up cost can be high relative to the machining cost. Figure 15.12 shows a unit cost curve as a function of quantity when set up lasts 2 hours and the unit machining time is 0.2 hour. If we can utilize the same setup to machine other items, the unit cost is considerably reduced. Graphically, it is shown by replacing the quantity scale (Q) by a (Q/N) scale that represents the division of quantity by number of items using the same setup. For example, the unit cost of an item produced in a batch size of five with a unique setup is 0.6 hour. If another item can use the same setup, the cost is reduced to 0.4 hour.

Hal takes advantage of this fact, and it can constrain the GPPP module to contrive a process plan that utilizes a given setup. This constraint results in deviation from the theoretical optimum process, but this devia-

15.4 Hal Benefits

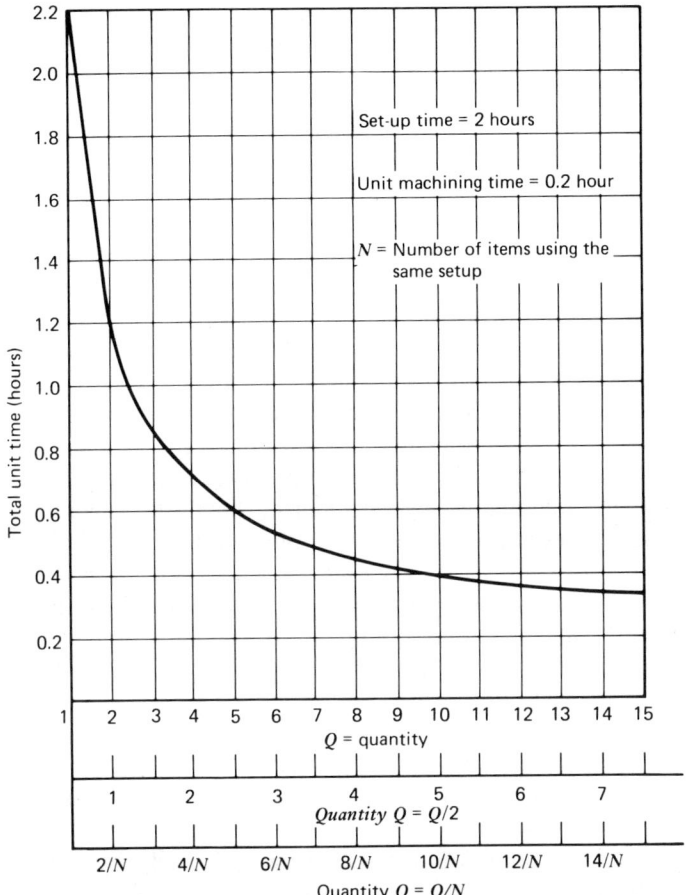

Figure 15.12 Unit cost as a function of quantity.

tion pays off on a product mix optimization basis. To demonstrate it, let us use the previous example. Suppose that the set-up constraints results in a 50% increase in machining cost—from 0.2 hour to 0.3 hour. From a product mix standpoint, the total unit cost is reduced from 0.6 hour (0.2 machining + 0.4 set up) to 0.5 hour (0.3 machining + 0.2 set up). In spite of a 50% deviation from the theoretical optimum, a 16% cost reduction is obtained.

Setups can be devised to minimize tool and fixture changes. For example, on a standard milling machine, a dividing head is permanently mounted at one end of the table, and a vise with interchangeable jaws is permanently mounted at the other end. On bar lathes, collet changing has been reduced by using only a restricted range of preferred bar sizes, even though it may mean making more swarf.

Standards

It is generally agreed that standard tooling, material, and so on should be used in order to reduce inventory and costs. However, in reality, this is seldom done, probably because there are no convenient tools to use and no way to make sure that designers use standards. Administrative tools, such as bulletins and verbal instructions, are not sufficient. Hal-Technology keeps company standards in its files, and from the design stage on draws the designer's attention to them and recommends their use. Furthermore, it would not accept any dimension or tool that was not standard without a special authorization. The benefits gained by this feature can be appreciated by considering the following reports. One company checked the number of holes they actually made in their parts. They found that 32 hole sizes were being made in their parts. They went through the design and either increased or decreased the hole sizes that were currently used without affecting the integrity of the design and standardized on 11 hole sizes. It does not sound particularly significant until you start to think of all the tools required to make 32 hole sizes as opposed to 11, such as drills, taps, and reamers. In addition, we commonly use fasteners where these holes occur. Again, a substantial reduction of fasteners inventory is obtained by stocking 11 sizes as opposed to 32.

At Herbert's Machine Tool Limited (England) similar steps and studies were carried out:

> Outcome of the exercise, and the subsequent redesign, was a reduction of 65% in the number of tools required and a reduction in floor-to-floor times of 10%. The design changes were mainly of ordinary dimensions, such as drain hole sizes and corner radii. The variety of drills and taps was reduced by adopting two preferred sizes of tapped hole, ¼ in. (British Standard Fine thread) BSF and ⅜ in. Whitworth standard thread. One striking fact revealed by the analysis was that there were 11 different tee slots in the group of top slides, requiring 26 different cutters. It was found that one H12 metric tee slot would suit every top slide.

The above demonstrates some of the benefits that can be realized by using Hal. In Chapter 16 the Hal features will be detailed.

Chapter Sixteen
Hal—in Engineering

The engineering phases of the manufacturing cycle include product design, process planning, and methods, time and motion study. The output of these phases is the bill of material, where information concerning product components, assemblies, and raw material is stored, and the routing, where information on how each of the above items is produced is stored.

The possible savings in manufacturing cost that can be brought about at the engineering stage are seldom studied in depth, and yet this is the area that probably holds the greatest potential for significant reduction in manufacturing costs. These potential savings can be obtained if engineering designs a product that, while performing its functions well (mechanical engineer or other), will be easy to handle (industrial engineer), easy to produce (process engineer), easy to assemble (production engineer), and made of inexpensive raw material (material engineer). Normally, there is a communication gap between the separate engineering professions, and hence the objective is not achieved. When the cost of an item or product becomes noncompetitive, management might apply value engineering in order to reduce cost. Usually, value engineering will not and cannot be applied to one-time small batch products or items. The objective of Hal is to close the communication gap between the engineering professions and thus apply value engineering to any design. This is preferably done by automatic design or else by interactive design, with the computer taking the role of safeguarding the design interest of the other engineering professions.

The other objective of Hal is to assist designers in their work by supplying reference data, strength analysis computations, and drafting; by eliminating any duplication of work; and by performing routine work that can save the engineer time and effort. These functions are discussed in this chapter.

16.1 Product Design—Preview

Product design is composed of two basic stages: the conceptual design and the detailed design, as discussed in Section 3.1.

Conceptual Design

In this stage designers employ their creativity. They can give free rein to their imaginations and come up with any wild idea. The more extravagant the ideas, the better designers they are. A computer is a machine and has no imagination at all. For this task a human is far superior to the computer, and thus humans should perform this task. It is possible to devise intelligent software with automatic design capabilities (to be introduced in the next subsection); however, using it in the conceptual design stage will limit and hinder technological progress. If they wish, designers can employ Hal for the following purposes:

- To remind them of secondary objectives.
- To help construct and solve a decision matrix. This might serve as a checklist and, more importantly, can be the method by which documentation of this stage is preserved.
- To assist in generating ideas. The Hal data bank contains many modules for the automatic design of elements or subsystems. Scanning and/or retrieving these modules according to purpose might trigger ideas and supply data required for the decision matrix. A request for a power source, for example, might retrieve electric motors of different types, hydraulic motors, hydraulic cylinders, pneumatic cylinders, internal combustion engines, solid or liquid propulsion rocket motors, springs, turbines, solar cells, solenoids, propellers, and so on.

Detailed Design

Detailed design is of a more technical nature; therefore, a computer can be employed to perform this task. In some fields automatic design programs have been constructed and successfully used. The most advanced field in this area is electronics, where products are being designed, board layout prepared, process established, and, in some cases, production carried out under computer control. We have also heard about a computer designing a new generation, better computer.

In construction planning, there are computer programs that, when given details about the number of floors, the load applied in the building,

16.1 Product Design—Preview

and the maximum wind velocity, give as output a detailed layout of construction steel and thickness of cement. The program also includes bill of material summaries.

Lately, a CAD program that generates hydraulic systems has been announced. It is used for the design of hydrostatic circuits in industry. Its main purpose is to calculate and check the stationary and dynamic behavior of circuits. The circuits can be composed of standard elements, special elements, or function groups.

In the mechanical field, automatic gear train design programs are available. Engineers type their special design requirements into the computer and get back, in minutes, all of the data they and the shop need—in easily readable, abbreviated English.

I am sure that many more programs of this nature exist. The common characteristic of these programs is the limited number of design alternatives. In the building design, for example, I was astonished to learn that there are only two alternatives of steel bar layout. The automatic design program decides which of the two alternatives to use and then adjusts dimensions to suit the particular building. It is possible to create a library of such programs, cataloged and indexed by function and by some more general names (purpose).

Automatic design programs should be used with care. One should be aware of their level of design optimization, that is, optimal design, rational design, or standard design. In addition, the date on which the program was prepared must be known. Designers must rely on their judgment and knowledge to make sure that technical developments and new materials introduced since the program was prepared do not outdate the available program.

Tool Automatic Design

Tool design for a specific part and given type of processing requires some degree of creativity and a great deal of technical know-how and experience. Therefore, it is better suited to automatic design than to product design. The main difficulty is that in many cases the process remains somewhat of an art, depending heavily on experience. A thorough understanding (similar to that of the GPPP) is needed in order to build up the scientific basis upon which an automatic design computer program is constructed. Many programs of this nature are available. Battelle Laboratories developed programs for the design of forging and extrusion dies. The extrusion die designer starts with the description of the extrusion shape. The cross section of the extrusion is expressed in terms of X, Y

coordinates and associated fillet or corner radii. These data are used to calculate such geometric parameters as cross-sectional area, perimeter, shape factor, location of centroid, and the size and location of the circumscribing circle. The user specifies press characteristics, such as capacity, container diameter, maximum billet length, and runout length (these data can be stored and retrieved by press number); the alloy to be used; and the extrusion temperature. On the basis of this information, load and yield calculations are made, and the number of openings required is established. Positioning of the holes is done by using the centroid technique; that is, the center of gravity of the hole opening is positioned to coincide with the center of gravity of the billet segment feeding that opening. The opening is also rotated so that its greatest distance is parallel and as close as possible to the chord of the segment. Figure 16.1 shows the arrangement of the entire die as displayed on a graphic terminal. After the layout is complete, the program corrects the openings for die cave and die deflection. A tool strength analysis is then performed, and the need for die support is determined. Die deflection analysis also includes the estimation of "tongue" deflections, so that the dimensions of the die opening can be modified to obtain the desired tolerances in the extruded shape.

As an extension of the die design program, NC punched tapes are prepared for: (1) machining the template of the finished extrusion; (2) machining the die opening itself; (3) machining the EDM (electric discharge machining) electrode; and (4) machining the die bearings into the back of the die.

In the electronics field, automatic design programs for printed circuit board assemblies are available. The layout system arranges all the components on the board so that the routing of the signal connections is as

Figure 16.1 Arrangement of the entire Battelle extrusion die.

simple as possible. In doing so, all the manufacturing rules, such as special placements and special ground plane requirements, are fully taken into consideration. When the board design is complete, the following additional output functions are available: (1) generation of the NC drill tapes for both plated and unplated holes and of a list of instructions for the drill operator; (2) electrostatic check prints of layout; (3) generation of photo-plotter drive tapes for the master artwork for each layer of the board; and (4) generation of photo-plotter tapes for silk-screen on the board.

At Cornell University, a computer-aided injection molding system is being prepared. Injection molding is the most important process in producing simple, as well as intricate, parts to tight tolerances for numerous engineering applications. Their approach is to treat the problem as an integrated system of mold design, mold-making, and process control. The input to the program is the part geometry and material. Programs of this nature can be assembled, cataloged, and stored in a library that becomes part of the system.

Structural Analysis Programs

To assist designers in performing their tasks, many analysis computer programs have been prepared. One of the outstanding programs is NASA's general-purpose structural program—NASTRAN. This program is now available to industry. It is designed to analyze the behavior of elastic structures under a range of loading conditions. Its general-purpose character makes it usable for:

- Any size, shape or class of structure.
- Geometric representations referred to any convenient coordinate system.
- Elastic relations ranging from isotropic to general anisotropy.
- Nonlinear behavior that can be represented by piecewise-linear approximations.
- Structural modeling with general one- and two-dimensional elements and special three-dimensional cases.
- Vibration frequency and mode determination.
- Synthesis of parts of the structure by experimentally determined static or dynamic properties.
- Loading conditions that embrace static surface and body forces, buckling factors, static thermal profiles, enforced deformations, steady sinusoidal excitations, time-varying surface and body forces, and stationary Gaussian random excitations.

- Dynamic combination with scalar force-producing systems.
- Solutions by most large second- and third-generation batch-mode or multiprocessing digital computers.
- Solution results plotted on any of three types of plotters.
- Complex as well as real matrix operations.

Analysts need no specialized knowledge of the computer because NASTRAN automatically takes over the direction of their problems.

Users regulate their problems on three separate levels. On the basic level, they describe their structural models in terms of bulk data. At the control level, they prescribe what portions of the bulk data are to be assembled for each subcase, the boundary conditions and loads to be applied, and the amount or type of output for printing or plotting. At the executive level, they set forth broad classifications covering the method of solution, the identification of information sources, and the provision for recovery.

Other structural analysis programs include:

- *GENESYS (general engineering system).* GENESYS is a master program containing an international library of civil and structural engineering subsystems. Subsystems include general-purpose finite-element analysis, frame analysis, soil mechanics, and project planning and design programs for highways, bridges, and concrete buildings.
- *PAFEC 75 (program for automatic finite element calculations).* PAFEC includes a library of 53 general, applicable elements which can be used for structural and heat-transfer problems. Calculation options include linear statics, plasticity, creep, natural frequencies, steady-state and transient heat transfer, transient response, and certain cases of large displacement.
- *ASAS (linear stress analysis system).* ASAS is a general-purpose finite-element computer program for structural analysis. Its element library includes straight and curved beams, membranes, plates, shells, solids, and special crack elements. ASAS contains a variety of analysis options, including linear statics, steady-state and transient heat transfer, and frequency and dynamic response.
- *PSA 5 (pipe stress analysis).* PSA 5 is used for the static and dynamic analysis of piping systems. Static analysis capabilities include thermal, gravitation, wind, and pressure loading. Among the features of the static analysis are constraint decoupling, unidirectional constraints, and limit stops. Dynamic analysis is based on the consistent mass method. Included in this analysis is an automatic procedure that retains the important degrees of freedom and results in the most economical solution.

16.2 Design of Parts and Products

Product and part design must meet their primary objectives and represent a compromise between several secondary objectives. A list of secondary objectives and the method by which to arrive at a good subjective compromise were discussed in Section 3.1. Any designer or plant might attach a different degree of importance to any single secondary objective, thus different designers will end up with different designs. In a way, design is a work of art that carries the personality of the designer. There is nothing wrong in this, since there are many ways to achieve a goal. In the design of any part, there are usually a few functional dimensions and shapes. The other dimensions and shapes are fillers. However, the engineering drawing must include and specify all part details.

For example, Figure 16.2 shows a locking handle. The only functional feature is the hole size (d). None of the other shapes and dimensions would affect the performance of the handle. The 20 deg inclination can be changed, the lever rod (d_3) can be changed to a different shape (e.g., square), and the method of connecting the three shapes can vary (thread, set screw, weld, or one piece). However, in a drawing no options are allowed—the part has to be uniquely defined. Figure 16.3 shows a connecting rod. It is a complicated part and probably requires a lot of static and dynamic analysis. However, the only functional dimensions are the distance between centers (151) and the bearing sizes (36 and 12). Any of the other dimensions and shapes can be varied (one more and another less) if it is found to be profitable.

It is estimated that about 70% of the details in a drawing are fillers, and only 30% are functional. The functional details can be divided into three categories: (1) the detail is constrained by the function that must be performed; (2) the constraint is due to purchased items (bearings) that are assembled in the product; and (3) the constraint is due to a decision made previously on another component (number and size of holes in a cover plate). The last two categories can be changed by selecting a different standard item (the variety is large) or by revision of previous decisions.

Thus, from a performance standpoint, the designer has a great degree of freedom. This freedom is utilized in design for economical manufacture. It is important for the designer to be aware of the various design features that can affect manufacturing costs. A minute change in a filler dimension or shape can cause a large variation in cost due to ease of machining, of assembly, of handling, and of use of standards.

Figure 16.4 shows the machining of a rectangular pocket. Corner radii have a major effect on part cost. Sharp corners are very difficult to machine and require many operations. It is good practice to use a tool diameter that corresponds to the corner radii. However, if the radii (R')

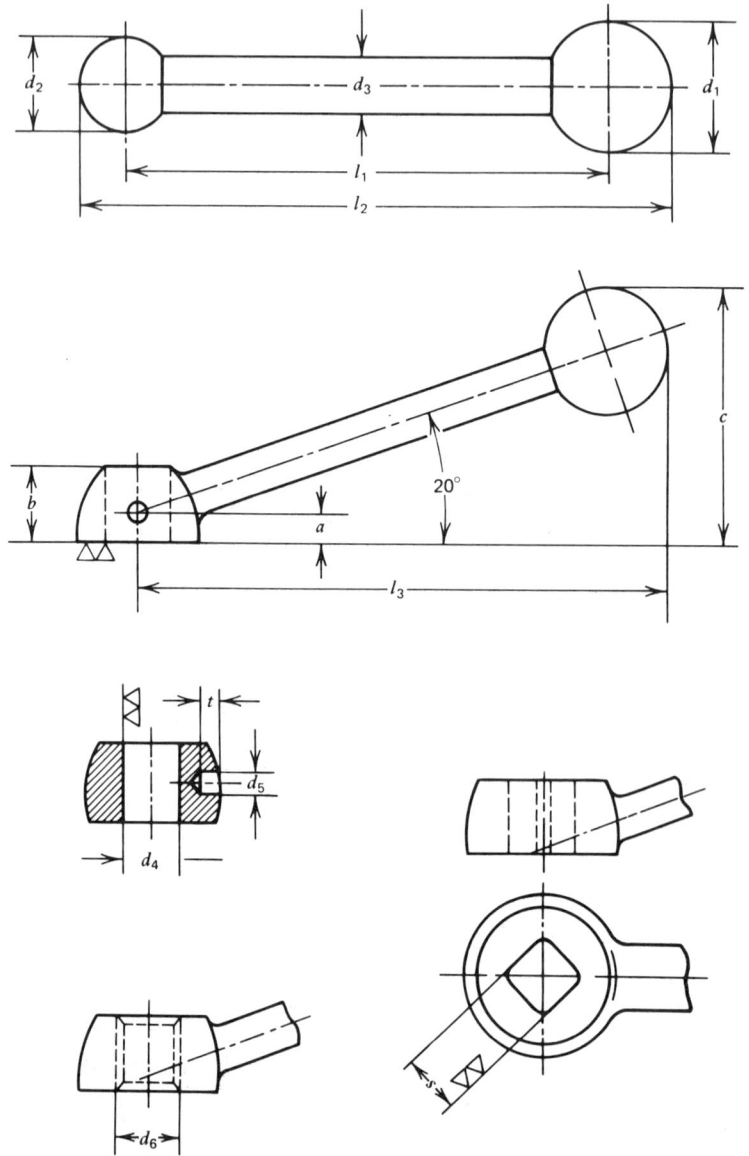

Figure 16.2 A locking handle.

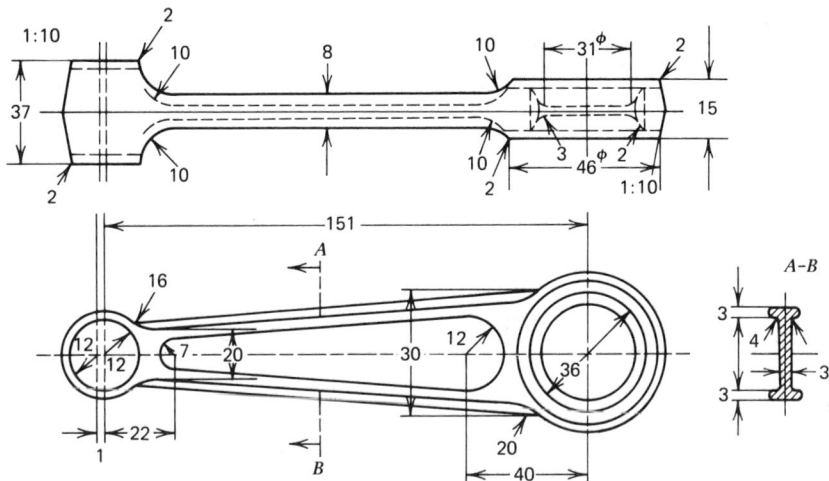

Figure 16.3 A connecting rod.

are small, machining the pocket requires a long tool travel path. It might be more economical to use a large tool diameter for material removal and then change cutters to machine the corners. If the radii match the economical tool diameter (R), the machining cost will be minimal. If the part is molded, appropriate corner radii and wall inclination will ease and reduce the cost of manufacturing.

Some of the factors that can increase the ease with which a product can be assembled are: (1) reduction in the number of parts; (2) provision of a suitable base component; (3) assembly in layer fashion from above along a vertical axis; (4) elimination of lengthy fastening techniques; and (5) design for ease of locating or aligning by providing chamfers, pilots, and so on.

Figure 16.5 shows the results of some experiments conducted to examine the effect of chamfers in the common process of inserting a peg into a hole, as presented by G. Boothroyd of the University of Massachusetts. It can be seen that the manual insertion time depends largely on the clearance between the peg and hole but can be significantly reduced for small clearances by the provision of suitable chamfers. Further work indicated that the use of an optimum curved chamfer can reduce insertion times to the minimum value for all clearances. This chamfer is a body of constant width (Figure 16.6) and ensures that, during assembly, no more than two contact points can arise between the peg and the hole, hence eliminating the possibility of jamming during the early stages of insertion.

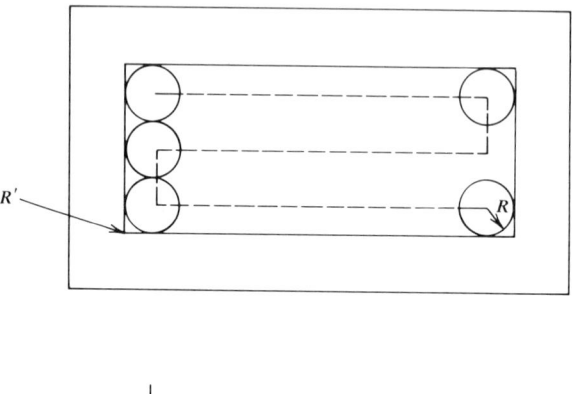

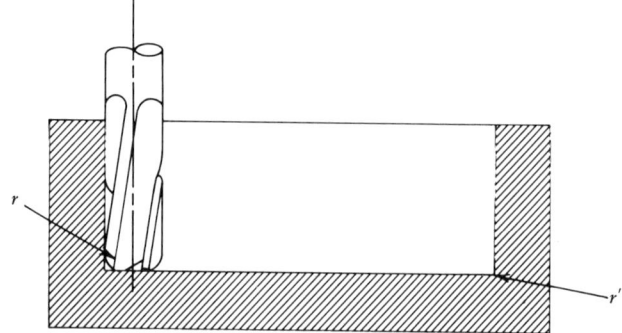

Figure 16.4 Machining of a rectangular pocket.

Figure 16.7 shows an example of the redesign of a box and lid assembly. The lid has been redesigned to eliminate asymmetry about the vertical axis and is self-locating. The screws have been produced with chamfered pilots and are also self-locating. The effects of various design changes were tested. The results show that the manual assembly time can be reduced from 28 to 14 seconds by using self-located/aligned screws and lid. The change of the symmetry of the lid had no measurable effect on the assembly time.

Design features for ease of assembly and ease of machining might be in conflict with one another. For example, in Figure 16.7 the lid without the self-locating shoulder is much easier to manufacture than the one with the shoulder, while the screw with a pilot might cost more than the screw without it. A compromise must be made between the conflicting demands.

In Part *a* of Figure 16.8 the design follows the ease of assembly

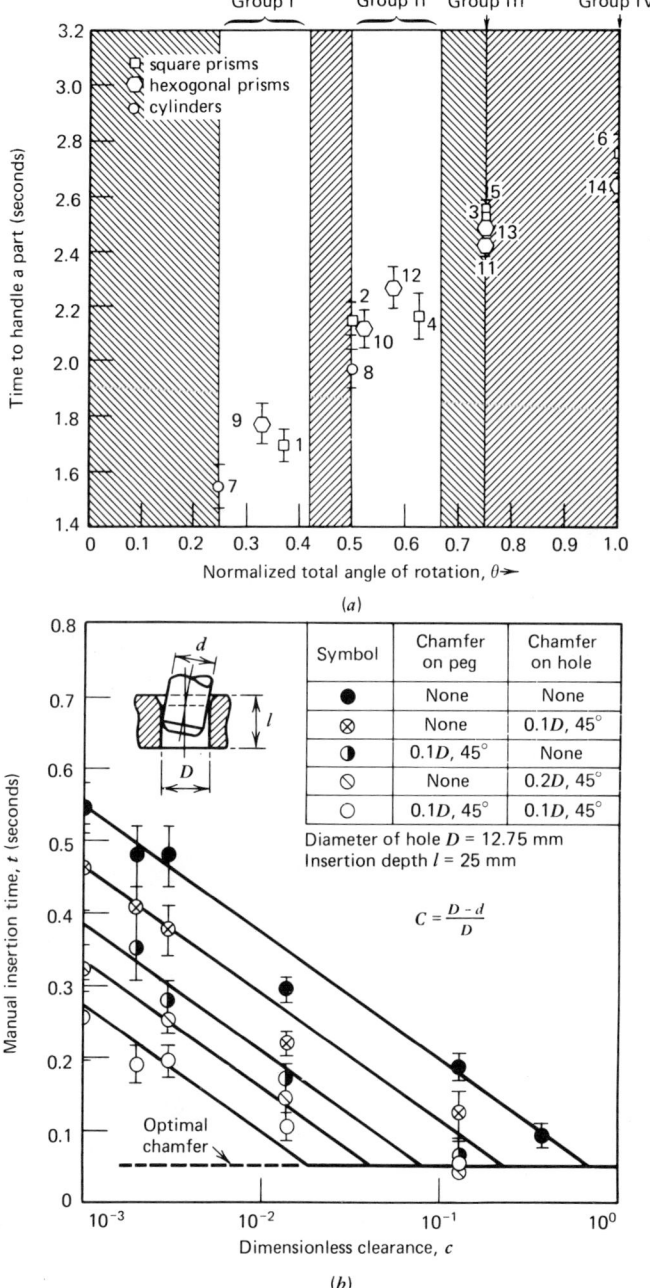

Figure 16.5 Results of experiments in assembly. (a) Relationship between the total angle of rotation and the time required to orient a part. The times are the average for two individuals. (b) Effect of clearance and chamfer design on manual insertion times.

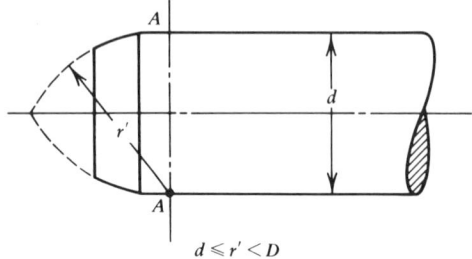

$d \leqslant r' < D$

Figure 16.6 Optimum chamfer design.

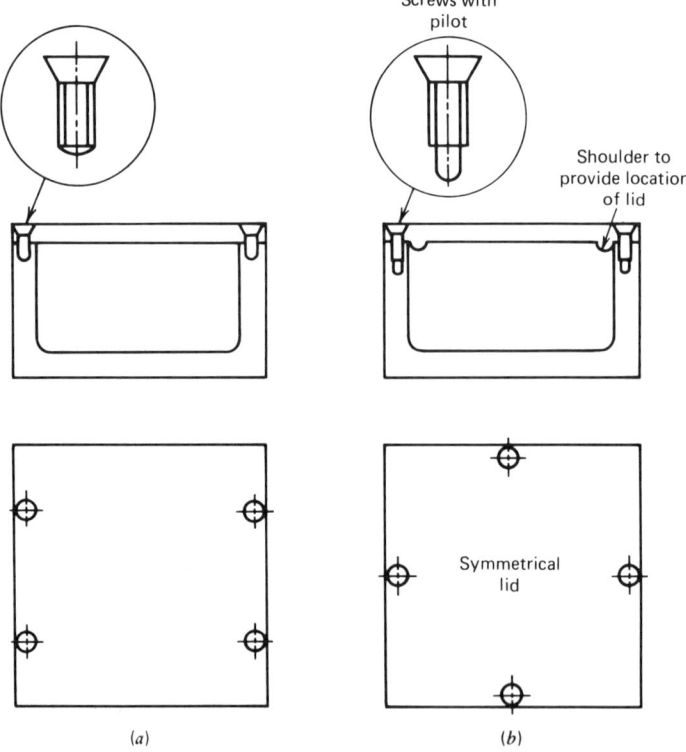

Figure 16.7 Redesign of a box and lid assembly. (a) Old design (assembly time = 28 seconds). (b) New design (assembly time = 14 seconds).

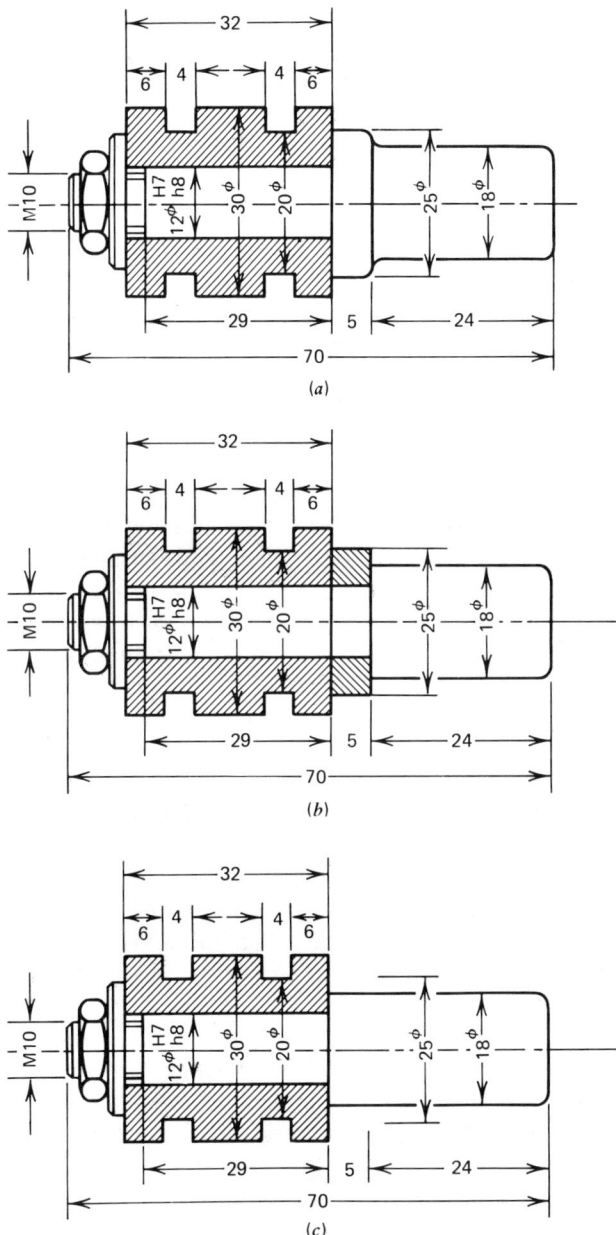

Figure 16.8 Design possibilities.

factor—reduction of the number of parts. The shoulder of 25 mm diameter and 5 mm thickness is a filler. Both dimensions can be varied without affecting the performance. In order to save material and reduce machining time, the shaft can be redesigned as shown in Parts *b* and *c*. These savings will increase assembly time and cost. Designers are professional engineers and hence know that they should consider all of the above factors. However, they are neither time and motion study experts, material or process engineers, nor economists. They will do their best, but, unfortunately, that will not be enough.

Hal suggests backing them up with computer programs that will draw their attention to all of the above factors and assist them in computing and selecting the best compromise.

Part Design

Under Hal, designers work in the interactive mode with a graphic display terminal, using the part description system as described in Section 15.3. They specify a shape or a body corresponding to the part description rules, and the shape appears on the terminal either as standard engineering drawing views or as a three-dimensional body that can be rotated to show the view from any desirable angle. As the designers proceed with the definition of the part, segment by segment, including special features, the display adds the separate segments to show a complete part drawing on the terminal.

A library of company standards is part of the Hal system. Designers can retrieve and display on the screen any feature they need and incorporate it into the design. On the other hand, the program keeps track of the design, and when the designers specify a feature (hole diameter, thread, keyway, etc.) that is not in agreement with the standards, a remark is displayed. The program will not accept nonstandard features unless a special authorization is given.

Standard tolerance system tables [e.g., the International Standards Organization (ISO)] are also part of the system. Designers can retrieve these tables and look up the recommended tolerance for the class of fitness they choose. A permanent remark about the effect of tolerances on cost will be displayed whenever tight tolerances are specified. At this stage the tolerances are accepted as specified.

At any moment designers can retrieve any of the structural analysis programs available in the Hal data base.

The designers review the design, change any detail if necessary, and when satisfied insert the "end of part" sign.

16.2 Design of Parts and Products

Hal then rearranges the part description data in the computer memory to allow the performance of Hal features.

A review program scans the part and identifies the most likely process for it. The scan function is performed in memory. No external data or classification and coding systems are required. Data organization and program logic are sufficient to scan a part in memory, to "see" part shape, wall thickness, and complexity of the geometric shape, and to determine the process.

Based on the selected process, a production cost estimate is made. The cost estimate is approximated by volume of chip removal in metal cutting, by tolerance–cost charts, material cost, die cost estimation, and so on. The estimated cost figure is displayed on the screen.

The cost analysis program then reviews the design and displays the messages that draw the attention of the designers to those features and dimensions that increase cost. In some cases change recommendations will also be displayed.

Next the material specified for the part is compared to other materials and to those that exist in the company inventory. The comparison is by specification and cost. The system might suggest the use of a less expensive material that possesses the required properties.

The system then reviews material utilization. In metal-removal processes the system computes the volume and cost of the removed material. The removed material is a waste of raw material cost, and of the time consumed in processing. If these values are high, the system scans the part to find out whether modifications in part design can reduce the cost. The system reasoning and recommended changes are displayed on the screen. Final decisions are reached by the designers. The program scans the part in search of profile sections. It checks whether a different profile (I, T, U, etc.) may possess a better moment to inertia to area ratio. If such a profile is encountered, a recommendation to change the design is displayed on the screen.

The program reviews all fillets and radii for ease of machining. It displays its recommendations on the screen for the designers' responses.

In case of castings or forgings, the program reviews such features as wall inclination, radius, wall thickness, and hidden corners for ease of production and displays its findings and recommendations on the screen.

Design features that can affect handling time, such as sharp corners, slipperiness, fragility, symmetry, balance, and gripping area, are examined and evaluated. Messages on the screen draw attention to those features that will present handling problems.

Machining aids are considered. The program scans the part, examining its shape for a reference machining area (*datum area*), easy gripping location, and ease of handling during production. If the company uses standard tooling, jigs, and fixtures, the program will consider their features. Recommendations to modify or even add features to the part for ease of machining will be displayed on the screen.

Miscellaneous functions are also carried out.

Many of the above features of the Hal system are not new. We all recognize their importance and try to achieve them. By its part description and file organization Hal enables designers to perform these functions in an automatic way. Once the logic is formulated, it can be used over and over again. The computer, with its data organization method, can see, scan, and manipulate part design. Any feature based on these capabilities, external data, and known logic can be added to the Hal system.

Product Assembly

Designers can retrieve any part available on file and display it on the graphic terminal. By the use of the assembly feature of the part description system, they define the location and orientation of another part and thus create an assembly drawing on the screen. Again, the presentation on the screen can be in an engineering drawing form or it can be a three-dimensional body that can be rotated, increased or decreased in scale, and presented according to any other standard feature in CAD systems.

The designers can check the assembly for interference, clearance, and motions until they are satisfied with the design.

The program scans the assembly in search of tight tolerances on non-mating segments and draws the attention of the designers to all such cases.

The program reviews the assembly in search of parts that are statically connected to one another. Such parts can be detected either by the information given on the GG entry line of the part description system or by the configuration of the assembly and the nature and shape of the parts in question. An attempt is made to reduce the number of parts in the assembly if it is profitable and practical. The program can compute the access space required for assembly and thus determine whether the suggested change is practical.

The program reviews the dynamic details of the assembly. These can be detected either by designer statement on the assembly entry or by the nature and shape of the part and the method by which it is secured in its

16.2 Design of Parts and Products

location. The program examines whether clearance, expansion, lubrication, adjustment, and so on have been considered by the designers. Any comments and recommendations will be displayed on the screen.

The program anticipates the sequence in which the product is to be assembled. It draws the attention of the designers to all the cases that are either impossible or very difficult to assemble because there is no access.

The program reviews the mating parts with respect to ease of assembly. A table containing common features for ease of assembly is part of the Hal data bank. While scanning the assembly, the program recognizes the existence of the prevailing conditions, looks up the table for recommendations, and checks whether these are incorporated into the design. A time table is utilized to approximate assembly time. Another module estimates extra machining cost of the features and optimizes the design. The resulting recommendations are displayed on the screen.

Company assembly standards, if available, can be stored in the Hal data bank. It might contain information on chamfers, clearances, reliefs, mating parts material restrictions, locking devices, and so on. The program will check whether the assemblies' details follow the standards.

Company policy and technological transfer can be considered. For example, the company might want to introduce a new technology, such as sealings, bondings (e.g., epoxies), or bearings. A description of such a technology, its capabilities and limitations, can become part of the system. During the design process, the system will draw the attention of the designers to the new technology. At the review stage the program will scan the assembly in order to detect conditions where such a technology can replace items of the original design. Recommendations for changes will be displayed on the screen.

In cases where the company applies standard assembly tools, handling devices, mechanical devices, or robots for automatic assembly, those standards and special features are kept in the Hal files. The program can detect environments where those standards are applied and draw the attention of the programmer to them. Recommendations for added features on the parts, such as gripping holes, locating surfaces, eye bolts, or any other special feature required, will be displayed on the screen.

During customization of the system, each individual company can add any miscellaneous feature it desires. It should be remembered that the system can "see" the assembly as a human designer can; however, being a machine, it does not possess imaginative power. The logic of the additional features should be supplied by the using company. This is done only once. Henceforth the system is able to apply this logic over and over again, whenever appropriate.

16.3 Other Design Features of Hal

In addition to direct assistance in design, Hal has many other features that can help with the daily work and reduce the cost at the design stage. These features are described in the following sections.

Drawings Management

All the company's drawings are kept on magnetic files. There is no need to keep hard copy drawings. Hal uses a condensed form by which part drawings are kept on file. It is estimated that 25,000 drawings can be stored on one tape reel (9 tracks, 1,600 BPI, 2,400 ft). With the new high-density tapes (6,250 BPI), about 100,000 drawings can be stored on one tape reel. Magnetic disk packs are available with a capacity of above 200 Mbytes and can store more than 100,000 part drawings. Any of the drawings stored can be retrieved in a matter of seconds, and a hard copy or microfilm can be prepared.

This type of drawing storage saves a great deal of floor space, provides easy back-up, secures the company's drawings against fire or other disasters, saves time in locating and retrieving a drawing, and eliminates the possibility of using an outdated drawing in manufacturing the items.

Retrieval by Key

Any drawing available in the system file can be retrieved by drawing, part number, or function key. The first is used when designers know exactly the part drawing number they need. It is applicable mainly in assembly design, revisions, and inquiries. A function key can be attached to each item, stating its normal purpose, such as brackets, springs, bearings, screws, cups, lids, or shafts. Each company can create its own function key. However, it is recommended that the next feature, retrieval by parameters, be carefully examined before using the function key, which reminds one of a classification and coding system. Function keys can be used for a limited number of purposes. A general-purpose key might retrieve too many drawings, thus defeating the purpose.

The designer might create standard subassemblies, such as lids, bearing housings, and hydraulic sealing units, and assign them a purpose name (function key) or a part number. In such cases the data are kept only by pointers to the separate parts, thus saving storage space.

Retrieval by Parameters

This feature is used to eliminate the redesign of existing parts and the need to design a new part if minor modifications on an existing drawing can

16.3 Other Design Features of Hal

create a new part drawing. This feature can also be used to create groups or families of parts, to be used for any desirable purpose.

Parts can be retrieved by a search argument that describes the desired features and the search limits. The inquiry request is not limited to any prearranged keys, but presented in a method similar to the part description system, with two exceptions: (1) Dimensions are not defined as a unique number but rather in a range, stating the lower and upper limits and (2) the part must not be uniquely defined, and portions of it can be omitted. This will indicate that the designer does not care about the shape in that particular section of the part.

The search argument can be defined in rough terms, such as: a part whose external dimensions are 100–150 by 70–90 by 40–60. In such a case, all parts, regardless of material, holes, threads, steps, and so on, will be retrieved.

The definition can restrict the search to only those parts made out of a specific material and possessing any special feature, such as a hole or taper or a combination of and/or/nor features. The search argument can be defined in fine terms, such as: a bracket of 80–90 by 20–25 by 5–7 that has a hole of ¼ to ⅜ in the middle and a 90-deg bend at 10–15 mm from the edge; or, a shaft has two bearings of known size 250–280 mm apart, and between them a pulley of inside diameter 1 in. is mounted and secured by a key—find an existing drawing that can be used. The search argument can also be defined in exact form, giving a sketch of the desired part.

Figure 16.9 shows the computer output for the case where several inquiries are entered as one batch. The inquiries are printed at the top of the page, followed by part numbers that correspond to the search request.

The inquiry is a free format; two letters designate the feature request, followed by the lower and upper limits. Figure 16.10 shows an inquiry in four steps, starting with a rough definition of only two parameters; 250 parts satisfy this request. A third parameter reduces the selected group to 50 parts. When a fourth parameter is defined, the number of parts complying with the inquiry is reduced to 5, and drops to only 1 part when a sixth parameter is added. This example was prepared using a file of 25,000 part drawings stored on a 3330 disk drive. It takes 9–15 seconds elapsed time per inquiry.

Automatic Change of Design

Whenever the company decides to adopt a new technology or a new design standard, an automatic change of all company drawings affected by it can be made. This feature is carried out for design standard (such as, hole sizes, threads, slots, and keys) by introducing a table listing all new standards and a change command. The above list will act as an inquiry for

SEARCH REQUEST LIST   16/01/79

GIDEON 4   01X   LN = 75000 , 77500,   BH = 180,190
GIDEON 4   01   OD=22850 , 25000,   ID = 1000 , 1100 , MC = 1,1
HL-1   GDD   01   OD=12000,20000,   LN = 2500,3300   MC=1,1,BH=280,320
HL-2   01   BH = 180,230,   OD = 13300,22000,   LN = 48000,55000
HL-3   ABC   01   ID = 610,660
HL-4   TST   01   OD = 600,650,   OD =   35000,38000   BH = 220,280
HL-4   TST   01X   OD=34000,38000,LN=5500,60000,MC=1,1,ID=800,950
HL-5   2345   01   BH=180,280
HL-6   ABCDEF 01X   MC = 2,3   LN =20000,25000   OD = 55300 ,60000
HL-6   ABCDEF 01   BH =   260 , 280
                     LN   = 60000,80000   OD=45000,50000
                     BH = 120 , 180   ID   = 1000, 1300

REQUEST NO.	PART NUMBER	REQUEST NO.	PART NUMBER	REQUEST NO.	PART NUMBER
HL-1 GDD	6005	HL-1 GDD	6010	HL-1 GDD	6015
HL-1 GDD	6025	HL-1 GDD	6030	HL-1 GDD	6035
HL-1 GDD	6045	HL-1 GDD	6050	HL-2	6773
HL-2	6873	HL-2	6923	HL-2	6973
HL-2	9323	HL-2	9373	HL-2	9423
HL-3 ABC	12524	HL-3 ABC	12574	HL-3 ABC	12624
HL-3 ABC	12724	HL-3 ABC	12774	HL-3 ABC	12824
HL-3 ABC	12924	HL-3 ABC	12974	HL-3 ABC	13024
HL-3 ABC	13124	HL-3 ABC	13174	HL-3 ABC	13224
HL-3 ABC	13324	HL-3 ABC	13374	HL-3 ABC	13424
HL-3 ABC	13524	HL-3 ABC	13574	HL-3 ABC	13624
HL-3 ABC	13724	HL-3 ABC	13774	HL-3 ABC	13824
HL-3 ABC	13924	HL-3 ABC	13974	HL-3 ABC	14124
HL-3 ABC	14124	HL-3 ABC	14174	HL-3 ABC	14224
HL-3 ABC	14324	HL-3 ABC	14374	HL-3 ABC	14424
HL-3 ABC	14524	HL-3 ABC	14574	HL-4 TST	14533
HL-3 ABC	14624	HL-3 ABC	14674	HL-4 TST	14724
HL-3 ABC	14824	HL-3 ABC	14874	HL-3 ABC	14924
HL-6 ABCDEF	19787	HL-6 ABCDEF	19792	HL-6 ABCDEF	19837
HL-6 ABCDEF	15887	HL-6 A3CDEF	19892	HL-6 ABCDEF	19937
HL-6 ABCDEF	19987	HL-6 ABCDEF	19992	HL-5 2345	20804
HL-5 2345	20814	HL-5 2345	20819	HL-5 2345	20824
HL-5 2345	20834	HL-5 2345	20839	HL-5 2345	20844
HL-5 2345	20854	HL-5 2345	20859	HL-5 2345	20864
HL-5 2345	20874	HL-5 2345	20879	HL-5 2345	20884
HL-5 2345	20894	HL-5 2345	20899		

Figure 16.9 Several inquiries to retrieve parts by parameters.

parts having the standard features. The retrieved part drawing will be changed automatically to correspond with the new standards. A report of all changes made will be prepared. The designer might ask to see any or all such drawings on the screen.

This mode of change can handle most of the mating parts automatically, since the part description system regards a feature as a neutral that becomes a male or female according to its location.

For new technology changes the user must specify what (and how) to alter in a given design. This definition is in general terms and not for a specific part. The above definition is the search argument by which the drawing file is to be searched. Such a definition can refer to an individual part or to an assembly. The retrieved parts and assemblies will be automatically changed as instructed.

Automatic Drawing Preparation

The drawings in the system file are part drawings. In many companies additional sets of drawings are prepared for inspection and production. These drawings deviate from the part drawings by tolerances and dimensions. The deviation is usually carried out by a set of rules. Such rules can be incorporated into the system, and the required additional set of drawings will be prepared on request.

This feature can also be used to prepare gauge drawings, special tools, forgings, castings, and any other drawings that can be derived by employing a set of rules on part drawings.

Miscellaneous

Any request based on uniquely defined rules requiring the ability to see, search, and retrieve drawings can be granted.

16.4 Process Planning under Hal

Part drawings, as defined by the part description system and stored in Hal files, are the input to the generative process planning program (GPPP). The GPPP was described in Part II of this book and is a part of the Hal system. Thus all the designer has to do is to specify a part number or drawing number and request a process plan. The process plan generated will be the practical optimum plan for the particular plant and will be used in the Hal manufacturing cycle, as shown in Figure 15.1. The Hal system keeps a dynamic environment in production, and the process plan phase

SEARCH REQUEST LIST

HL-600   01   LN = 35000,40030,   OD = 14000,18000   16/01/79

REQUEST NO.	PART NUMBER	REQUEST NO.	PART NUMBER	REQUEST NO.	PART NUMBER	REQUEST NO.	PART NUMBER
HL-600	6251	HL-600	6252	HL-600	6253	HL-600	6254
HL-600	6255	HL-600	6256	HL-600	6257	HL-600	6258
HL-600	6259	HL-600	6260	HL-600	6261	HL-600	6262
HL-600	6263	HL-600	6264	HL-600	6265	HL-600	6266
HL-600	6267	HL-600	6268	HL-600	6269	HL-600	6270
HL-600	6271	HL-600	6272	HL-600	6273	HL-600	6274
HL-600	6275	HL-600	6276	HL-600	6277	HL-600	6278
HL-600	6279	HL-600	6280	HL-600	6281	HL-600	6282
HL-600	6283	HL-600	6284	HL-600	6285	HL-600	6286
HL-600	6287	HL-600	6288	HL-600	6289	HL-600	6290
HL-600	6291	HL-600	6292	HL-600	6293	HL-600	6294
HL-600	6295	HL-600	6296	HL-600	6297	HL-600	6298
HL-600	6299	HL-600	6300	HL-600	6301	HL-600	6302
HL-600	6303	HL-600	6304	HL-600	6305	HL-600	6306
HL-600	6307	HL-600	6308	HL-600	6309	HL-600	6310
HL-600	6311	HL-600	6312	HL-600	6313	HL-600	6314
HL-600	6315	HL-600	6316	HL-600	6317	HL-600	6318
HL-600	6319	HL-600	6320	HL-600	6321	HL-600	6322
HL-600	6323	HL-600	6324	HL-600	6325	HL-600	6326
HL-600	6327	HL-600	6328	HL-600	6329	HL-600	6330
HL-600	6331	HL-600	6332	HL-600	6333	HL-600	6334
HL-600	6335	HL-600	6336	HL-600	6337	HL-600	6338
HL-600	6339	HL-600	6340	HL-600	6341	HL-600	6342
HL-600	6343	HL-600	6344	HL-600	6345	HL-600	6346
HL-600	6347	HL-600	6348	HL-600	6349	HL-600	6350
HL-600	6351	HL-600	6352	HL-600	6353	HL-600	6354
HL-600	6355	HL-600	6356	HL-600	6357	HL-600	6358
HL-600	6359	HL-600	6360	HL-600	6361	HL-600	6362

Figure 16.10 Retrieval of parts by parameters in four steps.

6363	HL-600	6364	HL-600	6365	HL-600	6366	HL-600
6367	HL-600	6368	HL-600	6369	HL-600	6370	HL-600
6371	HL-600	6372	HL-600	6373	HL-600	6374	HL-600
6375	HL-600	6376	HL-600	6377	HL-600	6378	HL-600
6379	HL-600	6380	HL-600	6381	HL-600	6382	HL-600
6383	HL-600	6384	HL-600	6385	HL-600	6386	HL-600
6387	HL-600	6388	HL-600	6389	HL-600	6390	HL-600
6391	HL-600	6392	HL-600	6393	HL-600	6394	HL-600
6395	HL-600	6396	HL-600	6397	HL-600	6398	HL-600
6399	HL-600	6400	HL-600	6401	HL-600	6402	HL-600
6403	HL-600	6404	HL-600	6405	HL-600	6406	HL-600
6407	HL-600	6408	HL-600	6409	HL-600	6410	HL-600
6411	HL-600	6412	HL-600	6413	HL-600	6414	HL-600
6415	HL-600	6416	HL-600	6417	HL-600	6418	HL-600
6419	HL-600	6420	HL-600	6421	HL-600	6422	HL-600
6423	HL-600	6424	HL-600	6425	HL-600	6426	HL-600
6427	HL-600	6428	HL-600	6429	HL-600	6430	HL-600
6431	HL-600	6432	HL-600	6433	HL-600	6434	HL-600
6435	HL-600	6436	HL-600	6437	HL-600	6438	HL-600
6439	HL-600	6440	HL-600	6441	HL-600	6442	HL-600
6443	HL-600	6444	HL-600	6445	HL-600	6446	HL-600
6447	HL-600	6448	HL-600	6449	HL-600	6450	HL-600
6451	HL-600	6452	HL-600	6453	HL-600	6454	HL-600
6455	HL-600	6456	HL-600	6457	HL-600	6458	HL-600
6459	HL-600	6460	HL-600	6461	HL-600	6462	HL-600
6463	HL-600	6464	HL-600	6465	HL-600	6466	HL-600
6467	HL-600	6468	HL-600	6469	HL-600	6470	HL-600
6471	HL-600	6472	HL-600	6473	HL-600	6474	HL-600
6475	HL-600	6476	HL-600	6477	HL-600	6478	HL-600
6479	HL-600	6480	HL-600	6481	HL-600	6482	HL-600
6483	HL-600	6484	HL-600	6485	HL-600	6486	HL-600
6487	HL-600	6488	HL-600	6489	HL-600	6490	HL-600
6491	HL-600	6492	HL-600	6493	HL-600	6494	HL-600
6495	HL-600	6496	HL-600	6497	HL-600	6498	HL-600
6499	HL-600	6500	HL-600				

Figure 16.10 (*Continued*)

SEARCH REQUEST LIST 16/01/79

HL-700   01   LN = 35000,40000,   OD = 14000,18000   MC = 1,1

REQUEST NO.	PART NUMBER	REQUEST NO.	PART NUMBER	REQUEST NO.	PART NUMBER	REQUEST NO.	PART NUMBER
HL-700	6251	HL-700	6252	HL-700	6253	HL-700	6254
HL-700	6255	HL-700	6256	HL-700	6257	HL-700	6258
HL-700	6259	HL-700	6260	HL-700	6261	HL-700	6262
HL-700	6263	HL-700	6264	HL-700	6265	HL-700	6266
HL-700	6267	HL-700	6268	HL-700	6269	HL-700	6270
HL-700	6271	HL-700	6272	HL-700	6273	HL-700	6274
HL-700	6275	HL-700	6276	HL-700	6277	HL-700	6278
HL-700	6279	HL-700	6280	HL-700	6281	HL-700	6282
HL-700	6283	HL-700	6284	HL-700	6285	HL-700	6286
HL-700	6287	HL-700	6288	HL-700	6289	HL-700	6290
HL-700	6291	HL-700	6292	HL-700	6293	HL-700	6294
HL-700	6295	HL-700	6296	HL-700	6297	HL-700	6298
HL-700	6299	HL-700	6300				

SEARCH REQUEST LIST 16/01/79

HL-800   01X   LN = 35000,40000,   OD = 14000,18000   MC = 1,1
HL-800   01   ID = 800, 1000

REQUEST NO.	PART NUMBER	REQUEST NO.	PART NUMBER	REQUEST NO.	PART NUMBER	REQUEST NO.	PART NUMBER
HL-800	6281	HL-800	6282	HL-800	6283	HL-800	6284
HL-800	6285						

Figure 16.10 *(Continued)*

```
                                          SEARCH REQUEST LIST              16/01/79

                        HL-900     01X         LN = 35000,40000,    OD = 14000,18000        MC = 1,1
                        HL-900     01                    ID =  800,  1000            BH= 180,220

REQUEST  NO.  PART  NUMBER         REQUEST  NO.  PART  NUMBER         REQUEST  NO.  PART  NUMBER
HL-900                 6283

                                          SEARCH REQUEST LIST              16/01/79

                        HL-1000    01X          LN = 33000,38000,    OD= 16000,16000,       MC = 1,1
                        HL-1000    01                ID =  900, 900         BH = 200,200

REQUEST  NO.  PART  NUMBER         REQUEST  NO.  PART  NUMBER         REQUEST. NO.  PART  NUMBER
```

Figure 16.10 *(Continued)*

should be able to generate processes to comply with instantaneous shop load, as will be discussed in Chapter 17. Hal's aim is to design and plan anything only once, let the computer system acquire that intelligence, and then use it over and over again in an automatic fashion. Although GPP is not a must, it is strongly recommended. If it is impossible or it takes too long to prepare a generative program, there are two alternate methods to treat process planning under Hal. Both alternatives utilize the concept, used in the GPPP, of separating the solution into engineering and mathematical stages. The engineering stage defines a theoretical process. It does not consider the specific facilities available. The mathematical stage will transform the theoretical process into a practical one, in a dynamic way. It will consider quantities, available machines, and load conditions. The mathematical stage is universal, that is, it can be used to handle any type of production. It can work with engineering data generated by a computer program or in any other way. The alternatives are, therefore, for the engineering stage and should be specified in general terms, not in fixed rigid terms. It should not refer to any specific machine, but rather to a type of machine and to the characteristics required.

The engineering data required are thus based only on the fixed elements of process planning, namely, geometric shape of the part, tolerances and surface finish, and raw material. The mathematical stage will handle the variable parameters, namely, quantity required, optimization criterion, and available facilities.

Retrieval Process Planning

Retrieval process planning is based on a family of parts, where a standard process plan exists for each family. The standard plan is the user's responsibility. In the existing systems (see Section 11.4), a classification and coding system is needed in order to retrieve a process. Under Hal, the data description system replaces the classification and coding system. The standard process plan changes its meaning and it is the theoretical process plan (the first module of GPPP). Thus the elements required for standard process definition and retrieval are geometric, and the conflict between geometric-, process-, or resource-oriented classification and coding systems is removed.

An existing retrieval process planning program (such as CAM-I, CAPP, TNO, AUTOPROS, and TAUPROG) can be used and become part of the Hal system. The standard process plan will define a theoretical process for a master part. The master part is defined by a range of dimensions and tolerances. This range definition can take place at the creation of the file, at the retrieval stage, or at both, where the definition at the retrieval stage supersedes that of the creation stage. In a way, the part description, among all its uses, also serves as a classification and coding system. The use of such classification and coding is internal in the computer program, and users do not even have to know that classification and coding are used. They just state, in a common engineering language, which kind of group or family of parts they want, and the system will retrieve it.

Since the classification and coding used as the interconnection of the retrieval process planning system and Hal are transparent to the user and automatically created, there are no error hazards in using a long code number (30 or more characters). Thus the classification is more refined without effort and expense and can be changed whenever wished.

The system is operated as follows:

1. Users specify a part number and state that they want to store a standard process plan for it. The system checks whether a standard plan already exists for this part or for a similar part. If such a process does not exist, the system creates the code number required by the retrieval system and accepts the standard plan. If a standard plan exists, the users can retrieve it, review it, and decide which action they wish to take. The method by which the standard plan is stored and the interactivity of the retrieval system used are maintained. At this stage Hal supplies an automatic classification and coding system and no more.

2. Users can retrieve any process stored by stating the exact part number or by issuing an inquiry to create a family of parts (see Section 16.3). The creation of a family of parts by inquiry is a dynamic grouping and can vary from one inquiry to another. In the retrieval process planning program the family of parts is rigidly connected to the standard process plan. Thus it is possible that more than one standard process will be retrieved. It is up to the users to decide which action they wish to take. They can use it as a study tool—it is a very strong tool. Comparison of the several standard plans retrieved, when using dynamic grouping, can be used to optimize the standard process plan and, even more importantly, to use it as a research tool in the preparation of a GPPP.

Manual Process Planning

The Hal system is not affected by the method used to derive a theoretical process plan for each part. It can be manually derived if the user wishes.

Users can start by specifying a process plan for each part manually. They can later use the dynamic grouping of parts feature of Hal to create a standard process for a family of parts, thereby moving to a retrieval system. In doing so, optimization of processes takes place. At a later stage, users continue their study and research and move to GPP.

Chapter Seventeen
Hal—in Production

Production planning is a developed field. A great deal of research work has been done in an attempt to optimize production. We find fascinating scheduling and sequencing theories, inventory systems, forecasting, dispatching rules, and so on. They all use advanced mathematical techniques in order to solve production problems and arrive at optimum operations. The engineering phases, that is, product design and process planning, are the fixed input data to which production planning applies the mathematical modules.

The main goals of production planning are to meet due dates and to keep capital tied down in production to a minimum. Sophisticated computer programs are available with built-in algorithms to decide when to split a job between two or more machines, when to apply overlap of operations, when to reduce transfer time, when to use an alternate machine, how to resolve competition between operations for facilities, and so on. These programs are backed up by on-line job recording and updating systems. Nevertheless, with all these, the goal is not usually achieved. The inherent fault is in the basic logic, not in the programs or the scheduling theories. In a plant there are, and will always be, "better machines" that will be overloaded and other machines that will usually be underloaded. Overload means that there are larger queues waiting for the machines, and these queues increase work-in-process. They are usually resolved by pulling jobs forward, which might result in not meeting due dates. Underload means that in order to balance the load and supply work to operators, the jobs are pulled backward, thus increasing work-in-process.

Requirement planning improves the investment in inventory. It issues a list of available items which are not required by any of the customer orders. Such items will always exist. They are leftover from assembly, since the actual rejection rate was not exactly the anticipated one. They are due to cancellation or change of orders or change in design. The requirement planning system merely indicates the existence of such items. It does not regard the resolving of production problems to be its duty. Dead stock and slow-moving items exist in inventory. The inventory system recognizes such items and issues a special report listing them. Resolving the situation is outside its scope.

It is characteristic of present-day technology (IMS, PICS, or any other) that whenever it encounters a situation that cannot be resolved mathematically, the system simply ignores the problem or turns to the interactive mode of operation. Users, however, are helpless. They have a problem but no tools to solve it with. Solving some of the problems in the interactive mode requires a great amount of data or computations. These data are unavailable to users, and computing them manually takes too long. Consequently, users are thrown back to relying on their intuition and memory.

Thus the optimum reached will be mathematically correct, but questionable from an overall manufacturing standpoint. This situation occurs because the production phases accept the engineering data as fixed and untouchable. They do not question whether the product design or the process planning is efficient or not, but might, and often do, optimize inefficient processes. To improve this situation, the overall (Hal) manufacturing process is looked on as one unit. The input data from the engineering phases, as discussed in Chapter 16, are regarded as the theoretical optimum. It is a starting point for the production phases, but can be modified if needed. It thus introduces additional courses of action for resolving problems, and enables a true optimum to be reached.

Hal can be employed as an extension of the existing production system or as a reorganization of the production phases. Both methods will be discussed in the following sections.

17.1 Adaptive Production-Process Planning

In the production planning and scheduling phase, the Hal concept means that the machining process advocated is not a constraint. In present-day practice, whenever a waiting line queues up for a certain machine, the jobs and product network are pulled forward to a later date. It is proposed that in such cases, instead of shifting networks, the machining process of the part be changed by predefined rules. In this approach, due dates will be met (unless they have been established arbitrarily) and the workload on all machines made smooth, without pulling operations forward or backward; therefore, work-in-process is not increased.

This goal seems to introduce a loop into the system. As long as the overall load profile is unknown, the machine loading profile is unknown, and no information on overloaded machines is available. Thus the process cannot be defined. As long as the process is not defined, job lead time is unknown, and due dates for items and operations cannot be set. Thus the overall load is unknown. To study this problem, an investigation of alternative machining processes was conducted. It is of interest to note that the dispersion of the results for maximum production was on the order of 10%, while the increase in cost was above 200%. Since this outcome is a function of the facilities introduced, tests were made with several varieties of machines. As long as the machines were of the same type, such as the universal lathe, the results were consistent. Only where a mixture of machine types was introduced (such as universal lathe, turret lathe, automat, and NC machines), and the alternatives were between machine types, did the time dispersion grow to over 200% and the cost to almost 300% (see Figure 17.1).

To continue with the research, one should set an acceptable limit for the time dispersion between alternatives. The question of which time should be assigned to operations for scheduling purposes is always a puzzle. Should standard or actual time be used? The incentive rate is usually between 25–35%. Should these figures be used? The other problem is how many working hours per day, or period, should be used as the available load. Should the actual time be assigned or should allowances be made for machine and tool breakdown, operator absenteeism, rush jobs, and so on? Another decision concerns the lead time to be used in requirement planning. In capacity planning, overlap, split, and reduced transfer time can be employed to reduce lead time. These are unknown at the requirement planning stage.

All of the above demonstrate that today's system tolerates a dispersion of more than 50% in machining time. If this figure is adopted, the loop problem can be overcome. The following simplified example will illustrate the basic idea of how the adaptive production-process planning is going to function.

It is required to produce 500 pieces of part 1105 and 500 pieces of part 2105. The advocated machining process with five alternatives can be seen in Figure 17.2.

One can note that machine 704 is the first choice for both parts, which in present-day practice means that one of the parts has to wait in queue.

In the suggested approach, the second alternative is examined, that is, machine 704 is not assigned to either of the parts. The algorithm will have to decide which part process will use machine 704 and which one will have to use an alternative machining process. One type of algorithm can use

17.1 Adaptive Production-Process Planning

Quantity (pieces)	Alternative	Maximum production		Minimum cost	
		Side 1	Side 2	Side 1	Side 2
1,000	1	500 NC	500 NC	520 TL	520 TL
		5.23 minutes		6.95 minutes	
		$8.84		$2.92	
	2	510 A	510 A	530 U	530 U
				510 A	606 U
				606 U	
		5.54 minutes		11.85 minutes	
		$5.10		$3.24	
	3	520 TL	520 TL	530 U	530 U
				510 A	510 A
					560 U
		7.19 minutes		9.11 minutes	
		$3.09		$3.38	
	4	530 U	530 U	510 A	510 A
				560 U	560 U
		14.28 minutes		7.18 minutes	
		$3.60		$4.60	
50	1	500 NC	500 NC	520 TL	520 TL
		5.31 minutes		8.47 minutes	
		$12.92		$3.93	
	2	510 A	520 TL	530 U	530 U
		8.90 minutes		15.13 minutes	
		$6.62		$4.46	
	3	510 A	510 A	606 U	606 U
		9.30 minutes		20.73 minutes	
		$9.12		$4.83	
	4	530 U	530 U	510 A	510 A
		15.23 minutes		9.30 minutes	
		$4.49		$9.13	

Figure 17.1 Summarized proposed process for part EXAPT 2-P106. Notation: U, universal lathe; TL, turret lathe; A, automat; NC, numerical control machine.

cost as a parameter and work in the following logic: Applying the alternative process for part 1105 will result in an added part cost of $0.5, while the added cost for part 2105 will be $2.41; therefore, it is recommended to let part 2105 use machine 704 and change the machining process of part 1105 to the second alternative, where machine 707 is recommended.

In production planning, it is the practice—in order to meet due dates or to smooth workload—to work on several lines (machines) in parallel (to split the quantity). The proposed approach can very easily handle this

	Maximum production				Minimum cost			
	Part 1105		Part 2105		Part 1105		Part 2105	
Alternative	Side 1	Side 2	Side 1	Side 2	Side 1	Side 2	Side 1	Side 2
0	717	724	724	724	704	724	704	714
		718				704		704
	22.04 minutes		39.55 minutes		22.94 minutes		42.39 minutes	
	$9.95		$11.93		$5.24		$9.60	
1	718	724	725	725	707	724	724	714
		718				707	707	724
								707
	22.04 minutes		39.55 minutes		23.59 minutes		43.32 minutes	
	$14.03		$23.80		$5.74		$12.01	
2	709	724	717	717	709	724	707	714
		705				709	705	707
								705
	22.62 minutes		40.77 minutes		23.47 minutes		48.71 minutes	
	$6.48		$13.62		$6.07		$12.55	
3	710	724	718	718	714	724	709	714
		705				714	705	705
								709
	22.62 minutes		40.77 minutes		24.51 minutes		45.29 minutes	
	$9.62		$27.25		$6.70		$12.98	
4	707	724	704	714	724	724	705	714
		705		705		717	714	705
	22.71 minutes		41.66 minutes		22.84 minutes		42.90 minutes	
	$6.30		$10.83		$7.14		$13.22	

Figure 17.2 Summarized proposed process for parts 1105 and 2105 with five alternatives. It is of interest to note that from the 12 available machines, only 10 have been selected for all five alternatives.

problem. The data required are on hand (see Figure 17.2), and the appropriate algorithm can be used. For example, one can note that for part 1105 the time and cost deviations between the second and third alternatives are negligible; hence, the quantity can be split between machines 707 and 709.

Naturally, this simplified example is not the solution to adaptive production-process planning. It merely serves as an illustration of the logic that should be used.

The adaptive production-process planning system is shown in Figure 17.3. The traditional routing file is replaced by an "operations sequence" generator. Machine loading is carried out in the customary way, and the

17.1 Adaptive Production-Process Planning

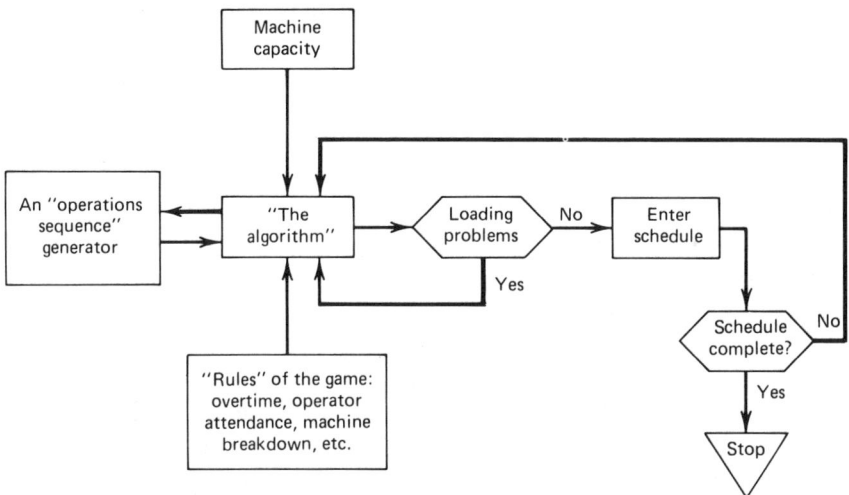

Figure 17.3 Adaptive production-process planning.

jobs are handled one at a time. The algorithm module turns to the generator with a request for a process. The generated process will be the optimum, that is, the first alternative. This process is transferred to the load module. Whenever a competition for facilities occurs, control is transferred back to the algorithm module, which examines the situation and requests additional alternatives for the jobs on hand from the operations sequence generator. The algorithm attempts to resolve the competition and forward a different set of processes to the load module. The load module will pull jobs forward or backward only as the last resort. The algorithm module can take any form, depending on the user's needs. It may consider item cost, lateness penalties, customer rating, special machine treatment, and so on. The operations sequence generator is the GPPP, as discussed in Part II of this book.

Process alternatives must be available and handled in efficient computer programs. The alternatives in the GPPP are derived by solving the matrix of the machine selection phase when selected machines are pulled out of it. At present, an alternative is defined as replacing the machine with the longest machining time. This is done by a special module that determines which machine to pull out of the matrix.

To adjust the programs for the adaptive production-process planning concept, only this module has to be changed.

17.2 Forced Process Planning

In adaptive production-process planning, the operations sequence generator played a passive role. It generated a process on request from the algorithm module. The power and capabilities of the GPPP exceed that task. The GPPP can play an active role and, based on a product mix, initiate beneficial changes and processes that will reduce machining cost and lead time. Such a feature is called forced process planning, and is discussed in the following sections.

Similar Setup

Setup is a costly and time-consuming activity. Production is a continuous process. At any instant or at the end of each job, a certain setup is available on each machine. Cost and time will be saved if the same setup is used for the following job. This fact is appreciated today, and in some scheduling systems a job with a similar setup will have loading priority.

The philosophy of forced process planning is not to wait for a chance before putting similar setup loading into practice, but rather to plan a process that will use a given setup whenever profitable. In today's practice, this is not done by the system. Good foremen might do it unofficially. Their intentions are good and locally they get good results, but in doing so they might sever scheduling and production planning from reality. The logic of forced process planning is to minimize process cost under the existing shop conditions rather than optimizing the cost of each item independently.

The GPPP normally assumes a stripped machine; thus

$$C_o = C_p + C_s$$

where C_o = optimum item cost
C_p = theoretical optimum process cost
C_s = set-up cost of the optimum process.

Considering existing shop conditions, the item cost is

$$C = C_p' + C_s'$$

where C = item cost
C_p' = actual process cost
C_s' = actual set-up cost.

When an existing setup is used,

$$C_s' = 0 \quad \text{or} \quad C_s' \ll C_s$$

17.2 Forced Process Planning

Thus

$$C < C_o$$

However, if the machine with the existing setup does not correspond to the machine and setup of the theoretical optimum process plan, it is profitable to deviate from the optimum process as long as $C < C_o$, that is, not to use optimum process, $C_p' > C_p$. The program attempts to select a process (C_p') that results in minimum cost. The steps taken are as follows:

1. Compute C_o by using the GPPP.
2. Prepare a list of all the machines that appear in the matrix of the GPPP (see section 13.4). These can be used to machine the item.
3. Check the existing setup of the machines on the list by using the job recording system.
4. Superimpose the existing setup as constraints of the machine on the list (the matrix).
5. Use the GPPP, together with the above machines and constraints, to generate a process and compute its cost C.
6. If C is less than C_o, use the process recommended in step 5.

Manufacturing Quantity Increase

Experiments conducted in process planning clearly point to the fact that machining cost is reduced as quantities increase. Each of the different operations needed to produce an item requires a different power and speed. With large quantities, each operation is performed by a suitable machine. At low quantities, the machine selected is a compromise between the requirements of the different operations.

The quantity can be increased by grouping parts whose machining operations are identical or similar. This feature can be applied in two manners: normal or forced grouping. The normal grouping is carried out by examining the processing of items required at a certain time; whenever the processing of an item or a portion of it is similar to that of other items, they are grouped to be machined as one batch. The forced grouping does not wait for a chance to group items in machining, but instead plans the process in such a manner that groups are formed.

The logic of the forced grouping mode is that a group of items, each having its own specific dimensions, can be produced from one common shaped raw material. Thus the required quantity of the shaped raw material is the sum of the quantities required by the individual items.

The part description system can automatically arrive at the common shaped raw material by the following steps:

1. Transform the item drawing into a minimum raw material item drawing. This is done by increasing the contour of the item by minimum and maximum depth of cut limits, as discussed in Section 12.3 and shown in Figure 12.5. The minimum raw material item drawing does not specify unique dimensions, but rather defines boundaries.
2. Establish the item chucking type and location. This is done by using the GPPP model, as discussed in Section 12.4.
3. Superimpose one item drawing on top of the others and adjust boundaries to suit as many items as possible. The programming technique used is similar to that used in determining the length of cut in the GPPP (see Section 12.3).
4. If step 3 fails to produce a sufficient quantity, increase the boundaries by a dimension equal to one rough cut permissible depth and repeat step 3.
5. By using the lower boundary limits, prepare an imaginary item drawing. This drawing represents the shaped raw material for the group of items. Pointers to all individual items are preserved. An order for this imaginary item, with a quantity equal to the sum of the quantities of the individual items in the group, is entered into the open job files. The process plan for each of the original items is arrived at by the GPPP when the imaginary item is defined as the shape of the raw material.

Flow of Components in Plant

One of the factors that controls manufacturing lead time is the flow of components in the plant. Group-Technology suggests solving this problem by introducing work cells, where each family of parts is started and finished in one cell. Users either enthusiastically report impressive saving figures or speak of failure and heavy investments.

The forced process can create a logical work cell while not changing the physical layout. Naturally, the expected result would not be as high as in the physical work cell, but the investment will be negligible.

The logical work cell can be formed by grouping families of parts and restricting their process to a predefined path or sequence of selected machines. The grouping of families of parts can be done by the part description system and retrieved by parameter option (see Section 16.3). All parts belonging to the same family will impose additional constraints on the machine preselection module of the GPPP (see Section 13.4) or replace the third step in machine selection. Thus the resulting process plan will conform to the desired logical work cell. A more refined method might be used with no family of parts at all. By this method additional constraints are imposed on the machine selection module of the GPPP. The constraints act on the machine selection sequence in general. Each machine points to a group of machines allowed (or not allowed) to perform any of the following operations. The allowable machines are determined by their location in the plant and form a smooth flow path.

The method works as follows. The third step in machine selection (GPPP) is omitted or the predetermined number of machines in the matrix is set to around 20. The constraint of sequence of machines is treated by assigning a value of "all '9'" in the transfer time table to the forbidden sequence. Assume, for example, that machine j does not belong to the same logical work cell as machine i; then the value of $R_{ij} = R_{ji} = 9,999$. If machine j belongs to the same logical work cell, but the path is only from machine i to machine j, then $R_{ij} = R$ and $R_{ji} = 9,999$.

Selection of the machine and sequence of operations is carried out in the normal way (see Section 13.6). The advocated process will comply with the desired path.

Miscellaneous

The last three features (i.e., similar set-up, quantity increase, and flow of components) demonstrate the ease with which any desired feature or option can be added to the system by using GPPP techniques and the part description system.

Any additional feature the user wishes for and can define can be added to the system.

17.3 Stock Utilization Features

For many justified reasons, stock usually builds up to include dead stock, slow-moving and rejected items, and used and leftover pieces. Although information about such items carried by the inventory system is available, they are usually not utilized by such production systems as MRP (material requirement planning). Due to system logic, such items cannot be treated automatically by a system and are thus treated manually (if treated at all).

The main difficulty in applying an automatic system approach is the rigidity of the systems. Items are classified and named by catalog or inventory number. These names are carried through and are the connecting link between the separate manufacturing stages. Bill of material, inventory, purchasing, and production all use the name of the items and disregard the item itself. They operate in a purely clerical way. Rejected items, for example, do not have a name in the system. As was discussed in Section 7.2, many tricks and immediate solutions are employed, but they serve mainly inventory accounting and not inventory management.

By using the Hal technique and capabilities, it is possible to treat the items as such and not as names or catalog numbers; that is, the items can be treated in an automatic system manner. The following sections will describe these options.

Dead Stock and Slow-Moving Items

The output of the MRP is a list of materials which should be purchased and a list of the unused materials in inventory. Both lists refer to the inventory catalog number. If pegging is employed (see Section 6.4), the source of the requirement of the materials to be purchased is known. If pegging is not used, it is possible to arrive at the source of the requirement (parts) for the materials to be purchased by using single-level where-used data in the bill of material file. It is recommended that, before proceeding with the purchase, an additional phase be added in an attempt to utilize dead stock.

The additional phase uses the part description system. It retrieves a part by name and transforms its drawing into a minimum raw material part drawing, as shown in Figure 17.4 (see Section 17.2). The new drawing is scanned for maximum values, and a circumference shape that is a minimum raw material dimension for the part is established. A rough estimation based on the volume of material removal and an economical algorithm is used to determine the maximum raw material dimensions.

The dead stock material is retrieved, unit by unit, and checked to see whether its dimensions fall within the minimum and maximum raw material dimensions range of the parts required. If it falls within the range, an accurate cost computation is made taking into account extra machining cost, material cost, and the time the material has been in stock. The material resulting in the most favorable cost ratio is selected.

If no material falls within the established range, a second pass with enlarged maximum limits can be made. The decision whether to use such a material or not is based on the accurate cost computations.

Stock in Cut Pieces

The sizes by which stock is purchased are not equal to those issued for production. Cut pieces are left over in inventory. They are recorded in the

17.3 Stock Utilization Features   459

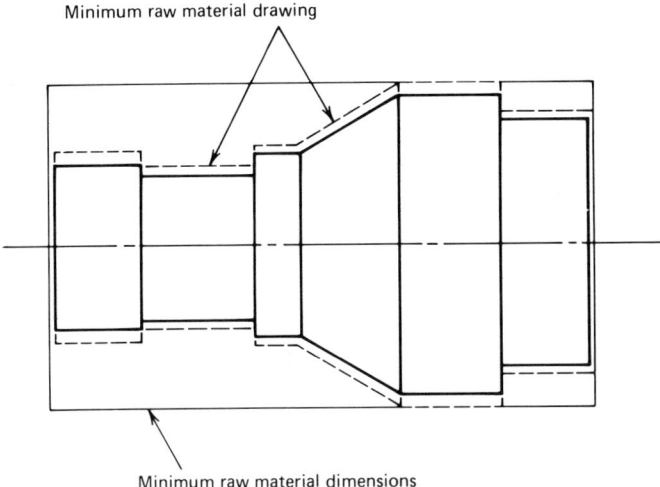

Figure 17.4 Transformation of a drawing.

inventory accounting and must be distinguished from the uncut material. In Section 7.2 it was suggested that a status code be used for that purpose. It solves the inventory accounting problem, but at the same time it prevents these pieces being considered for use by the MRP system.

The method used to utilize dead stock as previously described can also be used to handle this group of materials. Gripping allowances should be added to the minimum raw material dimensions.

Rejected and Leftover Items

In production planning, an anticipated rejection factor is used to increase the required quantity. If the actual rejection rate is not as anticipated, leftovers will be accumulated from assembly. The rejected parts are not considered for use by any of the automatic systems. The leftovers are considered by the requirement planning system. If no assembly requires these parts, they become dead stock. From the costing standpoint, the rejected and most of the leftover parts are covered and have no registered value. If such items can be used, it is practically all savings. In order to consider the use of such items in an automatic system manner, their geometric shape must be captured. Capturing these data is done as follows. The operation in which the rejection took place is known. Inspection reports specifying the dimensions causing rejection are available.

The GPPP keeps track of part geometry from raw material to the last operation. A machining drawing for each operation is easily prepared.

Thus the geometrical shape of the part before and after the operation that caused the rejection is available. Superimposing inspection reports on the original geometrical shape results in the geometrical shape of the rejected item.

The requirement planning system output is a list of orders to be released to the shop and a list of the unused items in inventory. It is recommended that an attempt be made to utilize the unused items and the rejected items by remachining prior to the issuing of shop orders. This is done by the use of the part description system. Each of the required items is matched against the surplus items.

In the first stage á test is made of the geometrical equality and inclusion of the required item within the surplus item. Required items that do not pass the test are released to shop and are excluded from further consideration. The remaining required items are compared not only for inclusion, but also according to the required special features and their locations, thereby reducing the number of alternatives. All alternatives are examined with respect to the conversion cost from the surplus item to the required item. This is done by defining the surplus items as shaped raw material to the GPPP. The output of the GPPP is an optimum process, including time and cost. An economic algorithm analysis decides which of the alternatives to choose, that is, which surplus item is to be remachined in order to obtain a required item.

More than one alternative can be chosen if the quantity of the surplus items is insufficient. Special shop orders are issued for such machining operations. If the economic module decides that it is not economical to transform items, a normal shop open order is issued for such items. Figure 17.5 shows the block diagram of this feature.

17.4 Hal—Master Production Planning

The Hal system is shown in Figure 15.1. As was discussed in Chapter 15, its purpose is to plan a realistic optimum production schedule that will serve management and the shop floor. It combines the traditional master production schedule, requirement planning, capacity planning, and the features and options previously discussed into one phase. Consequently, Hal eliminates the need to process the same data three times, each time with a different accuracy and for a different purpose; it thereby supplies management with a single accurate and unambiguous datum.

One may argue that management can do without the accuracy and the detailed level achieved by the Hal master production planning. However, such accuracy is needed for production purposes, and management data

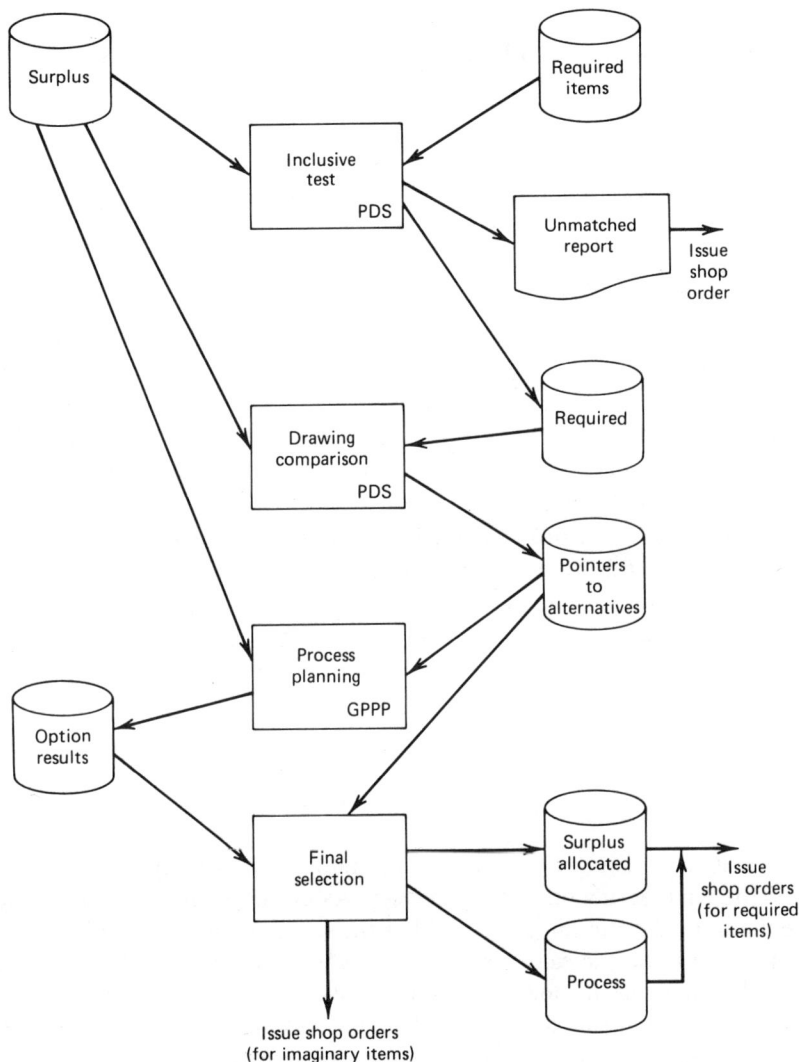

Figure 17.5 Utilization of surplus items module. Notation: PDS, part description system. GPPP, generative process planning program.

are just a by-product of the production system. Consequently, it costs less to supply management with accurate data than with rough estimates by the use of a separate application. The concepts and methods used in Hal master production planning are:

- The planning is carried out in a time period–product network matrix.
- The product network lead time is regarded as both elastic (i.e., it can be compressed or stretched within allowable boundaries) and rigid (i.e., it can be pulled forward or backward on a time scale as one unit).
- On-hand and on-order items are regarded as shaped raw material. They are allocated according to optimum requirements. The allocation is treated as a feedback completion entry.
- Load balancing is done on a work center and period basis. The tools used to achieve it are the product network properties and the allocation of available items period by period, rather than the allocation of an item through all the periods.
- The initial time periods are considered as the frozen zone. The allocation is fixed and the network planning is forward.
- In other time periods the allocation is temporary and the network planning is backward.

The master production planning is carried out as described in the following sections.

Basic Data Coefficient

The basic data required, that is, bill of material and routing, are retrieved from the engineering files of the system. As discussed in Chapter 16, they are theoretical optimum data. Neither the routing nor the machining time is expected to be the same, in practice, as the theoretical optimum values. Actually, one of the load balancing tools is to deviate from the item optimum in a controlled way in order to achieve a product mix optimization. However, data must be available for planning purposes. A coefficient is used to transform the theoretical values into practical ones:

$$T_p = CT_t$$

where T_p = practical time per item
T_t = theoretical time per item
C = coefficient.

The value of the coefficient (C) can be computed by using historical data and the Hal system logic. The load assigned is per period and department.

17.4 Hal—Master Production Planning

Thus C is computed by

$$C = \frac{L_p}{L_t} = \frac{\Sigma T_p}{\Sigma T_t}$$

The sum can be continuous over many periods, and the value of C can be varied from run to run. Initially, a value of $C = 2$, for example, can be used and adjusted after each run. The available capacity is the number of operators, or machines, multiplied by the number of hours per period.

If the user wishes, the coefficient can be applied to restrict the available capacity, the theoretical time values (T_t) being used for loading. The results will remain unchanged. However, since the length of the period is variable, it might be less confusing to work with actual available capacity and practical item machining time.

Time Period Length

The scheduling unit is the time period. The lead time for processing an item is known to be within the boundaries of the theoretical and the assumed practical times, but no exact and fixed process is known at this stage. This fits in with the basic concept of Hal. The exact process will be decided on only at the last minute, when the actual machining is to start. The process selected will be affected by the product mix required for the period. Similar setup, premachining, using alternatives to balance load, and so on will be considered according to the actual situation in the shop. The longer the duration of a period, the larger the product mix, and thus the probability of applying efficient machining methods increases (see Section 17.2). There is no way to predict and plan the completion of an item for any specific day within the period. The system guarantees that all the items planned for the period will be completed within that period and that they will be manufactured by an optimum method. Thus dependent items cannot be planned for the same period. An in-process inventory will be equal (at least) to the value of one production period. Product lead time is increased by at least half the duration of a period multiplied by the number of product levels. Consequently, the duration of a period should be minimal.

The duration of the period should be the best compromise between the conflicting parameters, as shown in Figure 17.6. The values are plant-dependent. They depend on the type of items, the probability of combining items into one machining run, the quantities, the unit theoretical machining time, set-up duration, plant layout, number of departments involved in machining, and so on.

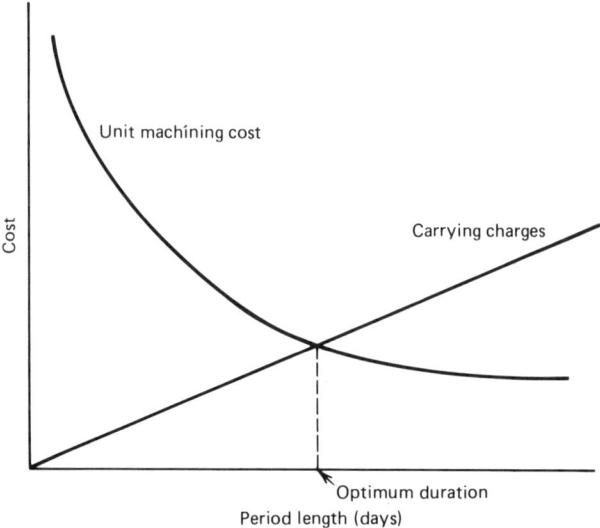

Figure 17.6 Time period length.

Since historical data are not available, we recommend 10–14 days for the initial 2–3 periods, 20–30 days for a period of 5–6 months ahead, and 6 weeks thereafter.

The construction of the product network is done on a calendar basis and thus is not affected by the length of the periods. The period serves two purposes. The immediate periods (2–3) are the frozen zone and serve as the basis for detailed process planning. Other periods serve for general information and for overall load balancing.

Production Lot Size and Due Date

Production planning is based on customer orders and/or forecasting. Such orders dictate the product, quantities, and delivery date. These data are external to the system and must be met. For planning purposes the orders on file will be treated as follows:

- *Yearly orders with gradual supply.* For these orders, existing economic lot size models will be used. One of the common lot size equations is

$$Q = \sqrt{\frac{2RS}{I}}$$

where Q = economic lot size
R = units per year
S = set-up cost
I = carrying cost per piece.

This equation considers only inventory carrying cost. When the purchasing expenses (F) are to be considered,

$$Q = \sqrt{\frac{2R(S+F)}{I}}$$

If one wishes to be more accurate and include warehousing cost (W), the equation becomes

$$Q = \sqrt{\frac{2RS}{I+2W}}$$

There are many equations. Each covers and considers different factors. It is important to use a model that fits the problem of the specific plant and not to memorize equations. The order will be split into several suborders according to economic considerations, each one having its own quantity and due date.

- *Confirmed orders for specific quantity and date.* For these orders, the economic lot size does not apply. It does not actually matter in which form the capital is tied down—in raw material, semifinished items, or finished products waiting to be shipped. The system will treat such orders as a unit to be produced and in the shortest lead time possible. During the production planning it is possible that orders will be split into several manufacturing lots if load and economic considerations make it necessary.

Product Network Planning

Each order on customer file, in any sequence, is planned from its due date backward. The lead time for each assembly, subassembly, and item is retrieved from the engineering file and multiplied by the practical coefficient. The system is shown in Figure 17.7. The lead time for purchased or subcontracted items is available in the engineering file and is either standard or specific for each item.

Product network planning is done twice for each product. The first planning utilizes the full value of lead time (theoretical time coefficient quantity), thus establishing the upper limit—the early start date (ES)—for

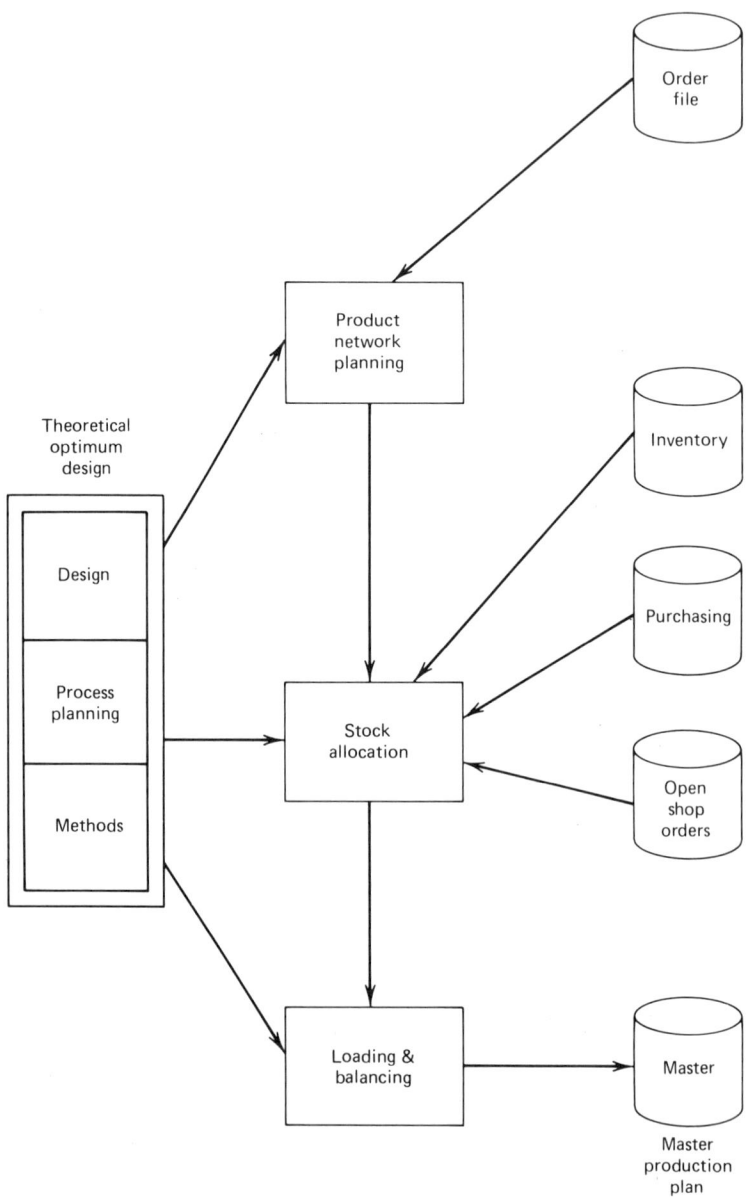

Figure 17.7 Master production planning module.

17.4 Hal—Master Production Planning

each item and assembly. The second planning employs the maximum overlap of components (see Figure 8.6 for the terminology).

The minimum overhang of the final assembly is 1–2 working days. All transfer of material between work stations is made at the end of a working shift. One can add another day for safety lead time. The quantity produced in the overlapped period is regarded as a separate lot for planning purposes. (In our case it is highly probable that one setup can serve many separate items. Thus the economic lot size previously mentioned is no longer applicable.) The starting date of each such lot is regarded as the latest start date (LS). The difference between the latest and the early start dates is the slack.

In overlap planning, four cases may be encountered, as shown in Figure 17.8:

1. The overhang time is greater than the lead time. In such a case, no overlap is used.
2. The lead time of component j is greater than or equal to that of component $j + 1$. In such a case, backward overlap is employed.
3. The lead time of component j is less than that of component $j + 1$. In such a case, forward overlap is employed.
4. The lead time of component j is significantly smaller than that of component $j + 1$ and $j - 1$. In such a case, component j will be produced in one lot, overlapping both component $j + 1$ and $j - 1$. Note that each consecutive component is separately overlapped and has its own lot sizes. Figure 17.8 demonstrates such cases.

Backward product network planning continues regardless of the current date so that activities can be planned for the past. Components that are in production, that is, in the frozen zone, are planned forward (from that network stage) with full lead time. If the completion date falls beyond the due date, the network is compressed symmetrically so as to make the completion date coincide with the delivery date. Figure 17.9 demonstrates this case. The compression should bring the new ES to somewhere between the old ES and the LS. If the new ES is later than the LS, the slack becomes negative, and the network is regarded as critical. An initial attempt to resolve this situation is made by rearranging lot sizes. If this is successful, the critical flag is removed.

Stock Allocation—First Step

Stock in inventory and on order is considered as free stock and can be allocated as needed to any open order. The allocation takes the form of a

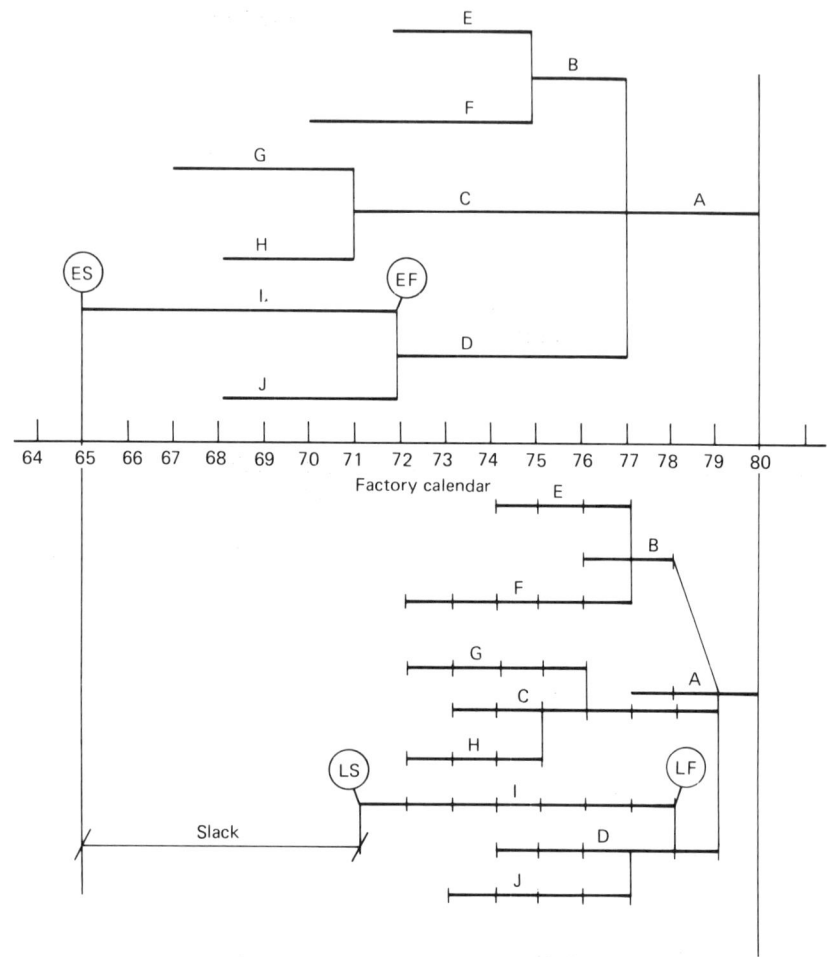

Figure 17.8 Overlapping of components. Notation: ES, early start date; EF, early finish date; LS, latest start date; LF, latest finish date.

"completion" feedback entry, which marks the job of producing the component "completed." The allocation model uses the minimum lead time product network, that is, the latest start dates are used for each component.

The allocation model examines the requirement of components for all orders per period, starting with the first and advancing consecutively to the last period.

17.4 Hal—Master Production Planning

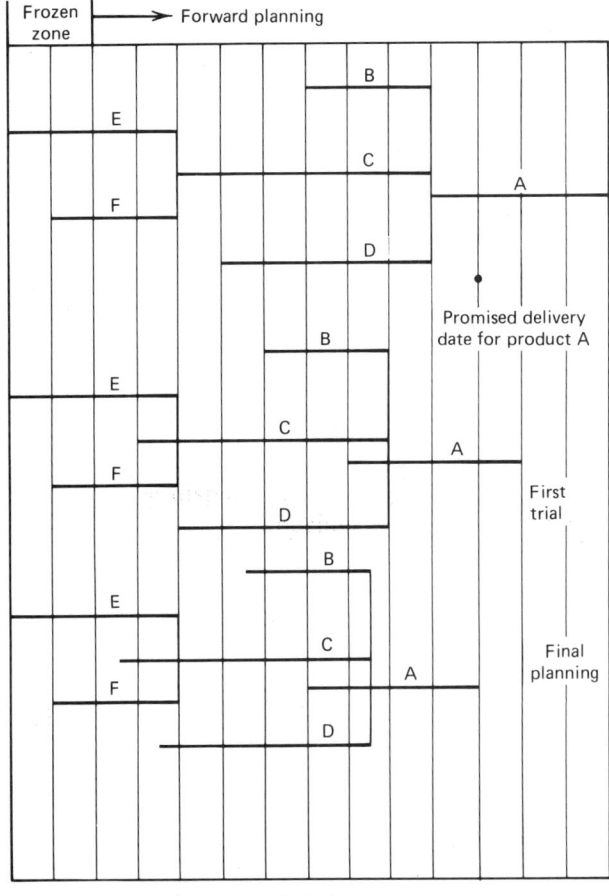

Figure 17.9 Forward network planning.

The system recognizes three types of periods, each one being treated differently (see Figure 17.10):

1. *Periods that fall in the past.* Components required in these periods present critical orders and get priority in stock allocation. The program scans the period column until it reaches a row with requirements (see row 4, line 1 in Figure 17.10). This row represents an order. The program walks through the row (the product network in the ascending period direction) until it reaches the top level of the order, that is, the product ordered (row

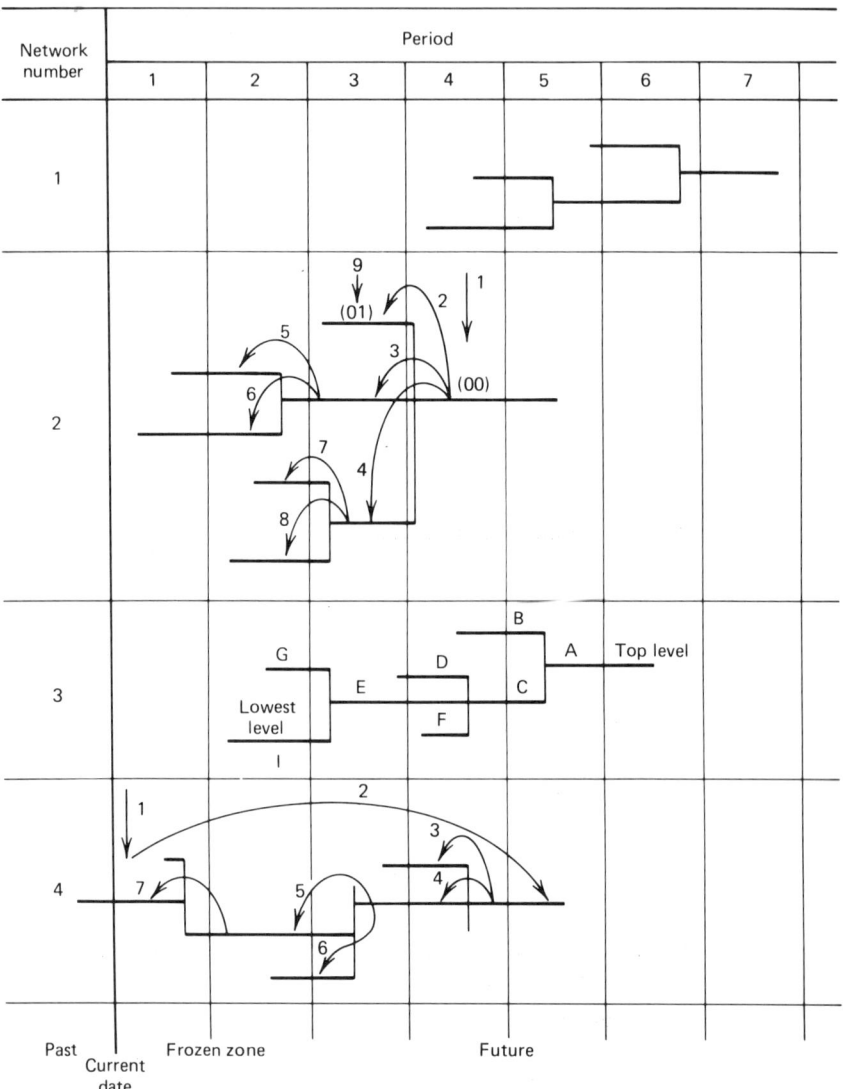

Figure 17.10 Allocation by periods.

17.4 Hal—Master Production Planning

4, line 2). If the above product is available in stock, the required quantity is allocated to that particular order. The product order is marked as completed. The program then walks through the row, in the descending period direction, marking all the branches of the network as completed (row 4, lines 3, 4, 5, 6, and 7). If only a partial quantity is available in stock, the order is split into two lots, one with the available quantity marked as completed and the other with the remaining quantity. The lead time and product network (now having a shorter duration) are adjusted to fit the new lot size.

In such cases and when no stock is available for the product, the program walks through the product network in the descending period direction until it reaches a product assembly level lower than the one handled (row 4, line 3). An attempt to allocate stock to this level is made as before. The entire above procedure is repeated until the whole network is marked "completed" or the initial period has been reached.

2. *Periods that fall in the frozen zone.* These periods can contain components of two types: components that are already in production, and components that are scheduled to be started. The components that are already in production (as a result of the previous run) are under the production (shop floor) control, and are ignored by the system. They are marked as completed in the product network matrix and need not be allocated. As previously described, the product network of such components is planned forward and does not include items that have already been produced. Other components that are present in the frozen zone periods are handled by the system. Such components are not critical. However, any delay in production will make them critical. If such components are available, they are going to be allocated instead of manufactured. The program scans the period column until it reaches a row (an order) with requirements. The row specifies the item number. If the item is available in stock, the required quantity is allocated, and the item is marked as completed. The requirements of the same order in the future periods are treated separately as belonging to future demands.

3. *Periods that fall in the future.* The allocation is carried out level by level, starting with low-level code 00 components (the low-level code indicates the lowest level at which a particular item is found in any bill of material), and period by period in ascending order. The program starts with the first period following the frozen zone. It scans the period column until it reaches a requirement for a component having a low-level code 00 (row 2, line 1). If stock is available, the required quantity is allocated. The product and all its components are marked as completed. If any of the

components is already marked as completed, it is transferred to the free stock list. This may occur if a component is in production or in the frozen zone and was allocated in the previous stage. If only a partial quantity is available in stock, the order is split into two lots: one for the available stock and marked as completed, and the other for the remaining quantity, with adjusted lead time to fit the new lot size. In such a case, and when stock is not available, the programs skip this order. It will be handled level by level in the following steps (row 2, line 9, etc.). When a component is required by several orders in the same period and the available stock is not sufficient to cover all requirements, the order with the longest lead time will be first served. The above steps are then repeated for the same low-level code and the subsequent periods. In the last period, the low-level code is increased by one, the period is set back to the initial period, and the allocation steps are repeated until the last low-level code and the last period have been dealt with.

Stock Allocation—Second Step

As a result of the first step, two lists are available: a list of the required items and components (purchased and produced) and a list of dead stock, where dead stock is defined as available stock not required by any of the open customer orders. If the dead stock is on order and has not yet been received, its order should be cancelled, even if it involves a reasonable penalty. The following steps are employed in an attempt to utilize the dead stock:

1. Examine whether assemblies can be torn down and whether it is profitable. If tearing down is possible, create a new free stock file listing the components derived from the single-level explosion of the available assembly. This process is shown in Figure 17.11 in a block diagram form. The information needed to decide whether an assembly can be torn down is available in the 'GG' entry of the part description system. The method of assembly is indicated for CAD purposes in the form of thread, rivet, press fit, weld, and so on. The procedures of the first allocation step are repeated by using the new free stock file. Whenever an allocation is made, a disassembly job is added to the time period–product network matrix.
2. The above steps are repeated until no more items are available in the new free stock file, that is, the assembly was single-level exploded, step by step down to its elementary items.

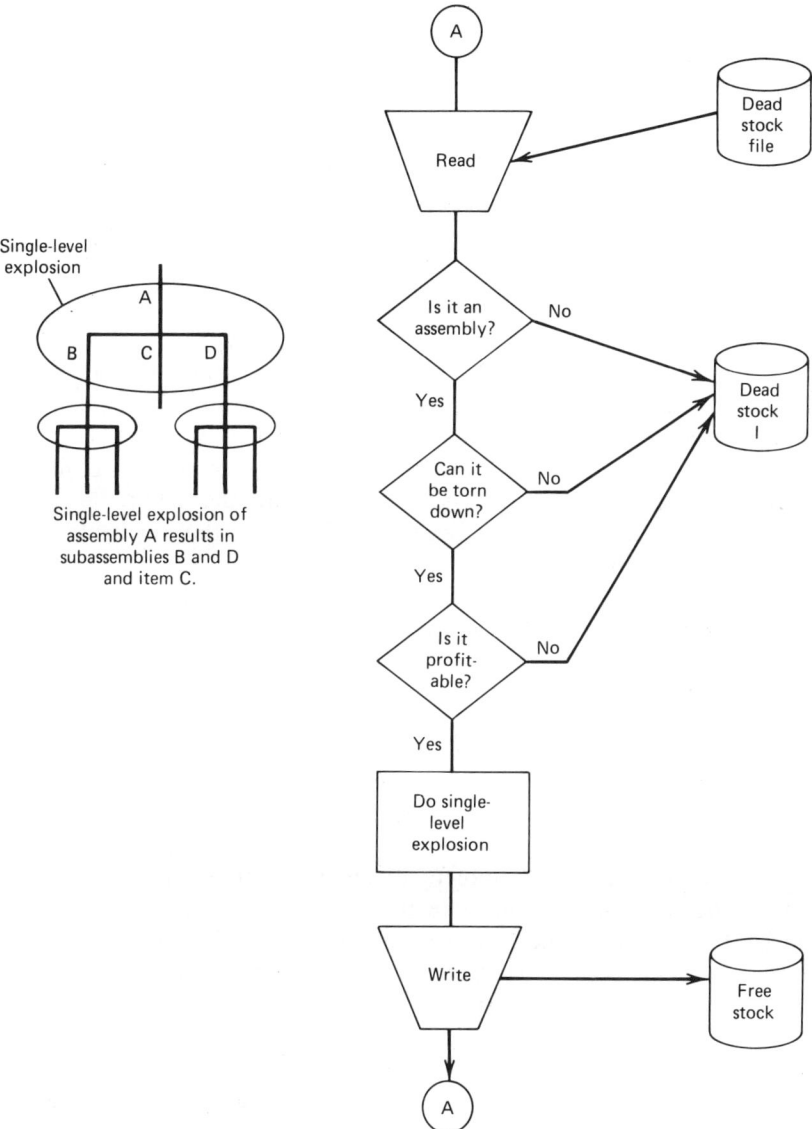

Figure 17.11 Utilization of dead stock by tear down of assemblies.

3. The remaining dead stock list includes three groups of items:
    a. Assemblies that cannot be torn apart.
    b. Elementary items that are in assemblies that can be torn apart.
    c. Elementary items.

The stock utilization feature, as described in Section 17.3, is employed for the last two groups.

Stock Allocation—Third Step

A critical order is defined as an order whose portion of the network falls in a past period. This means that if normal manufacturing practice is used, the delivery date cannot be met. The first attempt to solve this situation is made by utilizing slow-moving items, where a slow-moving item is defined as an available item that will not be required for a long time (six months plus its lead time). Such an item is transformed into the required item, and thus its lead time is reduced. The method used is the stock utilization feature, as described in Section 17.3. If this attempt fails, the use of a substitute design is examined. By using the part description system, the program checks whether any of the dead stock or slow-moving items can be substituted for the required item. This is done by replacing the required item with the available item and checking for possible assembly, clearance in movement, and functions.

Load Balancing

The workload required to complete all customer orders on time is arrived at by reviewing the present state of the time period–product network matrix. The total load and work center load are considered separately for each period. Components marked as completed or as purchased items are ignored. Figure 17.12 shows the matrix (Part *b*) and the load profile (Part *a*). The required load includes critical orders, that is, periods that are in the past, and disregards overloaded or underloaded periods. In order to arrive at a master production plan, the load should be balanced and the past period abolished. For load balancing purposes, past periods are considered to have zero available load. Any required load in these past periods transforms them into overloaded periods. A single technique can be used to resolve overloaded and past periods.

Overload cases can be resolved by network shifting if the average load is equal to or less than the available load. Load profiles reveal such information. If the required load over all periods is above the available

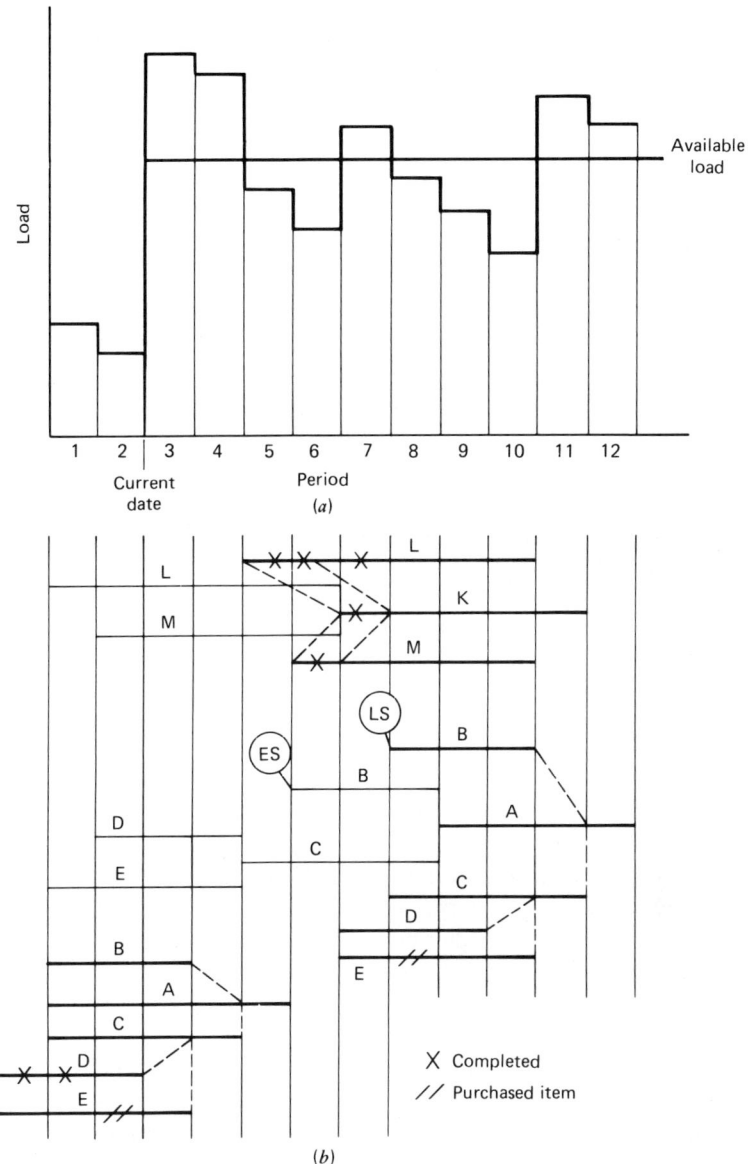

Figure 17.12 Load profile and matrix. (a) Load profile. (b) Matrix.

load, as shown in Part *a* of Figure 17.13, the system will supply information and recommendations to management as to which course of action to take, but the decision rests with the management. The available capacity can be increased by working overtime, extra shifts, or by the purchase of new equipment. The required load can be reduced by subcontracting work, turning down orders, or reducing order quantity. In such cases, the work center load profiles are used to pinpoint the bottleneck in production. A similar situation may occur if the initial periods are overloaded and not preceded by underloaded periods. Load profiles are prepared of the orders whose due dates fall within the overloaded zone. This is shown in Part *b* of Figure 17.13. If such a profile shows overload, or even 70% load, the overloaded zone should be treated by management as described before. An overloaded situation is resolved by shifting the planned dates of the components forward or backward to fill underloaded periods. The networks are shifted in a telescopic manner, where the order due date is fixed.

Load balancing is done in two steps: (1) resolving overload throughout all the periods, starting with the initial period, and (2) resolving underload throughout all the periods, starting with the initial period. When an overloaded period is not preceded by an underloaded period, forward shifting is used, while when an overloaded period is preceded by an underloaded period, backward shifting is used:

1. *Forward shifting.* Orders having the latest due date are handled first. Although the order network was planned on the basis of the latest start date (LS), as shown in Part *a* of Figure 17.14, several methods of shifting can be employed, such as:
   - *Splitting.* An initial attempt is made to shift forward items that do not affect any part of the network. If the lead time of an item is included in an overloaded and underloaded period, a split is used in the underloaded period, thus pulling the job forward. This is shown in Figure 17.15. Such a method used on an order for product A is shown in Part *b* of Figure 17.14. It could be applied only to components E, G, and C. By using this method, the overall lead time of product A was reduced from 20 to 18 time units.
   - *Splitting and shifting.* When a split is made on an assembly, all its components can be shifted forward. This situation is shown in Part *c* of Figure 17.14. Since assembly C was split, subassembly D can be shifted forward on the condition that a sufficient quantity is available for assembly. By using this method, the overall lead time of product A can be reduced to 14 time units.
   - *Order splitting.* When an overloaded zone is followed by a under-

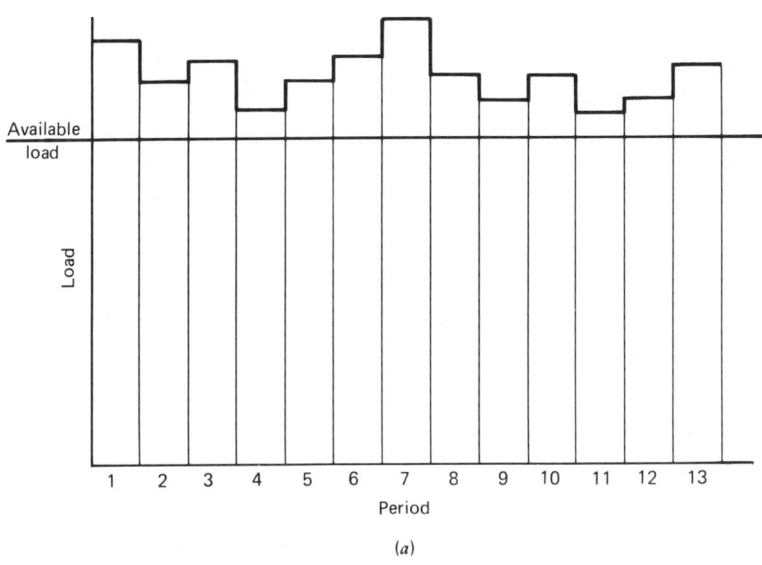

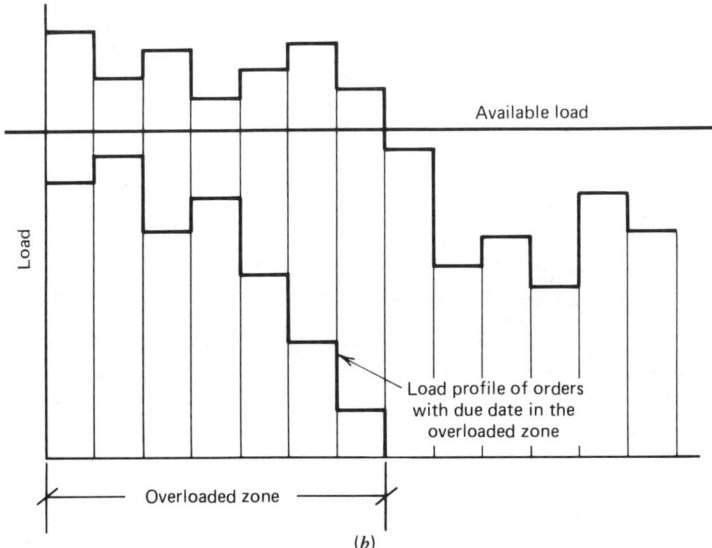

Figure 17.13 Load profiles.

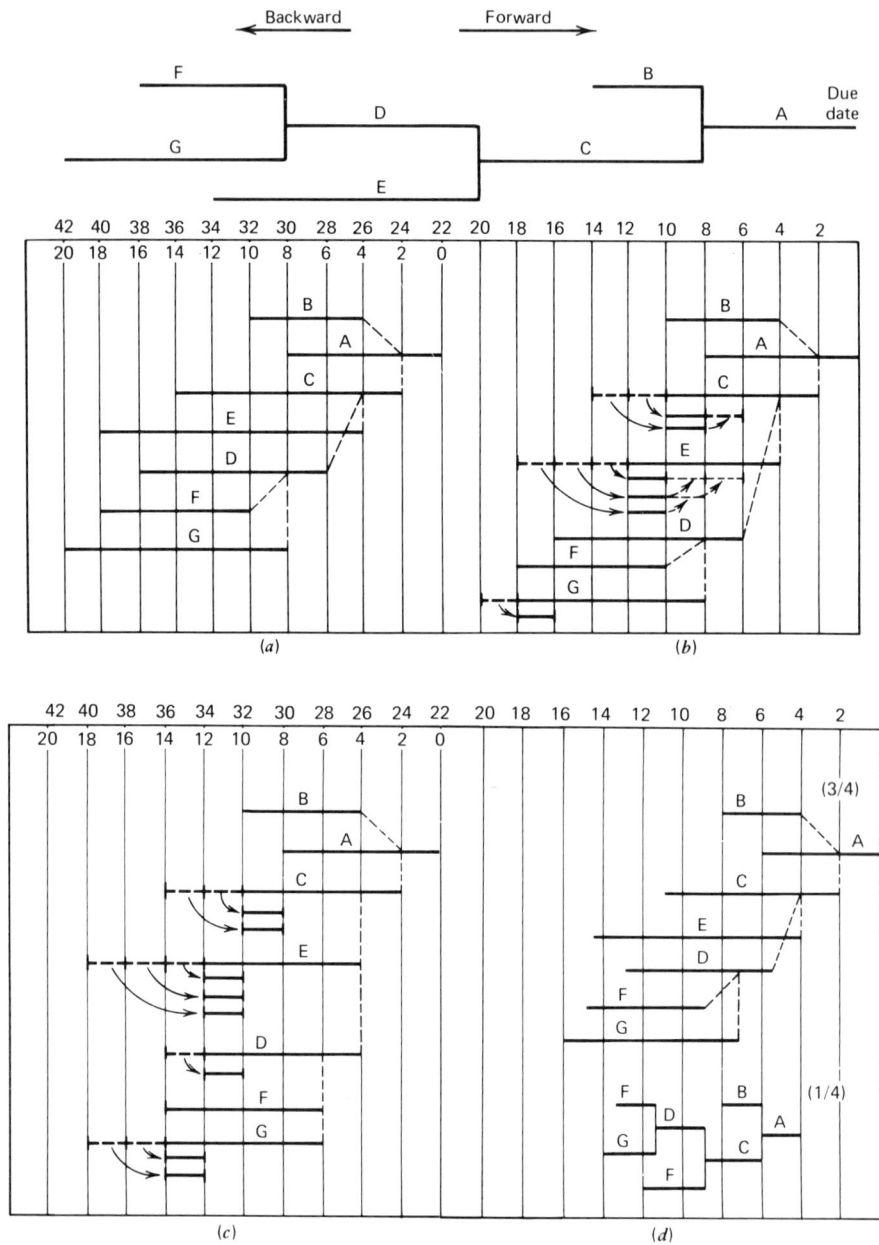

Figure 17.14 Forward shifting methods. (a) Forward shifting. (b) Splitting. (c) Splitting and shifting. (d) Order splitting.

17.4 Hal—Master Production Planning

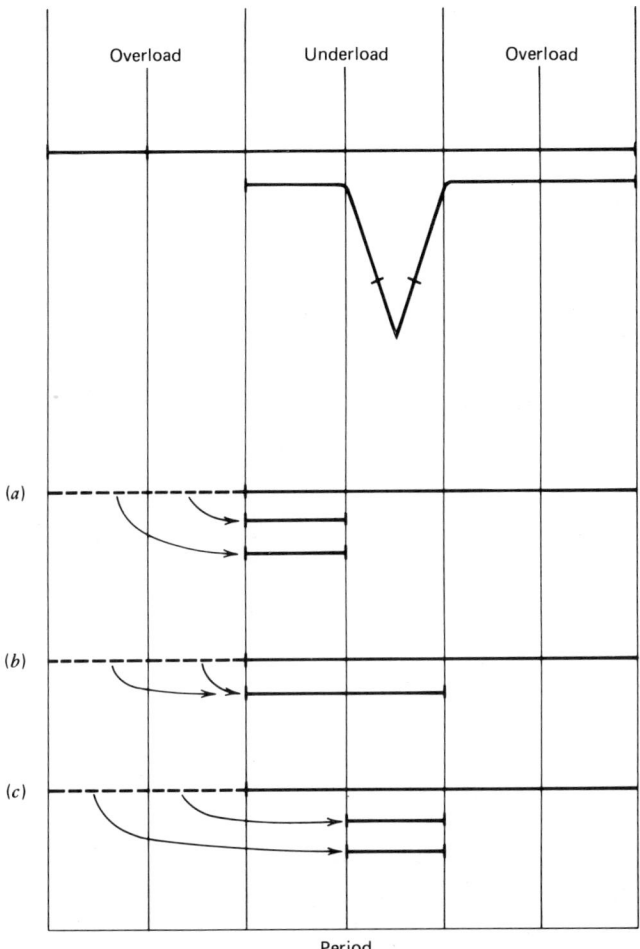

Figure 17.15 Pulling jobs to underloaded periods.

loaded zone that includes the order due date, the order quantity is split into two or more orders. This is shown in Part *d* of Figure 17.14.

- *Reducing lead time.* The system can absorb up to 30% of the lead time by improving the basic data coefficient, operation overlapping, and work efficiency. A network that ends in an overloaded period and starts (due date) in an underloaded period can be pulled forward by being recomputed with a three-step reduction in lead times of 10% each. Such orders will be marked for special treatment in the capacity planning phase.

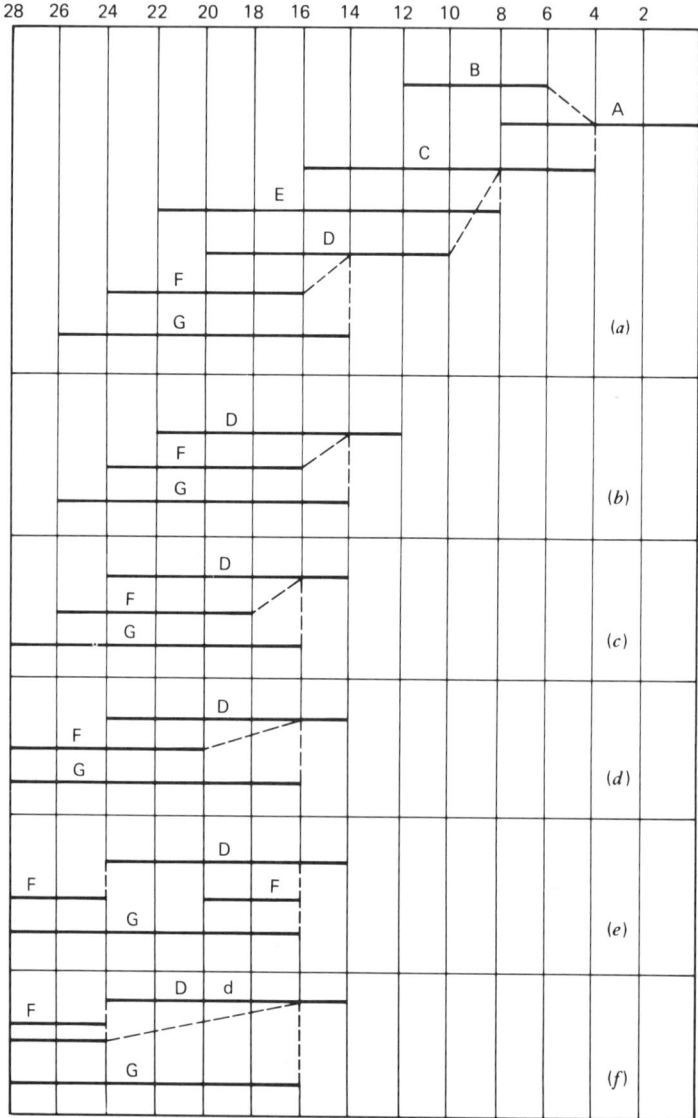

Figure 17.16 Backward shifting method.

17.4 Hal—Master Production Planning

2. *Backward shifting.* Orders having the earliest due date are handled first. Backward shifting is allowed within the range of early start (ES) and latest start (LS) dates. Several methods are used:

- *Network spreading.* When an overloaded period is preceded by substantial underloaded periods, the whole network is spread in a proportional way. This is done by increasing the overlap and overhang limiting factors. The increase is carried out in steps until no overlap is used at all, that is, the loading is by ES. Part *a* of Figure 17.16 shows this method (compare to Part *a* of Figure 17.14).

- *Assembly shifting.* When an overloaded period is encountered in only one or more work centers, it is sufficient to shift only those jobs that are in the overloaded work centers. An assembly is free to move as an individual item (i.e., not affecting any other part of the network) within the boundaries formed by the present overlapping factor and the minimum overlapping factor. Subassembly D in Part *b* of Figure 17.16 was shifted backward without affecting its components G and F. From this point the components are locked to their assembly and are moved as a unit. This is shown in Part *c* of Figure 17.16.

- *Item shifting.* An item can be shifted backward to any location between its ES and LS without affecting the network (see item F in

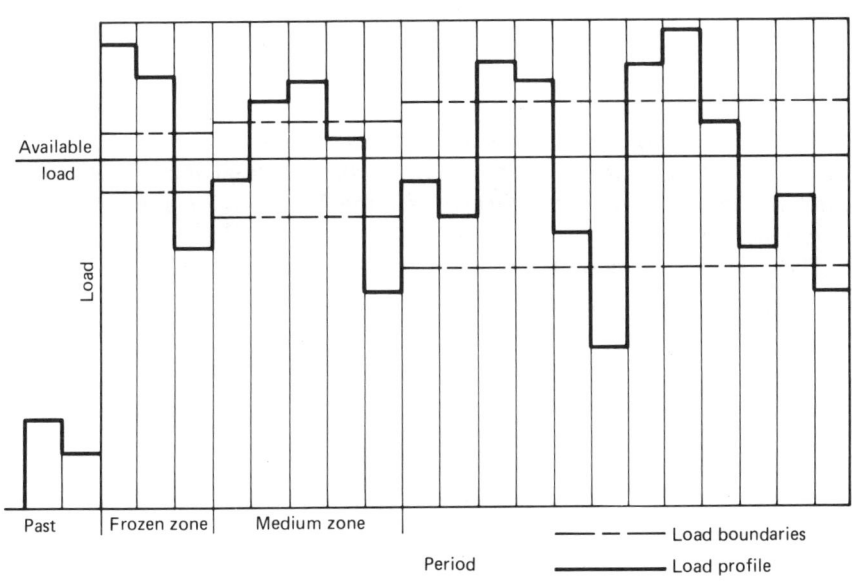

Figure 17.17 Loading boundaries.

Part *d* of Figure 17.16). Items B and E in Part *a* of Figure 17.16 can also be shifted.
- *Item shift and split.* When a narrow underloaded zone is encountered in a single work center followed by an overloaded period, the method of shifting and splitting the quantity shown in Parts *e* and *f* of Figure 17.16 can be used.

Summary

After the load balancing step, the matrix indicates the jobs and quantities to be manufactured in each period. It can present an order cross section or a load cross section for the whole plant or for each work center. This is the master production plan. The purchased items were ignored at the balancing step. At this stage, the purchasing requirements of items, quantities, and dates can be retrieved and transferred to the purchasing system.

Load balancing is a huge task, since the number of alternatives is large and the orders are interdependent. Available load is somewhat flexible. Its value depends on operator efficiency, machine breakdowns, and the actual process plan to be used. The use of boundary values for available load instead of a fixed value is therefore recommended. The boundaries (tolerances) are increased as the period is further away from the current date. The boundaries should not be symmetrically spread to both sides of the available load. The underload limit should be greater than the overload limit. This will eliminate overloaded periods and leave room for rush orders in the underloaded periods. It also enables us to quote early deliveries if necessary. Figure 17.17 shows the initial load profile, the available load and its boundaries, and the load profile after the load balancing step.

The frozen zone periods are transferred to the production phase (see Figure 15.1), where the specific process plan is established in the manner described in Sections 17.1 and 17.2.

Chapter Eighteen
Hal—in Industrial Management

The Hal manufacturing system is basically an engineering system. It can assist management by supplying the information and simulations needed to make decisions of an engineering nature, such as facilities planning, expansion of the manufacturing capabilities, and introduction of new manufacturing technologies. By using the absolute optimum manufacturing method as a fixed reference point, Hal can supply management with an objective measuring tool of plant operations. These applications of Hal will be discussed in this chapter.

Hal is not meant to cover the other systems that obtain in an industrial enterprise. The administrative functions, that is, bookkeeping, inventory, costing, personnel, purchasing, sales, and job recording, are not significantly affected by Hal and are to be carried out in the manner presented in Part I of this book. However, some adjustments must or may be made as a result of employing the Hal system:

- *Costing.* Costing should be based on overall actual performance per period rather than on the actual cost of an item. This is because the optimization is performed on a product mix per period and not on an individual item. An individual item might deliberately be machined by an inefficient method in order to increase overall product mix efficiency.
- *Inventory.* Inventory can do without a classification and coding system and use "running numbers" as catalog numbers. The full description of an item or raw material is kept in the part description file. A cross reference is used between the two. Thus the work status code and status code (see Section 7.2) are not needed. They are superseded by a new catalog number assigned to such items in inventory and pointing to a record in the part description file.

18.1 Plant Performance Measurements

Measuring and evaluating manufacturing performance is usually an ambiguous job, since no clear objective standards are available and there are many conflicting goals. By its nature and method of planning, Hal can supply objective measuring reference points. These can isolate the individual effects and be used as a universal scale for performance measurements. The reference points are:

• *Absolute theoretical optimum.* The absolute theoretical optimum is a fixed universal reference point. Its value is based on actual available technologies. It considers real strength and technical constraints. It assumes an imaginary machine, that is, no machine constraints are considered. Thus the absolute theoretical optimum process plan is practical from the engineering standpoint and theoretical from a specific shop standpoint. Its value does not include set-up cost. Consequently, it is free from sales, lot sizing, grouping, and scheduling effects. It is a theoretical value that will most probably never be achieved. However, it is a fixed value, representing the state of technology. It can be used as a fixed measuring reference point. The numerical value of the absolute theoretical optimum is a by-product of the engineering phase of the GPPP, as discussed in Chapter 12. It is expressed in time units. For comparison purposes, it is necessary to use a common denominator, that is, cost. The conversion from time units to cost is accomplished by multiplying the machining time by its hourly rate. An arbitrary hourly rate for the imaginary machine can be assumed. The lowest hourly rate used in the shop is recommended as the assumed value. This guarantees that the dispersion will be to only one side of the fixed reference point.

• *Theoretical optimum.* The theoretical optimum is a fixed specific shop reference point. Its value is based on the actual facilities available in a specific shop. The theoretical optimum is practical from the standpoints of technology and of available facilities and theoretical with regard to production and capacity planning, that is, the availability of the required machine at the required time. Moreover, it considers a theoretical optimum quantity and not the actual quantity. The theoretical optimum quantity is defined as that quantity which, when increased, will have a negligible effect on the item cost. The quantity may either be set arbitrarily at about 3,000 (for job shop) or be computed by the equation

$$Q = \frac{\text{TRN}}{(T_M/N) \times 0.05}$$

18.1 Plant Performance Measurements

where Q = theoretical optimum quantity
TRN = transfer cost (use the highest value practiced in the shop)
T_M = total machining time, that is, the sum of the machining times specified in the theoretical operations list report of the GPPP (see Figure 14.10)
N = number of required operations (see Figure 14.10).

- *Practical optimum.* The practical optimum is the planned cost of producing a known product mix in a given period. Its value is retrieved from the master production planning (see Section 17.4). The product mix planned for each department or work center is defined within the limits of the frozen zone. It is regarded as practical. By using any method the production phase must make sure that it is actually produced. Lot size is considered as a constraint; grouping, operation overlapping, and so on have been employed to balance the work center load and use optimum processing methods. The planned load is available from the work center profile of the master production plan. The total cost of producing the planned product mix is equal to the total department (or work center) expenses for the period.

- *Actual performance.* The actual total expenses of the department or work center and the deviation from the plan in terms of items and quantities constitute actual performance. The job recording system should supply data about completion of planned jobs and about accumulated cost in producing the product mix.

Ratios of the above reference points indicate the performance level of separate functions in the company.

Management Performance Level

The absolute theoretical optimum is a value that probably cannot be achieved. It is based on an imaginary machine that probably does not exist at all. The theoretical optimum depends heavily on the facilities available in a specific plant. The ratio R_M is given by

$$R_M = \frac{\sum_{i=1}^{i=PM} (TO_i \times TQ_i)}{\sum_{i=1}^{i=PM} (ATO_i \times TQ_i)}$$

where R_M = management performance ratio
PM = product mix

ATO$_i$ = absolute theoretical optimum cost of item i
TQ$_i$ = theoretical optimum quantity of item i
TO$_i$ = theoretical optimum cost of item i.

The lower the value of R_M, the better the performance obtained. It indicates that more suitable facilities are available for the production of the given product mix.

In a way, it is an absolute ratio, as it is independent of the product mix. It can be used to compare the level of management performance at different plants. It evaluates management decisions concerning facility planning. Section 18.2 will demonstrate how the Hal system and this ratio can assist management in the decision-making process.

Sales and Production Planning Performance Level

By definition, the theoretical optimum process results in the minimum item cost. Its value is specific to each plant and probably cannot be met. The product mix, that is, the customer orders and their due dates, affects performance and increases cost.

The ratio R_s is given by

$$R_s = \frac{PO}{\sum_{i=1}^{PM} (TO_i \times TQ_i)}$$

where R_s = sales performance ratio
PO = practical optimum (planned cost of a product mix PM)
PM = product mix
TO$_i$ = theoretical optimum cost of item i
TQ$_i$ = theoretical optimum quantity of item i.

The lower the ratio, the better the performance obtained. It indicates that the manufactured quantities are closer to optimum, due dates were set in such a manner that less competition for facilities occurs, product mix is such that it allows separate items to be grouped in a single manufacturing batch, and there are sufficient orders and flexible due dates to allow the work centers to be loaded to their full available capacity. These factors can partially be controlled. It is up to sales to accept a sufficient number of orders, to promise realistic due dates, and to direct their sales effort to orders that complement each other in terms of load. The master production plan can be used as a guide to sales effort and establishing realistic due dates. Some future periods are left underloaded in the master production plan. When needed, load can be pulled forward within the allowable range and can supply sufficient work for the frozen periods. This is not done to enable sales to promise realistic early deliveries of the unavoidable rush orders. The R_s ratio can also be used to study different loading methods employed by production planning. A dispatching rules study might use this ratio as an objective measuring scale of the performance of each rule.

Shop-Floor Performance Level

The ratio R_F indicates the performance level of the shop. It can be used to evaluate foremen. The ratio R_F is given by

$$R_F = \frac{PO + \sum_{i=1}^{LO} (TO_i \times Q_i)C}{PO}$$

where R_F = shop-floor performance level ratio
 PO = practical optimum cost
 LO = leftover (unfinished planned job)
 TO_i = theoretical optimum cost of item i
 Q_i = quantity of unfinished planned item i
 C = practicality coefficient (see Section 17.4).

The performance level is purely an engineering indicator. The shop is responsible for the completion of all planned jobs within the period. The ratio indicates whether this task was accomplished. The actual cost and department expenses are controlled by other systems, such as budget, costing, overhead, and personnel. A ratio of anticipated (PO) to actual (AC) work center expenses R_E,

$$R_E = \frac{AC}{PO}$$

can be used to evaluate and explain why or how the planned jobs were not completed.

18.2 Facility Planning

Management has to make decisions concerning investments in equipment. Such decisions establish plant level performance and thus the ability to compete on the market. The need to make such decisions frequently arises because of:

- *Replacement of old equipment.* The life of a machine is estimated as 10–20 years. This means that 5–10% of the equipment has to be replaced every year.
- *Added manufacturing power.* When the master production plan shows a continuous overload situation, management has to decide on expansion or to turn down orders. If it decides to expand, new equipment must be purchased.
- *Disposal of unsuitable, inefficient equipment.* When a machine is continuously underloaded, or not selected as the first alternative for any product, management has to decide whether to dispose of it. The master production plan and GPPP are used to draw management's attention to such machines.
- *New products.* New products might require machine capabilities unavailable in existing equipment.
- *Technological changes.* New design technologies and new materials might call for new types of equipment. Management can use the design features of Hal, as described in Section 16.3, to evaluate such new manufacturing technologies as precision casting or molding instead of machining and bonding instead of welding. When such technologies are adopted, an automatic change is made in all company product designs and drawings, the technology is added to the engineering stage of the GPPP, and new equipment is needed.
- *A new generation of machines is introduced.* Management has to keep track of technological developments. The new generation of machines should be evaluated from a technical and commercial standpoint.

Management relies on economic models and techniques (e.g., total value analysis) in making its decisions. The decisions are restricted to the engineering data fed into the economic model. Again, it is a chain of activities, with the accompanying constraints, that limits the number of alternatives and thus the quality of decisions. The use of Hal can improve decisions by introducing engineering into the economic model. The introduction of new machines might have a tremendous effect on manufacturing.

New machines usually possess more capabilities than the old ones. If optimum processes are to be used, all company routings should be examined and new process plans prepared. Merely replacing machine numbers in the routing file will result in inefficient manufacturing methods. However, it is impractical, by using today's techniques, to prepare a new set of

18.2 Facility Planning

process plans whenever a machine is added to the plant. It is a huge job and seldom done in general practice. Processes that might benefit remain unchanged. Thus the data fed to the economic model are incomplete. The Hal system does not suffer from such a degradation of manufacturing efficiency. Routings are not stored in files. The process plan is recomputed whenever needed by the GPPP with the present available machine list.

For decision purposes, Hal can be employed in one of two options: machine recommendation or machine evaluation.

Machine Recommendation Mode

The GPPP is divided into two major parts. The engineering stage, as was discussed in Chapter 12, is aware of the available technologies, that is, their capabilities and limitations. It computes a process to be carried out on an imaginary machine. The output of this stage is an operations list, as shown in Figure 14.10. The characteristics of the imaginary machine are precisely specified. The types of machining operations, such as turning, drilling, reaming, grinding, and milling, are specified in the column marked "SOG". The power, speed, and feed needed are specified in the appropriate columns. The special attachments required are specified by codes. The physical size of a machine is dictated by the part dimensions. The operations list can thus be regarded as a machine specification sheet. These specifications disregard economic considerations, that is, machine cost and its load forecast. An economic model should be used to transform the above ideal machine into a practical solution. Such a model should include machine cost, product quantities, and facilities loading. Any plant can use the above data in its own economic model. The following is an example of how machine specifications can be established. The steps are:

1. Select a typical plant product mix or use the master production plan as a test case.
2. Perform stage one of the GPPP. Get the operations list for each item in the product mix.
3. Separate the list into several lists, each including only one type of machining (lathe, milling, grinding, press, weld, etc.).

The following steps refer to each machine type separately.

4. Arrange the list in the descending order of power required per operation.

5. Select a representative machine from the operations list of each item. This is done as follows as a function of the quantity required:
   a. *For quantities of up to 50 pieces—one machine.* Compute C_j values by

   $$C_j \bigg|_{j=1}^{j=n} = \sum_{i=1}^{i=J} T_i \cdot \frac{P_i}{P_J} \cdot (P_J)^{0.67} + \sum_{i=J+1}^{n} T_i \cdot (P_J)^{0.67}$$

   where T_i = time of operation i
   P_i = power of operation i
   P_J = power of operation J
   n = number of operations on the list
   $J = j$ = number of operations (J starts at the top of the list, i.e., maximum power machine, and increases by one for each sum)
   C_j = sum value for limit set at operation J.

   Compute minimum C_j:

   $$C_K = C_j \text{ min}$$

   The representative machine is the one with a power as listed on line K of the list. Its speeds and feed range are the minimum and maximum values on the list. Its special attachments are as coded in any operation on the list. Its physical size is as per item requirements.

   b. *For quantities between 50–300 pieces—two machines.* Divide the list into two sections. The dividing line is operation K. It is characterized by

   $$\sum_{i=1}^{K} (T_i) \geq 0.4 \sum_{i=1}^{n} (T_i)$$

   Each section is computed as in step 5a and selects one machine.

   c. *For quantities between 300–1,000—three machines.* Divide the list into three sections and select one machine from each section. The dividing lines are operations K and L, determined by

   $$\sum_{i=1}^{K} (T_i) \geq 0.3 \sum_{i=1}^{n} (T_i)$$

   and

   $$\sum_{i=1}^{L} (T_i) \geq 0.6 \sum_{i=1}^{n} (T_i)$$

   Solve each section separately as in step 5a.

d. *For quantities above 1,000—unlimited number of machines.* Group the operations on the list by reasonable power ranges (e.g., 5 hp), and select a machine with the maximum values.
6. Compute item load on each representative machine by

$$L = T_{iL} \times Q \times \frac{1}{60} \times C$$

where L = load in hours
Q = quantity anticipated
T_{iL} = total machining time of operation to be performed on this machine (minutes).
C = accuracy constant (a value of 1.5–2 is recommended).

7. Arrange the list of the representative machines in descending power order.
8. Scan the list and combine machines whose power and size fall within a predetermined range.
9. Compare machines on the list to the existing ones and eliminate duplications.
10. Machines whose anticipated load is equal to the available load (number of hours per period) are selected as needed.
11. Representative machines, computed by a method similar to that described in step 5a, are selected for those machines having a partial loading.

Machine Evaluation Mode

The machines to be evaluated are introduced into the available machine list used by the GPPP. The program is used to generate processes for a selected typical plant product mix or for the products specified by the master production plan. One of the outputs of the GPPP is a summary report, as shown in Figure 14.5. The load of the evaluated machine is retrieved from this summary report. This load represents the first alternative. If the anticipated load is above the available load, the purchase of the machine is recommended. The possible savings can be estimated by the following method:

1. Scan the summary report and prepare a list of items using the evaluated machine.
2. Generate processes for items on the above list, once with the evaluated machine and then without it. The cost of the item in each case is given in the summary report.

3. Compute the difference in cost of the two alternatives and multiply it by the quantity.
4. The total of cost difference over all items results in total savings or loss. By using the alternative option, the GPPP allows the evaluation of several tested machines in one run.

The above recommendation and evaluation methods are not restricted to universal machines. The term "machine" can be expanded to cover any manufacturing technique, including NC machines, work cells, machining centers, and automatic factory.

NC, CNC, and DNC Machines

For the purpose of generating processes by the GPPP, this group of machines is no different from any other machine. The actual machining (metal removal) is unchanged. The difference lies in the handling times. Start–stop machine, engage feed, adjust tool, and so on are performed automatically and much faster than on universal machines.

On the other hand, programming and tape preparation are needed. The setup is different. All these differences are covered by the handling time table of the GPPP. It is permissible to use several columns in the table, where each machine points to the appropriate column. Thus evaluating machines of this group, in comparison with other types, can be done by exactly the method previously described.

Production Line (Transfer Line)

This type of manufacturing is characterized by having several types of machines laid out along a transfer line. Raw material, or initial body, is entered at the feed station, and a finished product emerges at the end of the line. All in-between operations are carried out automatically. The automation is achieved by mechanical means and controlled by switching circuits technology. Thus unless set-up work is done, the production line is capable of producing only one preplanned sequence of operations. The actual machining operations are carried out as on any other machine. The GPPP regards the production line as one machine, having an appropriate column in the handling time table. Recommendation or evaluation of this technique can be carried out for only one product, that is, the product mix used for evaluation is restricted to one product or limited to a selected group of products. The evaluation technique is as previously described. In addition, line balancing capabilities can be introduced.

18.2 Facility Planning

Machining Center

The machining center is basically equivalent to the production line. The difference lies in the method of control. In a machining center, the control is achieved by a computer and is therefore more flexible. The detailed operations, such as depth of cut, speed, and dimensions, may vary from one item to another. The sequence of operations and the shape may also differ from one item to another. Consequently, the product mix upon which the recommendation or evaluation is made can be increased. A general product mix can be introduced. The system will automatically select and evaluate all products that can be machined in the machining center.

Work Cell

The work cell technique can be looked on as a production line without the automatic controls. The lack of controls and transfer lines makes it more flexible and allows it to be used for small lot sizes. Operation times, however, increase. From the GPPP standpoint, the difference lies in the handling times that accompany the machines. The machine recommendation mode is used to construct a work cell, that is, determine which machines are to be selected for a cell. The 11 recommended steps (of this mode) specify the most suitable machines or those that are available for the production of the required product mix, so that now information is available about which machines are needed to produce which item, the sequence of machining operations that must be preserved (see the GPPP priority number), the quantity, and the required times. A components flow analysis (CFA) can thus be made as an extension of the 11 steps previously described.

The machine evaluation mode regards each machine located in a work cell individually, but reduces transfer time between machines in the same work cell, as described in Chapter 13. Thus appropriate values in the transfer time table allow the use of exactly the method previously described.

Automatic Factory

The automatic factory will consist of DNC machines, controlled transfer lines, or industrial robots. Process and load optimization are carried out by the central computer. A direct control over any device in the plant is possible through a computer hierarchy network. The network links the main central processor to the individual microprocessor controlling a single device. The automatic factory utilizes the benefits of all the facilities previously mentioned. It has the same reduced operation handling time as the DNC machine (handling time table), the same interoperation transfer time as in the machining center (transfer time table), the same chucking and gripping as in the production line (handling time table), and the same increased flexibility as in the work cell. Thus by assigning the appropriate values in the relevant table, the system can be used in decision-making concerning the automatic factory.

Index

Absolute check, and error detection, 41
Absolute theoretical optimum, 339, 484
Accuracy, 404
   machine constraint, 347
Action code, 45, 46
Activity planning, 151–155
Actual cost, 197, 264–268
Actual performance, 485
Adaptive production process planning, 449–453
Allocation of stock; *see* Stock, allocation
Allowance, 60
Analogy, and creativity, 52
Analysis
   dimensional, 307
   forecasting statistical regression, 131
   part design cost, 435
   structural programs, 425–426
Application approach to data processing, 7
Application-oriented data processing, 10
APT, 91
Arrangement of operations in matrix, 352
Artificial constraints, 303, 394
Assembly of product, 436–437
   demand file, 177
   ease of, 437
   shifting, 481
   standards, 437
Automated job shop, 85
Automatic design, 422
   change of design, 43, 439–441
   preparation of drawing, 441
Automatic factory, 86, 396, 493
Automatic tools design, 423–425

Backward scheduling, 204
Backward shifting, 481
Balancing load, 130, 474–482
Basic cutting conditions equation, 306–310
Basic data coefficient, 462
Basic dimensions, 63

Batch serial number, 189
Benefits of Hal, 414–420
Bilateral tolerance system, 60
Bill of materials, 104–113, 124, 163, 253, 259
   engineering, 114
   item master, 105
   low-level code, 107
   options, 117
   organization, 105, 113–121
   product
      definition, 104–113
      structure, 107
      variant, 117
   production, 114
Body, three-dimensional, 402
Body flat, 402
Body of rotation, 402
Bookkeeping, 236
Bounded geometry, 401
Brainstorming, 52
Budget, 22, 123, 125, 252, 259
   dynamic, 37
   indirect expenses, 147
   investment, 147
   and management control, 146–148
   research and development, 148
   zero base, 147
Built to order, 32, 199
Built to stock, 32, 199
Business, financial aspects of, 31

CAD (computer aided design), 86–93, 366, 401, 423
CAM (computer aided manufacturing), 86–93
   in group technology, 93
   NC programming, 91
CAM-I CAPP, 295, 446
Capacity planning, 18, 136, 141, 165, 199–219, 234, 343, 365, 399
   master production planning, 141

496 Index

Capacity planning *(Continued)*
  objectives, 200–203
  queuing problem, 343
  RPS (requirement planning), 174
  short-term, 221–226
  technique, 206–217
  terminology, 203–205
CAPP CAM-I, 295, 446
Cash-flow, 144–146
  profit planning, 36
Casting, 70
Causes of chatter and prevention methods, 321
Center lines, 402
Center of machining, 493
Certainty coefficient, 54
Chatter, 302, 315, 320
  causes and prevention methods, 321–322
  effects
    on tool wear, 320
    on surface finish, 315
Check digit, 40
Chip equivalent thickness (CET), 309
Chucking, 329–336
  location, change of, 330–333
  type, change of, 333–335
Classification and coding, 79–86, 96–104, 183, 294, 401
  code types, 97
  by computer, 93
  design-oriented, 79–82
  events of, 236
  external and internal, 103
  group, meaningful, 103
  mixed system, 103
  production-oriented, 82
  resource-oriented, 83–86
  running-numbers, meaningless, 97
  secondary objectives, 96
  semi-meaningful, 104
  type, meaningful, 97
CNC, NC, DNC machines, 492
Coding; *see* Classification and coding
Company, organization of, 22
Company standards, 66, 434
Components, flow of, 456
Computer
  classification by, 93
  evolution of applications in industry, 7
Computer-aided group technology, 93
Computer science, 304
Computerized vs. manual systems, 155–162
Concentricity, 404
Conceptual design, 49–56, 422
Constraints in design, 70–72
  artificial, 303
  machine accuracy, 347
  maximum torque, 347

  spindel bore, 347
Construction of machine-operation matrix, 343
Control reports, 247
Conversion cost, 253–259
  labor, 348
  material, 258
Coordinate system, 401
Cost planning and control, 252–273
  actual, 197, 264–268
  analysis, 435
  standard, 264
  tolerance of, 62
Costing system, 190, 252, 483
Count of stock, 192, 195
Creativity, 52
Cumulative average, 190
Current date, 203
Customer orders, 17, 125–129, 195
  forecasting, 18
  variations and option, 128
Cutting conditions, 299, 306–310
  feed rate and cutting forces, 310
  forces, 302, 311
  metal cutting technology, 302
  speed, 286, 307, 340
    economical, 343–344
    power adjustment of, 344–347
    and surface finish, 316
Cycle count, 195

Daily capacity planning, 221
  exception report, 247
  management report, 247
  performance report, 247
  release of jobs report, 247
Data base, 46
Data collection, 243
Data entry integration, 8–10
Data processing, 234
  application approach, 7
  application-oriented system, 10
  distributed, 11
  two-way, 44, 46, 197, 240
DDATE EDD (earliest due-date), 228
Dead stock, 180, 458
Decision-making process, 26–29
  matrix, 53, 56
Decision-rule priority, 226
Depth of cut, 302
  effect on feed rate, 312–315
  finish cut, 323
  maximum, 347
  non-symmetrical split of, 315
  surface finish, 318–319
Design cost analysis, 435
Design engineering, 16, 48–56
  automatic, 422, 439

Index

classification system, 79–82
concepts of manufacturing system, 24–32
conceptual, 49–56, 422
detailed, 56, 422
features of Hal, 415–417, 438–441
material utilization, 435
of parts and products, 427–437
product preview, 422–426
product reliability, 56–64
for production, 67–72
for reliability, 37–46, 240–246
of inventory system; 192–195
standards, 434
tolerance, 434
tools, automatic, 423–425
Detection error methods, 39, 40, 42, 43, 129, 174
Dimensional analysis, 307
Direct-to-indirect ratio report, 248
Direct standard hourly rate, 255
Discipline needed for requirement planning, 173–178
Discrepancies in inventory, 194
Dispatching order-release, 19
Dispatching rules, 226–234
Distributed data processing, 11
DNC, NC, CNC machines, 492
Drawing, engineering, 16, 401
automatic preparation, 441
file management, 438
retrieval
by key, 438
by parameters, 438
Due date, 203, 464
DDATE EDD, 228
of order, 164
Dynamic budget, 272
for production, 37
Dynamic standard cost, 272

Early finish, 204
Early start, 204
Ease of assembly, 437
Economic order quantity, 159, 167
Economical cutting speed, adjusting for, 343–344
Economics of the basic turning operation, 284–289
Effect of depth of cut on feed rate, 312
Effect of chatter on tool wear, 320
Efficiency, periodic report, 248
Elapsed manufacturing time, 205
Engineering data control, 95–124
design, 14–16, 48–56
drawing, 16, 401
generative process planning, 305–337
in Hal, 421–447
in manufacturing process, 47–94

value engineering, 421
Error detection methods, 39, 40, 42, 43, 129, 174
job identification, 242
reporting quantity, 242
requirement planning, 173
Exception daily report, 247
Expediting, 19
Expert-opinion forecast, intuition in, 131
Exponential forecast smoothing model, 135
External and internal classification system, 103

Fabrication by joining parts, 69
Facilities planning, 123, 143, 487–494
machine evaluation mood, 491
machine recommendation mood, 489
Factor of safety, 57–59
Factory, automatic, 86, 396, 493
Failure, mode of, 56
Failure mechanism, 56
Family of parts, 294
FCFS (First-Come-First-Served), 228, 233
Feed rate
cutting forces, 311–312
effect on depth of cut, 312
maximum, 311
minimum, 310
product of cutting speed, 310
surface finish and tool-nose radius, 316
File organization, 46, 105
Finance, management control, and planning, 142–149
aspects of the business, 31
commitment report, 196
and personnel systems, 36
and production, 76
Finite capacity, 205
First-Come-First-Served (FCFS), 228, 233
First-In First-Out (FIFO) method of inventory control, 190
Flat body, 404
Flow of components, 456
Forced process planning, 454–457
Forces in cutting, 302, 311
Forecasting, 17, 125, 129–135, 155
customers' orders, 17
exponential smoothing model, 135
financial control, 130, 146
intuitive expert opinion, 131
manufacturing lot-size, 130
master production planning, 131
profit, 146
seasonal load balancing, 130
statistical regression analysis, 131
Forging, 68
Forming, from solid, 67–69
liquid casting, 67
metal removing, 69

Index

Forward shifting, 476
  and splitting, 476
Free stock, 164, 177
Frozen-zone capacity planning, 141

General matrix solution, 358
Generative process planning, 277-389, 441
  engineering, 305-337
  example, demonstration of, 367-389
  mathematics, 338-365
  prerequisite, 277-304
  system flow, 367
Geometric shape, 289
Geometry, bounded, 401
Gross requirements, 151, 153, 164
Group, meaningful classification system, 103
Group technology, 6, 76-86, 93, 294, 394

Hal technology, 393-494
  benefits, 414-420
  concepts, 393, 395-397
  design, 415-417
    features of, 438-441
  engineering, 421-447
  industrial management, 483-494
  process planning, 441-447
  production, 417-420, 448-482
  standards, 420
  system architecture, 397-400
  work cell, 417
Handling time, 348
Hash totals, 41
Heat treatment, 70
Hourly machine rate, 340
Hourly standard, direct rate, 255
Hypothesis and creativity, 52

Idle time, 222
IMS (Integrated Manufacturing System), 5, 12, 13-46, 47, 96, 125, 141, 174, 233, 255, 268, 394
Incentive system, 238
Indirect expenses budget, 147
  direct-to-indirect ratio report, 248
  lead-time, 206-211
  operations, 122
Industrial manufacturing and management, 21-24, 483
Information, 29, 248
Innovation, 52
Inspection, 20, 122, 175, 196, 243
Internal and external classification system, 103
Interoperation time, 206, 237
Intuition, in expert-opinion forecast, 131
Inventory management and control, 20, 34, 45, 153, 155, 168, 179-198, 247, 265, 483
  accounting, 179
  design for reliability, 192-195
  discrepancies, 194
  level, 159, 174
  objectives, 180-182
  price of items, 190
  service-level, 155
  statistical control, 156
  system as management-control tool, 195
  system technique, 183
Inversion and creativity, 52
Investment budget, 147
Item
  allocation of, 162
  bill-of-material item master, 105
  leftover, 459
  rejected, 459

Job completion, 237, 243-246
Job recording, 19, 174, 198, 234, 240, 248
Job release, 219, 247
Job-shop simulation, 233

Keypunch and verify, 39

Last-In First-Out (LIFO) method of inventory control, 190
Last price, 190
Latest finish, 203
Latest start, 203
Lead time, 124, 479
  indirect, 206
  manufacturing, 153
  production, 78
  safety, 164
Leftover items, 459
Load balancing, 130, 474, 482
Location of storage, 189
Look-ahead look-back feature, 166
Loss and profit reports, 36, 196
Lot size, 156, 160, 180, 484
  forecast, 130
  master production schedule, 140
Low-level code, 107, 163

Machine
  accuracy constraint, 347
  hourly rate, 340
  operation matrix, 343
    construction, 343-348
  periodic status report, 247
  selection, 348, 352
  single machine solution, 356-358
Machine evaluation mood, 491
Machine recommendation mood, 489-491
Machines, NC, CNC, and DNC, 492

Index

Machining center, 493
Management, and finance systems, 36, 143–149
  budget and control, 146
  controls, 22, 179–198, 483–494
  daily reports, 247
  of inventory system, 179, 195
  manufacturing, 21
  performance level, 485
  periodic information reports, 248
Manpower planning, 123, 143
Manual vs. computerized systems, 155–162
Manufacturing
  activity planning, 151–155
  elapsed time, 205
  and industrial management, 21–24
  integrated control system, 12, 13, 21, 29, 32, 43, 96, 125, 174, 393
  lead time, 153
  lot size, 130
  process, 3, 14–20, 31
  routing, 121–124, 153
  system design concepts of, 24
Margin of safety, 57
Market standards, 65
Marking and non-machining treatment, 409
Master production planning, 18, 22, 125–149, 150, 163, 168, 179, 399, 460
  capacity profile, 139
  under Hal, 460–482
  lot size, 140
  scheduling, 18, 22, 123, 135, 163, 168, 179, 199
Master routing, 123
Material
  conversion cost, 258
  handling, 20
  pricing, 190–192
  utilization, 435
Mathematical review, 341–343
Mathematics for generative process planning, 338–365
Matrix
  decision for, 53, 56
  general solution, 358–364
  machine operation, 343
    construction, 343–348
  operation, arrangement in, 352–353
  solution, 353–364
Maximum depth of cut, 347
  feed rate, 311
  torque, 347
Measure, unit of, 183, 404
Measurement of plant performance, 484–487
Metal-cutting technology, 302
  forming, 69
  removal processes, 306

Methods for chatter prevention, 321–322
  error detection, 39, 40, 42, 43, 129, 173, 174, 242
Minimum feed rate, 310
  overhang, 212
  overlap time, 213
  time before overlap, 212
Mixed-code classification system, 103
Mode of failure, 56
Modulus, 10, 11, 40
Motion and time study, 17
MRP; *see* Requirements for facilities planning

National standards, 65
NC, CNC, and DNC machines, 492
NC programming, 91
Net change requirement planning, 168
Net requirements, 153, 164
Network product planning, 465
Non-machining treatment and marking, 409
Non-symmetrical split of depth of cut, 315

Objectives
  of capacity planning, 200–203
  of classification, 96
  of inventory, 180–182
  primary, 51
  secondary, 51
On-hand free stock, 164
Operation
  arrangement in matrix, 352–353
  establishment, 323–329
  indirect, 122
  machine matrix, 343–348
  priority, 221
  research, 341
  sequence of, 361
  time, 206
Optimum
  absolute, 339
  absolute theoretical, 484
  practical, 485
  theoretical, 397, 484
Options and variations, 117, 128
Order
  customers', 125, 195
  due date, 164, 464
  economic order quantity, 159, 167
  point of, 156
  release, 19, 199–219, 226
  variations and options, 128
Organization of bill of materials, 113–121
Organization of company, 22
Overhang, 213
Overlap, 212

Overload, 205, 216
Overtime, 226

Parameters, analysis, 310
Part
  cost analysis, 435
  definition, 367
  description system, 304, 400–414
  design, 427–437
  family, 294
  machining aids, 436
  material utilization, 435
  programmer, 281
  standards, 434
  tolerances, 434
Part-period, 165
Pegging, 169–173
Performance
  actual, 485
  daily report, 247
  evaluation, 56–64
  management level, 485
  plant level, 484–487
  production planning level, 486
  sales level, 486
  shop floor level, 487
Periodic reports
  direct-to-indirect ratio, 248
  efficiency, 248
  machine status, 247
  management information, 248
  scrap, summarized, 248
Personnel, 37, 123, 143
PERT, 343
Plant performance measurement, 484–487
Power adjustment, 344–347
Practical optimum, 485
Practical process plan, 303
Preparation of automatic drawings, 441
Preventing chatter, 321–322
Preview product design, 422–426
Price
  last, 190
  unit of, 197
Pricing methods, 190–192
Primary objective, 51
Priority
  decision rule, 226
  number, 361
  rating, 216, 227
Process
  adaptive production-process planning, 449–453
  decision making, 26–29
  engineering in manufacturing, 305–337
  forced process planning, 454–457

generative process planning, 277–389
  example of, 367–389
  manufacturing, 3, 14–20, 31
  metal removal, 306
Plan, 123
  practical, 303
  theoretical, 303, 338
Planning, 17, 72–76, 78, 92, 234, 289–294
  deviation, 291
  generative, 299–304, 366–389
  under Hal, 441–447
  retrieval, 294–298, 446
Product
  assembly, 436–437
  bill of material, product structure, 107
  definition of, 104
  design, 48–56, 89–91, 427–437
    preview, 422–426
    reliability for, 56
  network planning, 465–467
  structure, 151
  variants, 117
Production, 400, 417–420
  adaptive process planning, 449–453
  bill of materials, 114
  budget, 147
  capacity planning, 141
  control, 78
  design for, 67–72
  dynamic budget, 37
  and finance, 76
  forecast, master, 131
  in Hal, 400, 417–420, 448–482
  lead time, 78
  line, 492
  master planning, 17–18, 22, 123, 125–149, 150, 162, 168, 179, 199, 252, 399, 460–482
    capacity profile, 139
    control, 84, 253, 294, 448
    -oriented classification system, 82–83
    performance level, 486
Profit and loss, 36, 196
  cash flow, 36
  forecasting, 146
Programs for structural analysis, 425–426
Purchasing, 19, 34, 45, 173, 176, 196

Quantity, economic order, 160, 167
  reporting transaction, 242
  required, 289
Queuing problem and capacity planning, 343
Quotation, 128

Random, 229
Rating, priority of, 216, 227

supplier value, 196
Ratio, direct-to-indirect, 248
Raw material, 101, 289
Recording, job, 19, 198, 219, 234, 248
    design for reliability, 240–246
    requirement planning, 173
Reference lines, 401
Regeneration requirement planning, 168
Regression forecast, 131
Rejected items, 459
Reliability
    and design, 37–43, 43–46
        of inventory system, 192–195
        of products, 56–64
    job recording system, 240–246
Report
    control of, 247
    daily exception, 247
    daily management, 247
    daily performance, 247
    daily release of jobs, 247
    errors in quantity, 242
    financial commitments, 196
    periodic, direct-to-indirect ratio, 248
        efficiency, 248
        machine status, 247
        management information, 248
        scrap, summarized, 248
    profit and loss, 36, 196
    quantity transactions, 242
Required quantity, 289
Requirements for facilities planning, 143, 487–494
    manpower planning, 143
    planning system (RPS, MRP), 18, 123, 141, 150–178, 180, 199, 234, 399, 448
        and capacity planning, 174
        discipline for, 174
        errors in data, 174
        gross, 153, 164
        and inventory level, 174
        and job recording, 174
        net, 153, 164
        net change, 168
        and purchasing, 174
        regeneration, 168
        technique, 162
        time phase, 199
Research and development budget, 147
Resource-oriented classification system, 83–86
Retrieval
    of drawings
        by key, 438
        by parameters, 438
    process planning, 294–298, 446
Review mathematical methods, 341

Robots, 396
Rolling, 68
Rough classification of events, 236
Rough cut, 326
Routing, 17, 76, 121–124, 253, 259, 269
    master, 123
    work, 123

Safety factor, 57–59
    lead time, 164
    margin of, 57
    stock, 156
Salaries and wages, 36
Sales performance level, 486
Schedule, backward, 204
    master production planning, 18, 22, 123, 135, 163, 179, 199, 294
    capacity profile, 139
Scheduling; *see* Production planning
Scrap, periodic summarized report, 247
Seasonal load balancing, forecasting, 130
Secondary objectives, 51, 96
Selection
    of machine, 348
    process, 70–72
Self-checking number, 40
    detecting errors, 42
Semi-meaningful classification system, 104
Sequence of operation, 339, 361
Sequence work center, 224–226
Serial batch number, 189
Service level, 179
Setup
    and tear down, 237
    similar, 221, 454
Shifting assembly, 481
    backward, 481
    forward, 476
        and order splitting, 476
        and splitting, 476
Shipment file, 174, 196
Shop
    automated, 85
    floor, 20, 196
        activities, 264
        control of, 84, 220–251
        open order file, 196
        performance level, 487
Short process time (SPT), 229, 231
Short-term capacity planning, 221–226
Similar setup (SIMSET), 221, 229, 454
Simulation of job shop, 233
Slack, 204, 229, 232
Slow-moving items, 458
Smoothing exponential forecast model, 135
Solution for general matrix, 358–364

Spindle bore constraint, 347
Splitting, 211, 315, 476
SPT; *see* Short process time
Standard cost, 190, 259−264
  direct hourly rate, 255
  dynamic cost, 272
  plan, 295
Standardization, 64−67, 93
Standards, by Hal, 420
  for assembly, 437
  for part design, 434
Start of job transaction, 240−242
Statistical forecast regression analysis, 131
  inventory control, 156
Status code, 185−189
  machine report, 247
  work code, 188−189
Stock
  allocation, 162, 341, 467−474
  count, 194
  in cut pieces, 458
  utilization features, 457−460
Storage location, 189
Structural analysis programs, 425−426
Subcontracting, 176, 197, 217, 266
Summarized periodic scrap report, 248
Supplier rating value, 196
Surface finish, 289, 302, 306, 323, 404
  and chatter, 315
  and cutting speed, 316
  and depth of cut, 318−319
  and feed rate and tool-nose radius, 316
  and tolerances, 289
Surface integrity, 319
Systematic method and creativity, 52

Taylor equation, 306
Team work, 237
Tear-down and setup, 237
Technique for capacity planning, 206−217
  for inventory system, 183−192
  for requirement planning, 162−169
Technological improvements, 278−281
Technological transfer, 437
Technology, all-embracing, 6, 393
Theoretical optimum, 397, 484
  absolute optimum, 484
  process plan, 303, 338
Three-dimensional body, 404
Time
  and cost conversion, 348
  period length, 463

phase of requirement-planning, 199
  study and methods, 17
Tolerance, 59−62, 289, 302, 404, 434
  bilateral system, 60
  and cost, 62
  part design, 434
  and surface finish, 289
  unilateral system, 59
Tools
  automatic design, 423−425
  breakage, 306
  life, 307, 321
  nose-radius surface finish and feed rate, 316
  wear, 306
    chatter effect on, 320−321
Torque constraints, 347
Transaction code, 184, 195
  job completion, 243
  quantity reporting, 242
Transfer line, 492
Transfer time, 361
Transit code, 184, 188
Treatment and marking, 409
Two-way data processing, 44, 46, 197, 240

Underload and overload, 216
Unilateral tolerance system, 59
Unit of measure, 183, 404
Unit price of inventory items, 197
Utilization of technologies, 281−284
  of material in part design, 435
  of stock features, 457

Validation check, 193
Validation reliability and detection of errors, 129
Value engineering, 421
Variations and option in customers' orders, 128
Verify and keypunch, 39
Vibrations, 321

Wages and salaries, 36
Wear of tools, 306, 320−321
Work cell, 83, 493
  center sequence of, 224−226
  in Hal, 417
  in process, 153, 174, 198, 227, 272
  routing file, 123
  status code, 188, 197
  team, 237

Zero base budget, 147
Zero check, 42